Ingo Lieb
Joachim Michel

The Cauchy-Riemann Complex

Aspects of Mathematics

Edited by Klas Diederich

*A Publication of the Max-Planck-Institut für Mathematik, Bonn

Ingo Lieb
Joachim Michel

The Cauchy-Riemann Complex

Integral Formulae and Neumann Problem

vieweg

Prof. Dr. Ingo Lieb
Rheinische Friedrich-Wilhelms-Universität Bonn
Mathematisches Institut
Beringstraße 6
D-53115 Bonn, Germany

ilieb@math.uni-bonn.de

Prof. Dr. Joachim Michel
Laboratoire de Mathématiques Pures et
Appliquées Joseph Liouville
50, rue Ferdinand Buisson
F-62228 Calais Cedex, France

michel@lmpa.univ-littoral.fr

Prof. Dr. Klas Diederich (Series Editor)
Bergische Universität – Gesamthochschule Wuppertal
Fachbereich 7 – Mathematik
Gaußstraße 20
D-42119 Wuppertal, Germany

klas@math.uni-wuppertal.de

Die Deutsche Bibliothek – CIP-Cataloguing-in-Publication-Data
A catalogue record for this publication is available from
Die Deutsche Bibliothek.

First edition, March 2002

All rights reserved
© Friedr. Vieweg & Sohn Verlagsgesellschaft mbH, Braunschweig/ Wiesbaden, 2002
Softcover reprint of the hardcover 1st edition 2002

Vieweg is a company in the specialist publishing group BertelsmannSpringer.
www.vieweg.de

Cover design: Ulrike Weigel, www.CorporateDesignGroup.de

Printed on acid-free paper

ISBN 978-3-322-91610-5   ISBN 978-3-322-91608-2 (eBook)
DOI 10.1007/978-3-322-91608-2

ISSN 0179-2156

Hans Grauert gewidmet

Preface

This book presents complex analysis of several variables from the point of view of the Cauchy-Riemann equations and integral representations. A more detailed description of our methods and main results can be found in the introduction. Here we only make some remarks on our aims and on the required background knowledge.

Integral representation methods serve a twofold purpose: 1° they yield regularity results not easily obtained by other methods and 2°, along the way, they lead to a fairly simple development of parts of the classical theory of several complex variables. We try to reach both aims. Thus, the first three to four chapters, if complemented by an elementary chapter on holomorphic functions, can be used by a lecturer as an introductory course to complex analysis. They contain standard applications of the Bochner-Martinelli-Koppelman integral representation, a complete presentation of Cauchy-Fantappiè forms giving also the numerical constants of the theory, and a direct study of the Cauchy-Riemann complex on strictly pseudoconvex domains leading, among other things, to a rather elementary solution of Levi's problem in complex number space $\mathbb{C}^n$. Chapter IV carries the theory from domains in $\mathbb{C}^n$ to strictly pseudoconvex subdomains of arbitrary — not necessarily Stein — manifolds. We develop this theory taking as a model classical Hodge theory on compact Riemannian manifolds; the relation between a parametrix for the real Laplacian and the generalised Bochner-Martinelli-Koppelman formula is crucial for the success of the method. In Chapter V we describe the Neumann problem for the Cauchy-Riemann complex and prove, in particular, the fundamental density theorems due to Friedrichs and Hörmander. An analysis of this problem in our context leading to the main technical results of our work is given in Chapters VI and VIII, whereas Chapter VII develops the necessary machinery of integral estimates and function spaces. The book ends with applications of these technical results to complex analysis: Mergelyan's and Gleason's problem on complex manifolds, in the framework of Hölder spaces.

Prerequisites for reading this book are some acquaintance with the elementary theory of functions of several complex variables and a good knowledge of classical analysis, in particular of distributions and integration theory. The basic notions of analysis on manifolds are essential — even for domains in $\mathbb{C}^n$. There are very few instances where we rely more heavily on the theory of Stein spaces — mostly when we study strictly pseudoconvex domains in general manifolds. If one concentrates on subdomains of Stein manifolds, one can neglect these arguments.

As can be guessed from the above, we use and present many ideas going back to different mathematicians; we have tried to describe the historical development to the best of our knowledge but have probably failed at many instances. For this we apologize.

Particularly useful for us have been the textbooks by M. Range, G. de Rham and Ch. Laurent-Thiébaut; Chapters I and III follow their presentation at many places. Chapter V is based on Hörmander's work, in Chapter VII we use Krantz' paper [Kra 76], whereas the main results of Chapters IV and VI and many results of ChapterVIII are due to Lieb and Range (and to Michel in the analogous pseudoconcave case). (The original solution of the $\bar{\partial}$-Neumann problem is due to Kohn.) The systematic use of local integral formulae and its combination with the theory of compact operators, which allows to pass from $\mathbb{C}^n$ to arbitrary manifolds, can be retraced to Kerzman, Henkin and Range; Grauert's "bump method" plays a decisive role in this context, and the study of arbitrary — i. e. non-Stein — manifolds requires some classical tools which go back to Remmert, Cartan and K. Stein. Finally, the powerful analytical methods developed by E. M. Stein have been an essential help in estimating the kernels which we construct.

Our cooperation has been supported by the SFB 256 of the Deutsche Forschungsgemeinschaft, by the university of Bonn and the Université du Littoral – Côte d'Opale, and by the European network ANACOGA. We are very grateful to these institutions. Our particular thanks go to Dipl. Math. C. Wallat, Dipl. Math. Ch. Lampert and Dr. T. Hefer for competently establishing the TEX-file, and, moreover, suggesting many improvements and correcting many a mistake in the different versions of our text. Prof. K. Diederich has kindly invited us to publish our work in his "Aspects of Mathematics" series; his comments on our work have been particularly useful. Finally, Ms. U. Schmickler-Hirzebruch from Vieweg Verlag has patiently encouraged us to continue — and to come to an end.

The help of all these institutions, friends, and colleagues has been invaluable.

The first named author was a student of H. Grauert. It is a pleasure and an honour to dedicate this work to H. Grauert who, over more than 40 years, has immensely contributed to the development of complex analysis.

Bonn and Calais, January 2002   Ingo Lieb and Joachim Michel

Contents

Introduction

0.   Let f be a function of n complex variables $z_\nu = x_\nu + iy_\nu$, $\nu = 1, \ldots, n$. We introduce the *Wirtinger derivatives*

$$\frac{\partial f}{\partial z_\nu} = \frac{1}{2} \left(\frac{\partial f}{\partial x_\nu} - i \frac{\partial f}{\partial y_\nu} \right),$$
$$\frac{\partial f}{\partial \bar{z}_\nu} = \frac{1}{2} \left(\frac{\partial f}{\partial x_\nu} + i \frac{\partial f}{\partial y_\nu} \right),$$

(0.1)

and the differentials

$$\partial f = \sum_\nu \frac{\partial f}{\partial z_\nu} dz_\nu,$$
$$\bar{\partial} f = \sum_\nu \frac{\partial f}{\partial \bar{z}_\nu} d\bar{z}_\nu.$$

(0.2)

Therefore, $\partial f + \bar{\partial} f = df$ yields the total differential of f. The operators ∂ and $\bar{\partial}$ carry over to differential forms as follows: if

$$f = \sum_{\substack{i_1, \ldots, i_p \\ j_1, \ldots, j_q}} a_{i_1, \ldots, i_p, j_1, \ldots, j_q} dz_{i_1} \wedge \cdots \wedge dz_{i_p} \wedge d\bar{z}_{j_1} \wedge \cdots \wedge d\bar{z}_{j_q}$$

is a form of type (p, q), then

$$\bar{\partial} f = \sum_{\substack{i_1, \ldots, i_p \\ j_1, \ldots, j_q}} \bar{\partial} a_{i_1, \ldots, j_q} \wedge dz_{i_1} \wedge \cdots \wedge d\bar{z}_{j_q};$$

(0.3)

so $\bar{\partial} f$ is a form of type $(p, q+1)$. ∂ is defined accordingly such that we always have

$$d = \partial + \bar{\partial}.$$

(0.4)

The main elementary properties of $\bar{\partial}$ are:

$$A \text{ function } f \text{ is holomorphic if and only if } \bar{\partial} f = 0,$$

(0.5)

$$\bar{\partial} \circ \bar{\partial} = 0.$$

(0.6)

(Both statements are valid under mild differentiability assumptions.)

Definition 0.7 *f is $\bar{\partial}$-closed if $\bar{\partial} f = 0$; it is $\bar{\partial}$-exact if there is a form u with $\bar{\partial} u = f$. The operator $\bar{\partial}$ is the operator of the Cauchy-Riemann equations. The system*

$$\bar{\partial} u = f, \, \bar{\partial} f = 0$$

$(CR)_{pq}$

for a given $\bar{\partial}$-closed form of type (p, q) is the system of Cauchy-Riemann equations.

1

The system $(CR)_{pq}$ is, in view of (0.5), even defined on a complex manifold.

Let us denote the space of C^∞-smooth (i. e. infinitely differentiable) (p,q)-forms on a complex n-dimensional manifold X by $C^\infty_{pq}(X)$. In general, $(CR)_{pq}$ is not soluble. We therefore introduce the complex of vector spaces and linear maps

$$0 \to C^\infty_{p0}(X) \xrightarrow{\bar\partial} C^\infty_{p1}(X) \xrightarrow{\bar\partial} \cdots \xrightarrow{\bar\partial} C^\infty_{pn}(X) \to 0 ; \tag{0.8}$$

it is called the *Cauchy-Riemann* or *Dolbeault complex*, its cohomology spaces

$$H^{pq}(X) = \frac{\ker(\bar\partial : C^\infty_{pq}(X) \to C^\infty_{pq+1}(X))}{\operatorname{im}(\bar\partial : C^\infty_{pq-1}(X) \to C^\infty_{pq}(X))} \tag{0.9}$$

are the *Dolbeault cohomology spaces* (or *groups*).

So $H^{p0}(X)$ is the space of holomorphic p-forms, and for $q \geq 1$, the *vanishing theorem*

$$H^{pq}(X) = 0 \tag{0.10}$$

is equivalent to the solubility of $(CR)_{pq}$.

1.   Important facts of function theory in a complex manifold are encoded in the Cauchy-Riemann system. We state, first of all

Theorem 0.11 $(CR)_{pq}$ *is always locally soluble.*

This means that each point x of a complex manifold X has a neighbourhood U such that $H^{pq}(U) = 0, q \geq 1$.

From here one deduces the *Dolbeault isomorphism*

$$H^q(X, \Omega^p) \xrightarrow{\sim} H^{pq}(X) \tag{0.12}$$

between the q-th cohomology of X with values in the sheaf Ω^p of germs of holomorphic p-forms and the corresponding Dolbeault cohomology which allows to compute a holomorphic object by solving the Cauchy-Riemann equations.

Now suppose that X is a Stein manifold.

Definition 0.13 X *is a Stein manifold if it has the following properties:*

  i. *For any two points $x \neq y$ in X there is a holomorphic function f on X with $f(x) \neq f(y)$ (X is holomorphically separable).*

  ii. *For any sequence (x_j) in X without accumulation point there is a holomorphic function f on X which is unbounded on the sequence (x_j) (X is holomorphically convex).*

It follows from Cartan's Theorem B that on a Stein manifold

$$H^q(X, \Omega^p) = 0, \; q \geq 1. \tag{0.14}$$

This means, in view of (0.12):

Theorem 0.15 *On a Stein manifold, $(CR)_{pq}$ is always soluble.*

Theorem 0.15 is an important example of a *vanishing theorem* which expresses the solubility of the Cauchy-Riemann system in all cases. For many applications, *finiteness theorems* are almost as good: they state finite dimensionality of cohomology spaces. Here is the most important example (due to Grauert):

Theorem 0.16 *Let X be a strictly pseudoconvex manifold. Then*

$$\dim_{\mathbb{C}} H^q(X, \Omega^p) < \infty \text{ for } q \geq 1.$$

The Dolbeault isomorphism again translates this statement into the language of the Cauchy-Riemann equations. The concept of *strict pseudoconvexity* will be defined and carefully investigated in Chapters III and IV.

2. The above results (0.14)–(0.16) are "qualitative": even if one has additional information on the right-hand side of the Cauchy-Riemann system, for instance its boundedness, this information gets lost in the construction of the solutions. This is inevitable in view of the generality of the class of manifolds which are admitted in Theorem 0.15. On the other hand, for the more restricted class of strictly pseudoconvex domains or manifolds more precise "quantitative" results on the boundary behaviour of solutions should be and are possible, results which then have important applications to function theory. The present book is dedicated to such a quantitative theory.

3. The first important results in this direction are due to J. J. Kohn. He showed, in the early sixties,

Theorem 0.17 *Let Ω be a strictly pseudoconvex subdomain of a complex hermitian manifold X. Then there are constants C_q such that for any $(0, q)$-form f on Ω*

$$\|f\|_{L^2_{k+1/2}} \leq C_q(\|f\|_{L^2} + \|\bar{\partial} f\|_{L^2_k} + \|\bar{\partial}^* f\|_{L^2_k})$$

(for $1 \leq q \leq n = \dim X$).

The norms involved are the *Sobolev norms* of order k, $k + 1/2$; the operator $\bar{\partial}^*$ is the *Hilbert space adjoint* of $\bar{\partial}$ with respect to the scalar product on forms defined by the hermitian metric. All this will be carefully explained in Chapters V and VI.

Kohn's *basic estimate* above implies, in particular, existence and regularity theorems for the $\bar{\partial}$-operator:

Theorem 0.18 *Let Ω be as above. There is a linear operator K_q which associates with every $\bar{\partial}$-exact $(0, q)$-form f in the Hilbert space of square integrable forms a form*

$$u = K_q f$$

such that $\bar{\partial} u = f$ and $\|u\|_{L^2_{k+1/2}} \le C_q \|f\|_{L^2_k}$. C_q is independent of f. — In particular, if f is smooth up to the boundary, so is u.

Related theorems were somewhat later established by L. Hörmander.

4. Kohn's and Hörmander's methods and results can be extended to cover more general classes of pseudoconvex domains; there has been an active development in this direction marked by the work of Diederich, Fornæss, Kohn, Catlin, D'Angelo and others. On the other hand, it is only quite recently that their methods could be used to prove regularity results in function spaces different from the scale of L^2-Sobolev spaces. A more direct approach to establish, for instance, regularity theorems in Hölder norms, was started by Grauert/Lieb and Henkin, in 1969, and uses suitable integral representations which generalise the classical Cauchy integral. It is this method which we use in the present book — in fact, an extension of this method.

Originally, Henkin, Grauert/Lieb and Ramírez constructed integral kernels giving special solutions to the Cauchy-Riemann equations by combining the local geometry of strictly pseudoconvex domains with global results of function theory (as, for instance, Kohn's theorem above). Their solution operators can be easily controlled in L^p-norms or Hölder norms, but not in norms involving derivatives, like C^k-norms. We modify their approach in two ways. 1° *We work with locally defined integral kernels.* This allows to immediately work on a manifold, and it leads to finiteness theorems and, with some additional effort, to vanishing theorems which imply much of the classical theory of Stein manifolds. 2° *We analyse the $\bar{\partial}$-Neumann problem* which is basic for Kohn's theory *in terms of integral operators* and so arrive at satisfactory estimates in C^k-Hölder norms.

5. Here is a more detailed description of our work. We essentially use three fundamental integral formulae.

A. *The Bochner-Martinelli-Koppelman formula.* This formula refers to arbitrary bounded domains Ω in $\mathbb{C}^n$ (or even in a hermitian manifold). It represents a $(0, q)$-type differential form f on Ω as

$$f(y) = \int_{b\Omega} f(x) \wedge B_{nq}(x, y) - B_q \bar{\partial} f(y) - \bar{\partial} B_{q-1} f(y) , \tag{0.19}$$

with

$$B_q g(y) = \int_{\Omega} g(x) \wedge B_{nq}(x, y) ,$$

where the Bochner-Martinelli-Koppelman kernels B_{nq} are singular differential double forms on $\mathbb{C}^n \times \mathbb{C}^n$ of the appropriate double type which do not depend on the domain;

the formula holds under mild regularity assumptions on f. It implies a number of classical results on holomorphic extension and on local regularity of the Cauchy-Riemann operator which are exhibited in the first chapter. — On a hermitian manifold it includes an additional "error term".

B. *A basic homotopy formula,* and

C. *A basic integral representation for the $\overline{\partial}$-Neumann problem.*

Both B and C refer to strictly pseudoconvex domains or manifolds. They rely on the existence of *holomorphic support functions* (Levi polynomial, Ramírez-Henkin function) on such manifolds which satisfy precise lower estimates — see Chapters III and IV for the exact meaning of the above. These support functions are the essential building blocks of double differential forms $\mathcal{T}$, $\mathcal{S}$, $\mathcal{P}$ etc. which define integral operators T, S, P etc. with the properties described below.

B. The *basic homotopy formula* can be stated as

Theorem 0.20

$$f = Pf + T\overline{\partial}f + \overline{\partial}Sf.$$

Here f is a $(0,q)$-form on a strictly pseudoconvex manifold, $q \geq 1$. The integral operators P, T, S are L^2-bounded, and P is compact. A slightly different formula holds for $q = 0$.

This formula is proved in Chapters III and IV; it corresponds to a similar *parametrix formula* for the Laplace operator on compact Riemannian manifolds, and we shall indeed explain that the analogy is not only superficial — see Chapter IV. The formula almost immediately yields finiteness theorems for the Dolbeault cohomology and a solution to Levi's problem — the characterisation of Stein domains by local properties of their boundary. It also solves the $\overline{\partial}$-Neumann problem (without giving, however, the important regularity results valid in this situation) — see below.

C. *The basic integral formula for the $\overline{\partial}$-Neumann problem.*

The $\overline{\partial}$-Neumann problem consists, in its simplest version, in solving the system

$$\overline{\partial}u = f, \ \overline{\partial}^* u = 0.$$

The integral formula which we develop for its solution reads as follows.

Theorem 0.21 *Let f be a $(0,q)$-form, $q \geq 1$, on a strictly pseudoconvex manifold. Then*

$$f = T\overline{\partial}f + T^*\overline{\partial}^* f + R(f, \overline{\partial}f, \overline{\partial}^* f),$$

with linear integral operators T, T^* and R whose properties are described below.

The operators T, its adjoint T^* and the *error term* R are constructed similarly to Theorem 0.20; the action of R on $\overline{\partial}f$ and $\overline{\partial}^* f$ can be neglected as compared to T and T^* (this will be later specified), and the action of R on f can be similarly neglected as compared to the identity operator. In addition to the assumptions of Theorem 0.20 the underlying hermitian metric has to be carefully chosen (as a Levi metric). — A similar formula holds for $q = 0$.

The formula yields a large number of regularity results for the $\bar{\partial}$- and $\bar{\partial}$-Neumann problem, in particular *Hölder estimates* for derivatives of any order:

Theorem 0.22 *In the situation of Theorem 0.21,*

$$\|f\|_{C^{k+1/2}} \leq \mathrm{const}(\|f\|_{L^2} + \|\bar{\partial}f\|_{C^k} + \|\bar{\partial}^* f\|_{C^k}).$$

Let us define the norms in question:

Definition 0.23 *For $\Omega \subset\subset \mathbb{C}^n$ and $f \in C^k(\overline{\Omega})$,*

$$\|f\|_{C^k} = \sum_{\substack{D^l \\ l \leq k}} \sup_{x \in \overline{\Omega}} |D^l f(x)|,$$

where D^l stands for differentiations of order l.

Similarly, for $0 < \alpha < 1$,

$$\|f\|_{C^{k+\alpha}} = \|f\|_{C^k} + \sum_{D^k} \sup_{x \neq y} \frac{|D^k f(x) - D^k f(y)|}{\|x - y\|^\alpha}.$$

For subdomains of complex manifolds and for differential forms one works with a fixed finite coordinate cover and takes the sum of the norms of all the coefficients to obtain a coordinate dependent norm; different choices of the coordinate cover give rise to equivalent norms.

From Theorem 0.22 we deduce, in particular, the following complement to Kohn's Theorem 0.18.

Theorem 0.24 *The operators K_q which appear in Theorem 0.18 satisfy the estimates*

$$\|K_q f\|_{C^{k+1/2}} \leq \mathrm{const} \, \|f\|_{C^k}$$

(provided the metric is chosen as in Theorem 0.22).

The integral representation 0.21 finally allows to carry over the mechanism of the $\bar{\partial}$-Neumann problem to L^p-spaces for $p \geq 1$ and even to spaces of finite measures. Moreover, the solution operator N of the Neumann problem, the so-called Neumann operator, can be deduced from 0.21 as an operator whose principal part is a fairly simple integral operator.

6.   Our methods use, in an essential way, the strict pseudoconvexity of the manifolds; this is an in-built limitation which cannot easily be overcome.. There is, however, one further case where they apply, namely to strictly pseudoconcave manifolds. This case was studied in detail by the second author [Mic 92]; it requires variants of the methods which we present here, and we do not pursue this theory.

7.   For more general classes of pseudoconvex manifolds either the results as stated above, or the methods which we use here, break down. This has led to interesting and deep problems, partly solved, partly still open; we will describe some aspects of this development in our book.

Chapter I

The Bochner-Martinelli-Koppelman Formula

The most important integral representation of classical complex analysis in dimension 1 is Cauchy's integral formula with kernel $\frac{1}{2\pi i}\frac{d\zeta}{\zeta-z}$; it can be derived, by partial integration, from a corresponding integral formula for harmonic functions with kernel $\frac{1}{2\pi}\log\frac{1}{|\zeta-z|^2}$. The same procedure can be applied to a fundamental solution of the Laplacian in higher dimensions and yields a – partial – analogue of Cauchy's integral, the Bochner-Martinelli representation for holomorphic functions,

$$f(z) = \int_{b\Omega} f(\zeta) B_{n0}(\zeta, z) \tag{1}$$

The kernel $B_{n0}(\zeta, z)$ of this formula has good regularity properties but – unlike the Cauchy kernel – it is not holomorphic in the parameter z (except for $n = 1$ where it coincides with the Cauchy kernel). Although this restricts its use it does have important applications in complex analysis, some of which will be described in this chapter.

The BM formula has to be generalised and modified in several directions. 1° *One needs an integral representation not only for holomorphic functions but also for arbitrary smooth functions, corresponding to the Cauchy-Pompeiu formula in 1 dimension.* 2° *One has to deal with integral representations for differential forms of arbitrary degree.* Both generalisations will be established at the very beginning and (1) will appear as a special case of a more general Bochner-Martinelli-Koppelman (BMK) formula. 3° *The BMK kernels obtained according to 1° and 2° have to be decomposed into $\bar{\partial}_\zeta$- and $\bar{\partial}_z$-exact summands (up to certain well-behaved error terms).* This will replace the holomorphy of the Cauchy kernel, but this can only be done on special domains and requires, moreover, the calculus of Cauchy-Fantappiè forms; it will be dealt with in the following chapters.

This first chapter develops essential ideas which will be later taken up in more general situations; here they appear in a context where explicit and complete calculations are possible. A good part of the following paragraphs overlaps with the book of Range [Ran 86]; we include them to keep our exposition self-contained. The notations we use are as conventional as possible; most of them are explained in the first two paragraphs and in §5; §6 contains material which does not often appear in the literature (an exception is [Kyt 95]).

9

§1 Forms on Product Manifolds

This paragraph – besides fixing our notations – should help translating formulas from the language of exterior differential forms on a product manifold into the language of double differential forms.

Let W be an (in general open) subset of a product manifold $X \times Y$, where X and Y are differentiable manifolds of dimensions m resp. n. The spaces of exterior complex k-forms on W will be denoted by $E_k(W)$; they are modules over the ring $E_0(W)$ of $\mathbb{C}$-valued functions. The *degree* k can vary between 0 and $m + n$.

A double differential form f of double degree (r, s) associates with each point $(x, y) \in W$ an element

$$f(x, y) \in E_{r,x}(X) \otimes_{\mathbb{C}} E_{s,y}(Y),$$

where $x \in X, y \in Y$ and $E_{r,x}(X)$ is the vector space of exterior differentials of degree r in x, $E_{s,y}(Y)$ the corresponding space of s-differentials in y. The space of double (r, s)-forms on W will be denoted by $E_{rs}(W; X, Y)$; it is again a $E_0(W)$-module. If $W = X \times Y$ we write more simply $E_{rs}(X, Y)$.

Let $x^1, \ldots, x^m$ and $y^1, \ldots, y^n$ be local coordinates on neighbourhoods U of x_0 resp. V of y_0. The pull-back of the differential forms dx^μ resp. dy^ν to $X \times Y$ under the product projections defines exterior 1-forms on $U \times Y$ and $X \times V$ which will again be denoted by dx^μ and dy^ν. They are at the same time double forms of double degree $(1, 0)$ resp. $(0, 1)$ on $U \times Y$ resp. $X \times V$.

Now let $(x_0, y_0) \in U \times V \subset W$ and let f be a double form of double degree (r, s) on W. Then f can be uniquely written on $U \times V$ as

$$f = {\sum_{\substack{|I|=r \\ |J|=s}}}' f_{IJ}\, dx^I \otimes dy^J, \tag{1.1}$$

where the summation is over all strictly increasing multiindices $I \subset M = \{1, \ldots, m\}$ of length $|I| = r$ and $J \subset N = \{1, \ldots, n\}$ of length s. The f_{IJ} are functions on $U \times V$. We call f *measurable* (or *continuous*, *differentiable* etc.) if the f_{IJ} have these properties.

We shall always identify the (canonically) isomorphic spaces $E_{r,x}(X) \otimes E_{s,y}(Y)$ and $E_{s,y}(Y) \otimes E_{r,x}(X)$ and write accordingly

$$dx^I \otimes dy^J = dy^J \otimes dx^I = dx^I \cdot dy^J = dy^J \cdot dx^I;$$

so $\otimes$ will be replaced by a dot. The *exterior product* between two double forms is defined via their local representations by the rule

$$dx^I \wedge dy^J = dx^I \cdot dy^J. \tag{1.2}$$

So the dx-differentials anticommute among themselves, and so do the dy, whereas the dx commute with the dy. More generally we have

$$f \wedge g = (-1)^{pr+qs} g \wedge f \tag{1.3}$$

for $f \in E_{pq}(W; X, Y), g \in E_{rs}(W; X, Y)$.

Since exterior p-forms on X or Y define by pull-back double forms on $X \times Y$ (of double degree $(p, 0)$ resp. $(0, p)$) we get in particular exterior products

$$E_p(U) \times E_{rs}(U \times V; X, Y) \xrightarrow{\wedge} E_{p+r,s}(U \times V; X, Y)$$
$$E_q(V) \times E_{rs}(U \times V; X, Y) \xrightarrow{\wedge} E_{p,q+s}(U \times V; X, Y)$$

If $f \in E_{rs}(W; X, Y)$ is differentiable, we can take its differentials d_x with respect to x and d_y with respect to y:

$$d_x : E_{rs}(W; X, Y) \longrightarrow E_{r+1,s}(W; X, Y)$$
$$d_y : E_{rs}(W; X, Y) \longrightarrow E_{r,s+1}(W; X, Y)$$

We easily verify the rules

$$d_x^2 = d_y^2 = 0, \quad d_x \, d_y = d_y \, d_x \tag{1.4}$$

$$d_x(f \wedge g) = d_x f \wedge g + (-1)^r f \wedge d_x g \tag{1.5}$$
$$d_y(f \wedge g) = d_y f \wedge g + (-1)^s f \wedge d_y g$$

There is no canonical way of defining a total differential d invoking both d_x and d_y. One possibility is to set (for $f \in E_{rs}$)

$$df = d_x f + (-1)^r d_y f \tag{1.6}$$

In that case (1.4) implies

$$d^2 = 0; \tag{1.7}$$

On the other side no Leibniz rule like (1.5) holds for d. Moreover, $(-1)^s d_x + d_y$ would be another good candidate for d.

The *integral of a double form* $f \in E_{rs}(X, Y)$ *over an r-dimensional chain c in X* is well-defined:

$$\int_c f(x, y) = F(y);$$

it is an exterior s-form on Y (provided the coefficients of f have the right integrability properties). Stokes' theorems subsist, for instance:

$$\int_{\partial c} f(x, y) = \int_c d_x f(x, y)$$

under the usual assumptions.

Now consider an exterior k-form on $W \subset X \times Y$. If $U \times V \subset W$ is the product of local coordinate patches, U in X and V in Y, with coordinates x^μ and y^ν, respectively, f can be uniquely written, on $U \times V$, as

$$f = {\sum_{\substack{r+s=k \\ |I|=r \\ |J|=s}}}' f_{IJ} \, dx^I \wedge dy^J.$$

We associate with the term

$$f_{r,s} = \sum_{\substack{|I|=r \\ |J|=s}} f_{IJ} \, dx^I \wedge dy^J$$

the double form

$$\tau f_{r,s} = \sum_{\substack{|I|=r \\ |J|=s}} f_{IJ} \, dx^I \cdot dy^J$$

on $U \times V$. The corresponding form

$$\tau \widetilde{f}_{r,s} = \sum_{\substack{|I|=r \\ |J|=s}} \widetilde{f}_{IJ} \, d\widetilde{x}^I \cdot d\widetilde{y}^J$$

which we obtain on a different coordinate patch $\widetilde{U} \times \widetilde{V}$, $\widetilde{x}$, $\widetilde{y}$, is easily seen to coincide with $\tau f_{r,s}$ on the intersection of the two patches (note that we work with coordinate transformations in X and Y, separately, so x is a function of $\widetilde{x}$ alone and does not depend on $\widetilde{y}$). Consequently this procedure defines double forms $\tau f_{r,s}$ on all of W, and a vector space isomorphism

$$\tau : E_k(W) \longrightarrow \bigoplus_{r+s=k} E_{r,s}(W; X, Y).$$

τ is not an algebra homomorphism, of course. More precisely call f of *double degree* (r, s) if τf has this double degree. Then, for f of double degree (r, s) and g of double degree (p, q), we have

$$\tau(f \wedge g) = (-1)^{ps} \tau f \wedge \tau g. \tag{1.8}$$

We finally note

$$d \circ \tau = \tau \circ d. \tag{1.9}$$

Clearly (1.9) implies (1.7). – The easy but tedious verifications of (1.9) resp. (1.7) will be omitted. – Once again, the definition of τ is somewhat arbitrary; we can interchange the roles of X and Y to obtain a different isomorphism σ which commutes with the alternative candidate for d.

We now carry over these notions to the product of complex manifolds X and Y of complex dimensions m and n, respectively, whose points will be denoted by ζ and z. Forms on X or Y can be decomposed into summands of type (p, q) according to holomorphic and anti-holomorphic differentials, the differentials d_ζ and d_z on X and Y have the corresponding decomposition

$$d_\zeta = \partial_\zeta + \overline{\partial}_\zeta, \quad d_z = \partial_z + \overline{\partial}_z, \tag{1.10}$$

and consequently double forms on $X \times Y$ are sums of terms of *double type* $(p, q; r, s)$. The definitions of τ and d carry over; and so the double type of an exterior form is well defined. We write

$$d = d_{\zeta,z} = \partial_{\zeta,z} + \overline{\partial}_{\zeta,z} \tag{1.11}$$
$$= \partial_\zeta + (-1)^{p+q}\partial_z + \overline{\partial}_\zeta + (-1)^{p+q}\overline{\partial}_z$$

on forms of double type $(p, q; r, s)$. These definitions imply

$$\partial_{\zeta,z} \circ \partial_{\zeta,z} = 0, \quad \overline{\partial}_{\zeta,z} \circ \overline{\partial}_{\zeta,z} = 0,$$
$$\partial_{\zeta,z} \circ \overline{\partial}_{\zeta,z} + \overline{\partial}_{\zeta,z} \circ \partial_{\zeta,z} = 0,$$
$$\tau \circ \partial_{\zeta,z} = \partial_{\zeta,z} \circ \tau \tag{1.12}$$
$$\tau \circ \overline{\partial}_{\zeta,z} = \overline{\partial}_{\zeta,z} \circ \tau;$$

all these definitions are invariant under biholomorphic coordinate changes in X and Y separately.

§2 The Complex Laplacian

1. Complex number space $\mathbb{C}^n$ will be equipped with the Euclidean metric and the following orientation: if $z^j = x^j + iy^j$ is the decomposition of z^j into real and imaginary parts, then the $2n$-form

$$dV = dx^1 \wedge dy^1 \wedge \ldots \wedge dx^n \wedge dy^n = \left(\frac{i}{2}\right)^n dz^1 \wedge d\overline{z}^1 \wedge \ldots \wedge dz^n \wedge d\overline{z}^n \tag{2.1}$$

is positive. This form is called the *volume element*. The metric defines a *scalar product* $\langle , \rangle$ on the spaces $E_{pq,z} = E_{pq}$ of (p, q)-covectors in z; for

$$f = \sum_{|I|=p,|J|=q} f_{IJ}dz^I \wedge d\overline{z}^J, \quad g = \sum_{|I|=p,|J|=q} g_{IJ}dz^I \wedge d\overline{z}^J,$$

$$\langle f, g \rangle = 2^{p+q} \sum_{I,J} f_{IJ}\overline{g_{IJ}} \tag{2.2}$$

(The powers of 2 – here and in the sequel – are due to the fact that the dz^j do not have unit length.) Metric and orientation are combined to single out an isomorphism

$$* : E_{pq} \longrightarrow E_{n-q,n-p} \tag{2.3}$$

(more precisely

$$*_z : E_{pq,z} \longrightarrow E_{n-q,n-p,z}) \tag{2.3'}$$

with the following properties: * *is real*, that is

$$*\overline{f} = \overline{*f} \tag{2.4}$$

$$*1 = dV \tag{2.5}$$

$$f \wedge *\overline{g} = \langle f, g \rangle \, dV \tag{2.6}$$

$$** = (-1)^{p+q} \tag{2.7}$$

Properties (2.4) – (2.7) determine * uniquely; we have the explicit formula

$$*dz^A \wedge d\overline{z}^B \wedge (dz \wedge d\overline{z})^C = \alpha(a, b, c) dz^A \wedge d\overline{z}^B \wedge (dz \wedge d\overline{z})^{C'}. \tag{2.8}$$

Here A, B, C are pairwise disjoint increasing index sets in $N = \{1, \ldots, n\}$ of cardinality a, b, c, respectively, $C' = N - A - B - C$, and

$$\alpha(a, b, c) = i^n 2^{h-n} (-1)^{h(h-1)/2 + a} \tag{2.9}$$

$$h = a + b + 2c. \tag{2.10}$$

Let us further note:

$$\langle *f, *g \rangle = \langle f, g \rangle. \tag{2.11}$$

The operator * is called the *Hodge operator*. It carries over to forms on subsets $M \subset \mathbb{C}^n$ by pointwise application.

Definition 2.12 *A form f on a subset $M \subset \mathbb{C}^n$ is square integrable if its coefficients have this property. The scalar product for square integrable forms is defined as*

$$(f, g)_{L^2(M)} = \int_M \langle f, g \rangle \, dV = \int_M f \wedge *\overline{g},$$

the L^2-norm as

$$\|f\|_{L^2(M)} = (f, f)_{L^2(M)}^{1/2},$$

where the space of square integrable (p, q)-forms on M will be denoted by

$$L^2_{pq}(M).$$

It is a Hilbert space (by well known theorems of integration theory). In most cases M will be a domain Ω in $\mathbb{C}^n$; the index Ω can be omitted as long as this does not confuse things.

2.   From now on, spaces of differential (p, q)-forms with arbitrary coefficients will be denoted by E_{pq}, whereas $C^r_{pq}(M)$ is the space of r-times continuously differentiable (p, q)-forms, $C^\infty_{pq}(M)$ of infinitely often differentiable ("smooth") forms on M; compact support will be denoted by an index c. – In the remainder of this paragraph M will be $\mathbb{C}^n$; it will no longer be indicated in the notations.

Lemma 2.13 *There is exactly one linear operator*

$$\vartheta : C^{\infty}_{p,q+1;c} \longrightarrow C^{\infty}_{p,q;c}$$

with

$$(\overline{\partial} f, g) = (f, \vartheta g) \tag{2.14}$$

for smooth forms f and g with compact support.

For the proof we note the identity

$$d(f \wedge *\overline{g}) = \overline{\partial} f \wedge *\overline{g} + (-1)^{p+q} f \wedge \overline{\partial} * \overline{g} \tag{2.15}$$

for $f \in C^1_{pq}$, $g \in C^1_{p,q+1}$. Therefore, in case of compact supports, we have

$$
\begin{aligned}
(\overline{\partial} f, g) &= \int \overline{\partial} f \wedge *\overline{g} \\
&= (-1)^{p+q+1} \int f \wedge \overline{\partial} * \overline{g} \\
&= -\int f \wedge *\overline{(*\partial * g)};
\end{aligned}
$$

so

$$\vartheta = -*\partial* \tag{2.16}$$

satisfies (2.14). It is obviously uniquely determined and a differential operator of first order. Note

$$(\overline{\partial} f, g) = (f, \vartheta g) \tag{2.14'}$$
$$(\vartheta g, f) = (g, \overline{\partial} f) \tag{2.14''}$$

whenever f and g are continuously differentiable and at least one of them has compact support. Clearly $\vartheta f = 0$ if f is a function.

Definition 2.17

    i. $\vartheta = -*\partial*$ *is the formal adjoint of $\overline{\partial}$.*

    ii. $\Box = \overline{\partial}\vartheta + \vartheta\overline{\partial}$ *is the complex Laplacian.*

Thus, $\Box$ is a second order linear differential operator which carries (p, q)-forms into (p, q)-forms. From (2.14) we get

$$(\Box f, g) = (f, \Box g), \tag{2.18}$$
$$(\Box f, g) = (\overline{\partial} f, \overline{\partial} g) + (\vartheta f, \vartheta g), \tag{2.19}$$

if f and g are twice continuously differentiable and f or g have compact support.

A straightforward but tedious calculation yields explicit formulae for the two operators ϑ and $\Box$ in terms of the coefficients of the forms involved – cf. [Ran 86]. We state the results:

Lemma 2.20 *Let*

$$f = \sideset{}{'}\sum_{IJ} f_{IJ}\, dz^I \wedge d\bar{z}^J$$

be a differentiable (p, q)-form. Then

$$\vartheta f = (-1)^{p+1} 2 \sideset{}{'}\sum_{I,J,K,k} \varepsilon^J_{kK} \frac{\partial f_{IJ}}{\partial z^k}\, dz^I \wedge d\bar{z}^K.$$

The summation is extended over all strictly increasing index tuples I of length $|I| = p$, J of length $|J| = q$, K of length $|K| = q - 1$ and $1 \leq k \leq n$. The coefficient ε^J_{kK} is defined as follows:

$$\varepsilon^J_{kK} = \begin{cases} 0 & \text{if } \{k\} \cup K \neq J \\ 1 & \text{if } kK \text{ differs from } J \text{ by an even permutation} \\ -1 & \text{otherwise} \end{cases} \tag{2.21}$$

kK is the ordered set $\{k, k_1, \ldots, k_{q-1}\}$, where $K = \{k_1, \ldots, k_{q-1}\}$.

Lemma 2.22 *If f is a ($2\times$ differentiable) function, then*

$$\Box f = -2 \sum_{k=1}^n \frac{\partial^2 f}{\partial z^k \partial \bar{z}^k} = -\frac{1}{2} \sum_{k=1}^n \left[\frac{\partial^2 f}{(\partial x^k)^2} + \frac{\partial^2 f}{(\partial y^k)^2} \right].$$

For a (p, q)-form

$$f = \sideset{}{'}\sum f_{IJ}\, dz^I \wedge d\bar{z}^J$$

we have

$$\Box f = \sideset{}{'}\sum_{IJ} \Box f_{IJ}\, dz^I \wedge d\bar{z}^J.$$

The complex Laplacian is consequently a diagonal operator which acts coefficientwise. Note that the definitions of both ϑ and $\Box$ depend on the (in our case Euclidean) metric. We shall later on define ϑ and $\Box$ with respect to arbitrary metrics and we shall have to investigate, to what extent Lemma 2.22 subsists.

3. The operator ϑ occurs in the following – elementary but very important – formulae.

Lemma 2.23 (Green's formulae)

$$\langle \bar{\partial} f, g \rangle\, dV = \langle f, \vartheta g \rangle\, dV + d(f \wedge *\bar{g}) \tag{2.24}$$

$$\langle \vartheta g, f \rangle\, dV = \langle g, \bar{\partial} f \rangle\, dV - d(\bar{f} \wedge *g) \tag{2.25}$$

for $f \in C^1_{0,q-1}$, $g \in C^1_{0,q}$.

The proof is immediate. – As a corollary one gets Green's formulae in integrated form:

Lemma 2.26 *Let f, g be C^1-smooth forms of type $(0, q-1)$ resp. $(0, q)$ on the closure $\overline{\Omega}$ of a bounded domain with piecewise smooth boundary (of class C^1). Then*

$$(\overline{\partial} f, g)_\Omega = (f, \vartheta g)_\Omega + \int_{b\Omega} f \wedge *\overline{g} \qquad (2.27)$$

$$(\vartheta g, f)_\Omega = (g, \overline{\partial} f)_\Omega - \int_{b\Omega} \overline{f} \wedge *g. \qquad (2.28)$$

§3   The Fundamental Solution

1.   Let $n \geq 2$ and introduce in $\mathbb{C}^n \times \mathbb{C}^n$ (with coordinates ζ and z) the function

$$\Gamma_0(\zeta, z) = c_{n0} \frac{1}{\|\zeta - z\|^{2n-2}}. \qquad (3.1)$$

The real constant c_{n0} will be determined later. Γ_0 is smooth for $\zeta \neq z$ and everywhere locally integrable in ζ (for fixed z); moreover, it is symmetric in ζ and z. A straightforward calculation gives

$$\square_\zeta \Gamma_0(\zeta, z) = 0 \qquad \text{for } \zeta \neq z. \qquad (3.2)$$

If f is a C^2-function with compact support in $\mathbb{C}^n$, the integral

$$(\square f, \Gamma_0) \stackrel{\text{def}}{=} \int_{\mathbb{C}^n} \square f(\zeta) \wedge *\Gamma_0(\zeta, z) \qquad (3.3)$$

is well-defined. The operators $\square$, $*$ and $\int$ all act on the ζ-variable; z is a parameter.

Since both Γ_0 and $\overline{\partial}\Gamma_0$ have integrable singularities, (3.3) can be transformed, using Green's formulae (2.25) resp. (2.27), into

$$(\overline{\partial} f, \overline{\partial}\Gamma_0) \qquad (3.4)$$

A second partial integration (using (2.24) and (3.2)) gives

$$(\overline{\partial} f, \overline{\partial}\Gamma_0) = \lim_{\varepsilon \to 0} \left[\int_{\mathbb{C}^n - B_\varepsilon(z)} \langle f, \square\Gamma_0 \rangle \, dV - \int_{S_\varepsilon(z)} f * \partial\Gamma_0 \right] \qquad (3.5)$$

$$= -\lim_{\varepsilon \to 0} \int_{S_\varepsilon(z)} f(\zeta) * \partial\Gamma_0(\zeta, z),$$

with

$$B_\varepsilon(z) = \{\zeta : \|\zeta - z\| \leq \varepsilon\}, \quad S_\varepsilon(z) = \{\zeta : \|\zeta - z\| = \varepsilon\}. \qquad (3.6)$$

The limit is easily determined:

$$\lim_{\varepsilon \to 0} \int_{S_\varepsilon(z)} f(\zeta) * \partial \Gamma_0(\zeta, z) \tag{3.7}$$

$$= \lim_{\varepsilon \to 0} \left[\int_{S_\varepsilon(z)} (f(\zeta) - f(z)) * \partial \Gamma_0(\zeta, z) + \int_{S_\varepsilon(z)} f(z) * \partial \Gamma_0(\zeta, z) \right]$$

$$= f(z) \lim_{\varepsilon \to 0} \int_{S_\varepsilon(z)} * \partial \Gamma_0(\zeta, z).$$

Now

$$* \partial \Gamma_0(\zeta, z) = \frac{c_{n0}}{\|\zeta - z\|^{2n}} (1 - n) \alpha(1, 0, 0) \sum_{j=1}^{n} (\bar{\zeta}^j - \bar{z}^j) \, d\zeta^j \wedge (d\zeta \wedge d\bar{\zeta})^{N-j},$$

so by Stokes' theorem and (2.9),(2.10)

$$\int_{S_\varepsilon(z)} * \partial \Gamma_0 = \frac{c_{n0}}{\varepsilon^{2n}} (1 - n) i^n 2^{1-n} \sum_{j} \int_{B_\varepsilon(z)} (d\bar{\zeta}^j \wedge d\zeta^j) \wedge (d\zeta \wedge d\bar{\zeta})^{N-j} \tag{3.8}$$

$$= \frac{c_{n0}}{\varepsilon^{2n}} n(1 - n) i^n 2^{1-n} \left(\frac{2}{i} \right)^n \varepsilon^{2n} \int_{B_1(z)} dV$$

$$= -c_{n0} \frac{2\pi^n}{(n - 2)!}.$$

The integral (3.8) is -1 if we choose

$$c_{n0} = \frac{(n - 2)!}{2\pi^n}. \tag{3.9}$$

We obtain from (3.3)–(3.9)

Proposition 3.10 *Let*
$$\Gamma_0(\zeta, z) = \frac{(n - 2)!}{2\pi^n} \frac{1}{\|\zeta - z\|^{2n-2}}.$$

Then for each C^2-function f with compact support in $\mathbb{C}^n$ we have

$$f(z) = (\Box f, \Gamma_0) = \int_{\mathbb{C}^n} \Box f(\zeta) * \Gamma_0(\zeta, z).$$

This tells us that Γ_0 is a fundamental solution for the complex Laplacian (on functions). – We can easily pass to higher order forms using – e. g.– Lemma 2.22. To this end we set for

$q \geq 0$

$$\Gamma_q = c_{nq} \Gamma_0 \gamma^q \tag{3.11}$$

$$\gamma = \sum_{j=1}^{n} d\bar{\zeta}^j \, dz^j \tag{3.12}$$

where c_{nq} will be determined a bit later, and γ^q is the q-th exterior power of γ, i.e.

$$\gamma^q = q! \sideset{}{'}\sum_{|J|=q} d\bar{\zeta}^J \, dz^J. \tag{3.13}$$

Γ_q is a double form of double type $(0, q; q, 0)$. If

$$f = \sideset{}{'}\sum_{|J|=q} f_J \, d\bar{\zeta}^J$$

is a $(0, q)$-form with compact support in $\mathbb{C}^n$ of differentiability class C^2, the integral

$$\int_{\mathbb{C}^n} \Box f(\zeta) \wedge *\overline{\Gamma_q(\zeta, z)} \overset{\text{def}}{=} (\Box f, \Gamma_q) \tag{3.14}$$

is a well defined $(0, q)$-form on $\mathbb{C}^n$; the operators $\Box$ and $*$ refer to the variable ζ. We obtain from (3.10) and (2.2):

$$(\Box f, \Gamma_q) = 2^q q! c_{nq} \sideset{}{'}\sum_{|J|=q} (\Box f_J, \Gamma_0) \, d\bar{z}^J$$

$$= 2^q q! c_{nq} f.$$

Setting

$$c_{nq} = 2^{-q} q!^{-1},$$

we arrive at the principal result of this paragraph

Theorem 3.15 *Set for $q = 0, \ldots, n$*

$$\Gamma_q = \frac{(n-2)!}{q! 2^{q+1} \pi^n} \frac{1}{\|\zeta - z\|^{2n-2}} \left(\sum_{j=1}^{n} d\bar{\zeta}^j \, dz^j \right)^q. \tag{3.16}$$

   i. If $f \in C^2_{0q,c}(\mathbb{C}^n)$,

$$(\Box f, \Gamma_q)(z) = \int_{\mathbb{C}^n} \Box f(\zeta) \wedge *\overline{\Gamma_q(\zeta, z)} = f(z). \tag{3.17}$$

   ii. $\Box_\zeta \Gamma_q(\zeta, z) = 0$ for $\zeta \neq z$.

The second part of the theorem follows easily from the preceeding argument; similarly we obtain from (2.19)

Theorem 3.18 *Under the assumptions of 3.15 we have*

$$f = (\Box f, \Gamma_q) = (\overline{\partial} f, \overline{\partial} \Gamma_q) + (\vartheta f, \vartheta \Gamma_q).$$

The identity

$$f = (\overline{\partial} f, \overline{\partial} \Gamma_q) + (\vartheta f, \vartheta \Gamma_q) \tag{3.19}$$

is even true if f is only once continuously differentiable (and of compact support, of course), as can be easily seen by approximating f in C^1-norm by C^2-forms.

Definition 3.20

$$\Gamma_q = \frac{(n-2)!}{q! 2^{q+1} \pi^n} \frac{1}{\|\zeta - z\|^{2n-2}} \left(\sum_{j=1}^{n} d\overline{\zeta}^j dz^j \right)^q$$

is called "the" fundamental solution of $\Box$ for $(0,q)$-forms in $\mathbb{C}^n$, $n \geq 2$.

For $n = 1$ we have to define

$$\Gamma_0 = \frac{1}{2\pi} \log \frac{1}{|\zeta - z|^2}$$

$$\Gamma_1 = \frac{1}{2} \Gamma_0 \, d\overline{\zeta} dz \tag{3.21}$$

as the corresponding fundamental solutions; also, it is convenient to set $\Gamma_{-1} = 0$.

2.   The following relation between the Γ_q will be important in the sequel:

Proposition 3.22

   i. $\vartheta_\zeta \Gamma_q = \partial_z \Gamma_{q-1}$

   ii. $\partial_\zeta \overline{\Gamma}_{q-1} = \vartheta_z \overline{\Gamma}_q$

Proof   Clearly i) and ii) are equivalent, so we just prove the first equality, or, equivalently,

$$*_\zeta \vartheta_\zeta \Gamma_q = *_\zeta \partial_z \Gamma_{q-1}. \tag{3.23}$$

We have

$$*\vartheta_\zeta \Gamma_q = - * * \partial_\zeta * \Gamma_q$$

$$= (-1)^q \frac{(n-2)!}{q! 2^{q+1} \pi^n} \partial_\zeta \frac{1}{\|\zeta - z\|^{2n-2}} \wedge *_\zeta \left(\sum d\overline{\zeta}^j dz^j \right)^q$$

$$= (-1)^{q+1} \frac{(n-1)!}{2^{q+1} \pi^n} \frac{1}{\|\zeta - z\|^{2n}} \partial_\zeta \|\zeta - z\|^2 \wedge \sum_{|J|=q} *_\zeta d\overline{\zeta}^J dz^J.$$

The right hand side of (3.23) becomes

$$-\frac{(n-1)!}{2^q\pi^n}\frac{1}{\|\zeta-z\|^{2n}}\partial_z\|\zeta-z\|^2 \wedge \sum_{|K|=q-1} *_\zeta d\bar\zeta^K dz^K.$$

So we have to show

$$\frac{(-1)^q}{2}\partial_\zeta\|\zeta-z\|^2 \wedge \sum_{|J|=q} *_\zeta d\bar\zeta^J dz^J = \partial_z\|\zeta-z\|^2 \wedge \sum_{|K|=q-1} *_\zeta d\bar\zeta^K dz^K. \qquad (3.24)$$

Using (2.8) – (2.10) this amounts to showing

$$\sum_j (\bar\zeta^j - \bar z^j)\, d\zeta^j \wedge \sum_{|J|=q} d\bar\zeta^J \wedge (d\zeta \wedge d\bar\zeta)^{N-J} \cdot dz^J$$

$$= \sum_k (\bar\zeta^k - \bar z^k)\, dz^k \wedge \sum_{|K|=q-1} d\bar\zeta^K \wedge (d\zeta \wedge d\bar\zeta)^{N-K} dz^K. \qquad (3.25)$$

Let J with $|J| = q$ be fixed. The coefficient of dz^J on the right-hand side of (3.25) is

$$\sum_{\substack{k\in J \\ |K|=q-1 \\ J=k\cup K}} \varepsilon^J_{kK}(\bar\zeta^k - \bar z^k)\, d\bar\zeta^K \wedge (d\zeta \wedge d\bar\zeta)^{N-K}, \qquad (3.26)$$

on the left-hand side the corresponding coefficient is

$$\sum_{j\in J} (\bar\zeta^j - \bar z^j)\, d\zeta^j \wedge d\bar\zeta^J \wedge (d\zeta \wedge d\bar\zeta)^{N-J},$$

which can also be written as

$$\sum_{\substack{j\in J \\ |K|=q-1 \\ J=j\cup K}} (\bar\zeta^j - \bar z^j)\varepsilon^J_{jK}\, d\bar\zeta^K \wedge (d\zeta \wedge d\bar\zeta)^{N-K},$$

that is as (3.26).   □

§4 The Bochner-Martinelli-Koppelman Formula

In this paragraph, which is crucial for all that follows, we will use the fundamental solution Γ_q of the Laplacian in order to carry over Cauchy's integral formula to the case of several complex variables (and forms of higher degree). As a first step we generalise Theorem 3.15, admitting forms with arbitrary support.

So let $\Omega \subset\subset \mathbb{C}^n$ be a domain with piecewise smooth boundary and $f \in C^1_{0q}(\overline\Omega)$ be a continuously differentiable $(0, q)$-form on $\overline\Omega$. Choose a point $z_0 \in \Omega$ and a smooth patching function χ which is 1 in a neighbourhood U of z_0 and has compact support in Ω, with $0 \leq \chi \leq 1$. Let $z \in U$; then

$$f = \chi f + (1 - \chi)f = f_0 + f_1. \tag{4.1}$$

Theorem 3.15 gives

$$f(z) = f_0(z) = (\overline{\partial} f_0, \overline{\partial} \Gamma_q) + (\vartheta f_0, \vartheta \Gamma_q). \tag{4.2}$$

The form f_1 can be handled by Green's formulae (2.27), (2.28):

$$(\overline{\partial} f_1, \overline{\partial} \Gamma_q) = (f_1, \vartheta \overline{\partial} \Gamma_q) + \int\limits_{b\Omega} f_1 \wedge *\overline{\partial} \Gamma_q \tag{4.3}$$

$$(\vartheta f_1, \vartheta \Gamma_q) = (f_1, \overline{\partial} \vartheta \Gamma_q) - \int\limits_{b\Omega} \overline{\vartheta \Gamma_q} \wedge *f_1 \tag{4.4}$$

(Note that for $\zeta \in U$ the integrands vanish.) Adding up and taking into account that

$$\Box_\zeta \Gamma_q = 0,$$

we obtain

Theorem 4.5 *For $z \in \Omega$ and $f \in C^1_{0q}(\overline{\Omega})$ one has the integral representation*

$$f(z) = - \int\limits_{b\Omega} f(\zeta) \wedge *\overline{\partial \Gamma_q(\zeta, z)} + (\overline{\partial} f, \overline{\partial} \Gamma_q) + (\vartheta f, \vartheta \Gamma_q) + \int\limits_{b\Omega} \overline{\vartheta \Gamma_q(\zeta, z)} \wedge *f(\zeta)$$

Definition 4.6 *The double differential form*

$$B_{nq}(\zeta, z) = - * \partial_\zeta \overline{\Gamma_q(\zeta, z)}$$

is called the Bochner-Martinelli-Koppelman kernel for $(0, q)$-forms (in $\mathbb{C}^n$).

Let us describe the kernel in some detail. An immediate calculation gives

$$B_{nq} = \frac{(n-1)!}{2^{q+1}\pi^n} \alpha(q+1, 0, 0) \sum_{\substack{j \\ |J|=q \\ |K|=n-q-1}} \frac{(\overline{\zeta}^j - \overline{z}^j)}{\|\zeta - z\|^{2n}} d\zeta^j \wedge d\zeta^J \wedge (d\zeta \wedge d\overline{\zeta})^K \cdot d\overline{z}^J \tag{4.7}$$

On the other hand, introducing

$$\widetilde{\beta} = \sum (\overline{\zeta}^j - \overline{z}^j) \, d\zeta^j,$$

one obtains

$$\widetilde{\beta} \wedge (\overline{\partial}_\zeta \widetilde{\beta})^{n-q-1} \wedge (\overline{\partial}_z \widetilde{\beta})^q$$
$$= (-1)^q(-1)^{n-q-1}q!(n-q-1)! \sum_{\substack{j \\ |J|=q \\ |K|=n-q-1}} (\overline{\zeta}^j - \overline{z}^j) \, d\zeta^j \wedge d\zeta^J \wedge (d\zeta \wedge d\overline{\zeta})^K \cdot d\overline{z}^J$$

$$(4.8)$$

Comparison of (4.7) and (4.8) gives in view of the definition of $\alpha(q+1,0,0)$ – see (2.9) –

Proposition 4.9

i) $\quad B_{nq}(\zeta,z) = (-1)^{q(q-1)/2} \binom{n-1}{q} \left(\frac{1}{2\pi i}\right)^n \frac{1}{\|\zeta - z\|^{2n}} \widetilde{\beta} \wedge (\overline{\partial}_\zeta \widetilde{\beta})^{n-q-1} \wedge (\overline{\partial}_z \widetilde{\beta})^q$

$\qquad = \frac{1}{\|\zeta - z\|^{2n}} \widetilde{B}_{nq}(\zeta,z)$

ii) $\quad B_{nq}(\zeta,z) = (-1)^{q(q-1)/2} \binom{n-1}{q} \left(\frac{1}{2\pi i}\right)^n \beta \wedge (\overline{\partial}_\zeta \beta)^{n-q-1} \wedge (\overline{\partial}_z \beta)^q$

with

$$\beta(\zeta,z) = \sum_j \frac{\overline{\zeta}^j - \overline{z}^j}{\|\zeta - z\|^2} \, d\zeta^j. \tag{4.10}$$

Part ii) is an immediate consequence of i). The form (4.10) is called the *generating form* for the Bochner-Martinelli-Koppelman kernels.

Note that we consider β and the B_{nq} as double differential forms on $\mathbb{C}^n \times \mathbb{C}^n$. If one prefers working with exterior forms the signs have to be adjusted, using the rules of §1.

We now use 3.22 to eliminate the adjoint operator ϑ from Theorem 4.5.

$$(\vartheta f, \vartheta \Gamma_q) = (\vartheta f, \partial_z \Gamma_{q-1}) = \overline{\partial}_z(\vartheta f, \Gamma_{q-1})$$
$$= \overline{\partial}_z \left[(f, \overline{\partial}_\zeta \Gamma_{q-1}) - \int_{b\Omega} \overline{\Gamma_{q-1}} \wedge *f \right]$$
$$= -\overline{\partial}_z \int_\Omega f(\zeta) \wedge B_{n,q-1}(\zeta,z) - \int_{b\Omega} \overline{\partial_z \Gamma_{q-1}} \wedge *f$$
$$= -\overline{\partial}_z \int_\Omega f(\zeta) \wedge B_{n,q-1}(\zeta,z) - \int_{b\Omega} \overline{\vartheta_\zeta \Gamma_q} \wedge *f.$$

Inserting this result into 4.5 we get the main result of this chapter:

Theorem 4.11 (Bochner-Martinelli-Koppelman formula) *If $f \in C^1_{0q}(\overline{\Omega})$ then for $z \in \Omega$*

$$f(z) = \int_{b\Omega} f(\zeta) \wedge B_{nq}(\zeta,z) - \int_\Omega \overline{\partial} f(\zeta) \wedge B_{nq}(\zeta,z) - \overline{\partial}_z \int_\Omega f(\zeta) \wedge B_{n,q-1}(\zeta,z).$$

We finally note several important properties of the BMK kernels:

Proposition 4.12   *i.* $\bar{\partial}_z B_{n,q-1}(\zeta,z) = (-1)^q \bar{\partial}_\zeta B_{nq}(\zeta,z)$

 ii. $\Box_\zeta B_{nq}(\zeta,z) = 0;\ \Box_z B_{nq}(\zeta,z) = 0$

 iii. For $z \notin \overline{\Omega}$,

$$0 = \int\limits_{b\Omega} f(\zeta) \wedge B_{nq}(\zeta,z) - \int\limits_{\Omega} \bar{\partial} f(\zeta) \wedge B_{nq}(\zeta,z) - \bar{\partial}_z \int\limits_{\Omega} f(\zeta) \wedge B_{n,q-1}(\zeta,z)$$

Property (*i*) follows immediately from 3.22; (*ii*) follows from the definitions and (*i*); (*iii*) is derived from (*i*) and Stokes' formula. – In particular, B_{n0} is $\bar{\partial}_\zeta$-closed, and the BM-projection

$$\int\limits_{b\Omega} f(\zeta) \wedge B_{nq}(\zeta,z)$$

is harmonic.

It is worthwhile to write down the 1-dimensional version of the BMK formula:

$$f(z) = \frac{1}{2\pi i} \int\limits_{b\Omega} \frac{f(\zeta)}{\zeta - z}\, d\zeta - \frac{1}{2\pi i} \int\limits_{\Omega} \frac{f_{\bar\zeta}(\zeta)}{\zeta - z}\, d\bar\zeta \wedge d\zeta, \tag{4.13}$$

which is the Cauchy-Pompeiu formula. Unlike in this classical case the Bochner-Martinelli kernel is not $\bar{\partial}_z$-closed in the higher dimensional case: the boundary integral in 4.11 is in general not a $\bar{\partial}_z$-closed differential form. We will return to these questions later.

§5   Types of Kernels and Regularity Properties

Let $\mathcal{K}(\zeta,z)$ be a double differential form of double type $(p,q;r,s)$ on a product domain $\Omega \times \Theta$ in $\mathbb{C}^n \times \mathbb{C}^n$. $\mathcal{K}$ defines a map

$$K : E_{pq}(\Omega) \longrightarrow E_{sr}(\Theta)$$

by the formula

$$Kf(z) = \int\limits_{\Omega} f(\zeta) \wedge *_\zeta \overline{\mathcal{K}(\zeta,z)}, \tag{5.1}$$

provided the integral exists; this imposes regularity conditions on f and $\mathcal{K}$. $\mathcal{K}$ is called the *kernel of the integral operator* K. Formula (5.1) is also written as

$$Kf = (f,\mathcal{K})_\Omega \tag{5.1'}$$

The formal adjoint of K is defined as the integral operator K^* with kernel

$$\mathcal{K}^*(\zeta,z) = \overline{\mathcal{K}(z,\zeta)}; \tag{5.2}$$

it maps $E_{sr}(\Theta)$ into $E_{pq}(\Omega)$.

The most important example of these integral operators is the *Bochner-Martinelli-Koppelman transform*

$$B_q f(z) = \int\limits_{\Omega} f(\zeta) \wedge B_{nq}(\zeta, z) \tag{5.3}$$

whose kernel is

$$(-1)^{q+1} * \overline{B_{nq}(\zeta, z)} \overset{\text{def}}{=} \mathcal{B}_{nq}(\zeta, z); \tag{5.4}$$

it carries forms of type $(0, q+1)$ into forms of type $(0, q)$ on Ω. — We shall permit us, occasionally, to call B_{nq} itself the kernel of B_q.

In all that follows we shall denote by Δ the diagonal of a product manifold $X \times X$, that is

$$\Delta = \{(x, x) : x \in X\};$$

its intersection with a subset $W \subset X \times X$ is the diagonal in W.

The regularity properties of the BMK transform are essentially governed by the behaviour of its kernel near the diagonal. As we have to deal with more general situations later we are now going to introduce a fairly wide class of kernels and operators containing the BMK kernel as an important special case.

Let ρ be a non-negative function on $\mathbb{C}^n \times \mathbb{C}^n$ which is smooth outside the diagonal, whose square ρ^2 is smooth everywhere, and which satisfies

$$\rho(\zeta, z) = \rho(z, \zeta) \tag{5.5}$$
$$c\|\zeta - z\| \leq \rho(\zeta, z) \leq C\|\zeta - z\|, \ 0 < c \leq C, \tag{5.6}$$

uniformly on compact sets in $\mathbb{C}^n \times \mathbb{C}^n$.

A double differential form $\mathcal{K}(\zeta, z)$ on an open set $W \subset \mathbb{C}^n \times \mathbb{C}^n$ will be simply denoted as $\mathcal{E}_l(\zeta, z)$:

$$\mathcal{K}(\zeta, z) = \mathcal{E}_l(\zeta, z), \quad l \in \mathbb{Z}, \tag{5.7}$$

if the following holds:

  i. *for $l \geq 0$, $\mathcal{K}$ is smooth on W and satisfies*

$$|\mathcal{K}(\zeta, z)| \leq \text{const} \cdot \rho^l \tag{5.8}$$

  locally near the diagonal Δ;

  ii. *for $l < 0$, $\mathcal{K}$ is smooth outside Δ and has a representation*

$$\mathcal{K}(\zeta, z) = \frac{\mathcal{E}_m(\zeta, z)}{\rho^{2t}}, \quad m \geq 0, m - 2t = l, \tag{5.9}$$

  locally near Δ.

$|K|$ stands for the absolute value of each coefficient of K.

Definition 5.10 *Forms $\mathcal{E}_l(\zeta, z)$ will be called isotropic kernels of order l and type $j = l + 2n$, the corresponding integral operators*

$$E_l f = (f, \mathcal{E}_l)$$

isotropic operators of the corresponding order resp. type.

A kernel of order l is a fortiori of order $l - 1$, so we should call these kernels more precisely of order $\geq l$. Note that (5.9) and (5.8) imply, for $l \in \mathbb{Z}$,

$$\begin{aligned}
|\mathcal{E}_l(\zeta, z)| &\leq \text{const} \cdot \rho^l, \\
|\mathcal{E}_l(\zeta, z)| &\leq \text{const} \cdot \|\zeta - z\|^l,
\end{aligned} \tag{5.11}$$

locally near points of Δ, and that derivatives of kernels of order l are of order $l - 1$:

$$D\mathcal{E}_l(\zeta, z) = \mathcal{E}_{l-1}(\zeta, z). \tag{5.12}$$

D stands for an arbitrary first order derivative with respect to ζ, $\bar{\zeta}$ or z, $\bar{z}$, applied to the coefficients of $\mathcal{E}_l$.

Taking $\rho = \|\zeta - z\|$, we see that

$$B_{nq}(\zeta, z) = \mathcal{E}_{1-2n}(\zeta, z);$$

the Bochner-Martinelli kernel is (uniformly) isotropic of type 1 (or order $1 - 2n$).

Kernels of type > 0 are locally integrable (in ζ for fixed z, and vice versa); we shall say more in the last paragraph of this chapter. Here we want to restrict attention to the Bochner-Martinelli kernel. From

$$|B_{nq}(\zeta, z)| \leq \text{const} \|\zeta - z\|^{1-2n}$$

we immediately deduce

Proposition 5.13 *Let Ω be a bounded domain in $\mathbb{C}^n$. The BMK transform*

$$B_q f(z) = \int\limits_{\Omega} f(\zeta) \wedge B_{nq}(\zeta, z)$$

is a bounded linear operator from

$$L^1_{0,q+1}(\Omega) \longrightarrow L^1_{0q}(\Omega).$$

A more precise result will be established at the end of this chapter.

Proposition 5.14 *Let* $f \in C_{0,q+1}^k(\Omega) \cap L_{0,q+1}^1(\Omega)$. *Then*

$$\boldsymbol{B}_q f \in C_{0,q}^k \cap L_{0q}^1(\Omega).$$

Proof Let D^z be a derivative with respect to z or $\overline{z}$, D^ς the corresponding derivative with respect to ς or $\overline{\varsigma}$. We consider a point $z_0 \in \Omega$, a neighbourhood U of z_0 with $U \subset\subset \Omega$, and choose a smooth patching function

$$\chi(\varsigma) = \begin{cases} 1 & \text{on } U \\ 0 \le \chi \le 1 & \text{everywhere} \end{cases}$$

with compact support in Ω. Then

$$\boldsymbol{B}_q f(z) = \boldsymbol{B}_q(\chi f)(z) + \boldsymbol{B}_q(1-\chi)f(z). \qquad (5.15)$$

The second form on the right hand side of (5.15) is clearly differentiable on U. So we can assume f with compact support in Ω, dropping χ. A translation

$$w = \varsigma - z$$

yields

$$\boldsymbol{B}_q f(z) = \int_\Omega f(\varsigma) \wedge B_{nq}(\varsigma, z) = \int_{\mathbb{C}^n} f(\varsigma) \wedge B_{nq}(\varsigma, z)$$

$$= \int_{\mathbb{C}^n} f(w+z) \wedge B_{nq}(w+z, z).$$

But $B_{nq}(w+z, z)$ is independent of z or $\overline{z}$; consequently

$$D^z \boldsymbol{B}_q f(z) = \int_{\mathbb{C}^n} D^z f(w+z) \wedge B_{nq}(w+z, z)$$

$$= \int_{\mathbb{C}^n} D^\varsigma f(\varsigma) \wedge B_{nq}(\varsigma, z)$$

$$= \int_\Omega D^\varsigma f(\varsigma) \wedge B_{nq}(\varsigma, z).$$

The case of higher order derivatives is treated by induction. $\qquad \square$

Note that the proof has established the commutator relation

$$\boldsymbol{B}_q D^\varsigma f = D^z \boldsymbol{B}_q f \qquad (5.16)$$

valid for smooth forms with compact support.

The analogue of 5.14 for arbitrary isotropic operators is true but more difficult to prove – see §10.

§6   Derivatives of the BMK Transform

1.   Let D be a first order derivative with respect to z or $\bar{z}$; we want to study the transform DB, that is the derivative of the Bochner-Martinelli-Koppelman transform. This will lead, among other things, to a new proof of the BMK formula.

So let $B = B_{n,q-1}$ be the Bochner-Martinelli-Koppelman kernel for $(0, q-1)$-forms. For $f \in C^1_{0q}(\Omega) \cap L^1_{0q}(\Omega)$ we set

$$F(z) = \int\limits_{\Omega} f(\zeta) \wedge B_{n,q-1}(\zeta, z) = \mathbf{B}f(z)$$

and prove

Theorem 6.1

$$\frac{\partial}{\partial z^k} \int\limits_{\Omega} f(\zeta) \wedge B_{n,q-1}(\zeta, z) = \text{p.v.} \int\limits_{\Omega} f(\zeta) \wedge \frac{\partial}{\partial z^k} B_{n,q-1}(\zeta, z)$$

$$\frac{\partial}{\partial \bar{z}^k} \int\limits_{\Omega} f(\zeta) \wedge B_{n,q-1}(\zeta, z) = -\frac{1}{n} f_k(z) + \text{p.v.} \int\limits_{\Omega} f(\zeta) \wedge \frac{\partial}{\partial \bar{z}^k} B_{n,q-1}(\zeta, z)$$

The principal value integrals are defined as

$$\text{p.v.} \int\limits_{\Omega} \ldots = \lim_{\varepsilon \to 0} \int\limits_{\|\zeta - z\| \geq \varepsilon} \ldots \quad ;$$

the $(0, q-1)$-form $f_k(z)$ is given by the equality

$$f(z) = d\bar{z}^k \wedge f_k(z) + f^k(z), \tag{6.2}$$

where $f_k(z)$ and $f^k(z)$ are $(0, q-1)$ resp. $(0, q)$-forms which do not contain $d\bar{z}^k$. Both formulae can be uniformly expressed as

$$D \int\limits_{\Omega} f(\zeta) \wedge B_{n,q-1}(\zeta, z) = -\frac{1}{n} D \lrcorner f(z) + \text{p.v.} \int\limits_{\Omega} f(\zeta) \wedge D B_{n,q-1}(\zeta, z) \tag{6.3}$$

because

$$\frac{\partial}{\partial z^k} \lrcorner f = 0.$$

Since we always have

$$\bar{\partial} f = \sum_{k=1}^{n} d\bar{z}^k \wedge \frac{\partial f}{\partial \bar{z}^k}$$

$$qf = \sum_{k=1}^{n} d\bar{z}^k \wedge f_k, \tag{6.4}$$

we deduce from 6.1 the

Corollary 6.5

$$\partial \int_\Omega f(\zeta) \wedge B_{n,q-1}(\zeta,z) = \text{p.v.} \int_\Omega f(\zeta) \wedge \partial_z B_{n,q-1}(\zeta,z)$$

$$\bar{\partial} \int_\Omega f(\zeta) \wedge B_{n,q-1}(\zeta,z) = -\frac{q}{n} f + \text{p.v.} \int_\Omega f(\zeta) \wedge \bar{\partial}_z B_{n,q-1}(\zeta,z)$$

In particular for $q = n$, because of

$$\bar{\partial}_z B_{n,n-1} = (-1)^n \bar{\partial}_\zeta B_{n,n} = 0,$$

$$\bar{\partial} \int_\Omega f(\zeta) \wedge B_{n,n-1}(\zeta,z) = -f(z).$$

This can of course also be derived from the BMK formula for $(0,n)$-forms, but it explains the dimension coefficient $-\frac{1}{n}$. The theorem is well known for $n = 1$ – cf. [Vek 62] –, it appears in a similar context in [Kyt 95].

2. **Proof** of Theorem 6.1. We assume at first f with compact support in Ω. Then (5.16) yields

$$D^z \int_\Omega f(\zeta) \wedge B(\zeta,z) = \int_\Omega D^\zeta f(\zeta) \wedge B(\zeta,z).$$

Now we have to distinguish the two cases $D = \frac{\partial}{\partial z^k}$ and $D = \frac{\partial}{\partial \bar{z}^k}$. The first case is easier:

3. The kernel $B = B_{n,q-1}$ can be written as

$$B = d\zeta^k \wedge B_k,$$

and so

$$J = \int_\Omega \frac{\partial f}{\partial \zeta^k} \wedge B = \int_\Omega \frac{\partial f}{\partial \zeta^k} \wedge d\zeta^k \wedge B_k.$$

Now

$$d(f \wedge B_k) = d\zeta^k \wedge \frac{\partial}{\partial \zeta^k}(f \wedge B_k)$$

$$= (-1)^q \left(\frac{\partial f}{\partial \zeta^k} \wedge B + f \wedge \frac{\partial B}{\partial \zeta^k} \right). \tag{6.6}$$

Consequently

$$J = \lim_{\varepsilon \to 0} \int_{\|\zeta - z\| \geq \varepsilon} \frac{\partial f}{\partial \zeta^k} \wedge B \tag{6.7}$$

$$= (-1)^q \lim_{\varepsilon \to 0} \int_{\|\zeta - z\| \geq \varepsilon} d(f \wedge B_k) - \text{p.v.} \int_\Omega f \wedge \frac{\partial B}{\partial \zeta^k}$$

$$\overset{\text{def}}{=} (-1)^q \lim_{\varepsilon \to 0} J_{1,\varepsilon} + \text{p.v.} \int_\Omega f \wedge \frac{\partial B}{\partial z^k}$$

To determine the limit in (6.7) we compute

$$J_{1,\varepsilon} = - \int\limits_{\|\zeta-z\|=\varepsilon} f(\zeta) \wedge B_k(\zeta, z) \tag{6.8}$$

$$= - \int\limits_{\|\zeta-z\|=\varepsilon} [f(\zeta) - f(z)] \wedge B_k(\zeta, z) - \int\limits_{\|\zeta-z\|=\varepsilon} f(z) \wedge B_k(\zeta, z),$$

where

$$f(z) = f(\zeta)\big|_{\zeta=z} = \sum_{|J|=q} f_J(z) d\bar{\zeta}^J.$$

The first integral in (6.8) is $O(\varepsilon)$, thus:

$$\lim_{\varepsilon \to 0} J_{1,\varepsilon} = - \lim_{\varepsilon \to 0} \int\limits_{\|\zeta-z\|=\varepsilon} f(z) \wedge B_k(\zeta, z). \tag{6.9}$$

Using again the notation

$$B(\zeta, z) = \frac{1}{\|\zeta - z\|^{2n}} \widetilde{B}(\zeta, z)$$

$$B_k(\zeta, z) = \frac{1}{\|\zeta - z\|^{2n}} \widetilde{B}_k(\zeta, z),$$

we get

$$\int\limits_{\|\zeta-z\|=\varepsilon} f(z) \wedge B_k(\zeta, z) = \frac{1}{\varepsilon^{2n}} \int\limits_{\|\zeta-z\|=\varepsilon} f(z) \wedge \widetilde{B}_k(\zeta, z) \tag{6.10}$$

$$= \frac{1}{\varepsilon^{2n}} \int\limits_{\|\zeta-z\|\le\varepsilon} d_\zeta(f(z) \wedge \widetilde{B}_k(\zeta, z))$$

$$= \frac{1}{\varepsilon^{2n}} \int\limits_{\|\zeta-z\|\le\varepsilon} d\zeta^k \wedge f(z) \wedge \frac{\partial}{\partial \zeta^k} \widetilde{B}_k(\zeta, z)$$

$$= 0,$$

because

$$\frac{\partial}{\partial \zeta^k} \widetilde{B}_k(\zeta, z) = 0.$$

(6.7), (6.8) and (6.10) give the first part of Theorem 6.1 in the case of compact support.

4.   The case $D^z = \frac{\partial}{\partial \bar{z}^k}$ requires more work. In order to study

$$J = \int\limits_\Omega \frac{\partial f}{\partial \bar{\zeta}^k}(\zeta) \wedge B(\zeta, z),$$

we decompose f and B as in (6.2):

$$f(\zeta) = d\bar{\zeta}^k \wedge f_k(\zeta) + f^k(\zeta)$$
$$B(\zeta, z) = d\bar{\zeta}^k \wedge B_k(\zeta, z) + B^k(\zeta, z)$$

$$(6.11)$$

and obtain

$$\frac{\partial f}{\partial \bar{\zeta}^k} \wedge B = d\bar{\zeta}^k \wedge \frac{\partial f_k}{\partial \bar{\zeta}^k} \wedge B^k + \frac{\partial f^k}{\partial \bar{\zeta}^k} \wedge d\bar{\zeta}^k \wedge B_k,$$

since $f^k \wedge B^k = 0$. So

$$J = \int_\Omega d\bar{\zeta}^k \wedge \frac{\partial f_k}{\partial \bar{\zeta}^k} \wedge B^k + \int_\Omega \frac{\partial f^k}{\partial \bar{\zeta}^k} \wedge d\bar{\zeta}^k \wedge B_k \qquad (6.12)$$

$$\stackrel{\text{def}}{=} J_1 + J_2.$$

J_1 and J_2 will be treated separately.

5.   For J_2 we use the relation

$$d(f^k \wedge B_k) = d\bar{\zeta}^k \wedge \frac{\partial}{\partial \bar{\zeta}^k}(f^k \wedge B_k)$$

$$= d\bar{\zeta}^k \wedge \frac{\partial f^k}{\partial \bar{\zeta}^k} \wedge B_k + d\bar{\zeta}^k \wedge f^k \wedge \frac{\partial B_k}{\partial \bar{\zeta}^k},$$

this gives

$$J_2 = \lim_{\varepsilon \to 0} \int_{\|\zeta - z\| \geq \varepsilon} \frac{\partial f^k}{\partial \bar{\zeta}^k} \wedge d\bar{\zeta}^k \wedge B_k \stackrel{\text{def}}{=} \lim_{\varepsilon \to 0} J_{2,\varepsilon}; \qquad (6.13)$$

$$J_{2,\varepsilon} = (-1)^q \int_{\|\zeta - z\| \geq \varepsilon} d(f^k \wedge B_k) - \int_{\|\zeta - z\| \geq \varepsilon} f^k \wedge d\bar{\zeta}^k \wedge \frac{\partial B_k}{\partial \bar{\zeta}^k}. \qquad (6.14)$$

The first integral in (6.14) is

$$\int_{\|\zeta - z\| \geq \varepsilon} d(f^k \wedge B_k) = - \int_{\|\zeta - z\| = \varepsilon} f^k \wedge B_k \qquad (6.15)$$

$$= - \int_{\|\zeta - z\| = \varepsilon} [f^k(\zeta) - f^k(z)] \wedge B_k(\zeta, z)$$

$$- \int_{\|\zeta - z\| = \varepsilon} f^k(z) \wedge B_k(\zeta, z)$$

$$= O(\varepsilon) - \int_{\|\zeta - z\| = \varepsilon} f^k(z) \wedge B_k(\zeta, z)$$

$$= O(\varepsilon) - J_{3,\varepsilon}.$$

Let us write

$$B_k = \frac{1}{\|\zeta - z\|^{2n}} \sum_{l \neq k} \left(\bar{\zeta}^l - \bar{z}^l \right) b_{kl},$$

where the b_{kl} are forms with constant coefficients which do not contain $d\bar{\zeta}^k$. Consequently

$$f^k(z) \wedge B_k(\zeta, z) = \frac{1}{\|\zeta - z\|^{2n}} \sum_{l \neq k} \left(\bar{\zeta}^l - \bar{z}^l \right) a_{kl}, \tag{6.16}$$

where now the a_{kl} are constant coefficient forms of type $(n, n-1)$ not containing $d\bar{\zeta}^k$. The sum in (6.16) is d_ζ-exact, and we obtain $J_{3,\varepsilon} = 0$.

Inserting this information into (6.13) we get

$$J_2 = \lim_{\varepsilon \to 0} \left(- \int\limits_{\|\zeta - z\| \geq \varepsilon} f^k(\zeta) \wedge d\bar{\zeta}^k \wedge \frac{\partial B_k}{\partial \bar{\zeta}^k} \right). \tag{6.17}$$

But we can still add

$$0 = \int\limits_{\|\zeta - z\| \geq \varepsilon} f^k \wedge \frac{\partial B^k}{\partial \bar{\zeta}^k}$$

to (6.17) to obtain our final result for J_2:

$$J_2 = - \text{p.v.} \int\limits_{\Omega} f^k(\zeta) \wedge \frac{\partial B}{\partial \bar{\zeta}^k} \tag{6.18}$$

$$= \text{p.v.} \int\limits_{\Omega} f^k(\zeta) \wedge \frac{\partial B}{\partial \bar{z}^k}$$

6.   Let us now turn to J_1 in (6.12). We note

$$d(f_k \wedge B^k) = d\bar{\zeta}^k \wedge \frac{\partial f_k}{\partial \bar{\zeta}^k} \wedge B^k + d\bar{\zeta}^k \wedge f_k \wedge \frac{\partial B^k}{\partial \bar{\zeta}^k},$$

and therefore

$$J_1 = \lim_{\varepsilon \to 0} \int\limits_{\|\zeta - z\| \geq \varepsilon} d(f_k \wedge B^k) - \lim_{\varepsilon \to 0} \int\limits_{\|\zeta - z\| \geq \varepsilon} d\bar{\zeta}^k \wedge f_k \wedge \frac{\partial B^k}{\partial \bar{\zeta}^k} \tag{6.19}$$

$$= \lim_{\varepsilon \to 0} \int\limits_{\|\zeta - z\| \geq \varepsilon} d(f_k \wedge B^k) + \text{p.v.} \int\limits_{\Omega} d\bar{\zeta}^k \wedge f_k \wedge \frac{\partial B^k}{\partial \bar{z}^k}.$$

The principal value integral in (6.19) can be written as

$$\text{p.v.} \int\limits_{\Omega} d\bar{\zeta}^k \wedge f_k \wedge \frac{\partial B}{\partial \bar{z}^k}. \tag{6.20}$$

(The difference between the two terms is 0.) Adding (6.19) and (6.18) we arrive at

$$J = \lim_{\varepsilon \to 0} \int\limits_{\|\zeta - z\| \geq \varepsilon} d(f_k \wedge B^k) + \text{p.v.} \int\limits_{\Omega} f \wedge \frac{\partial B}{\partial \bar{z}^k}. \tag{6.21}$$

7.   Remains to compute the limit in (6.21). We proceed as before:

$$J(\varepsilon) = \int\limits_{\|\zeta - z\| \geq \varepsilon} d(f_k \wedge B^k) = - \int\limits_{\|\zeta - z\| = \varepsilon} f_k \wedge B^k \tag{6.22}$$

$$= - \int\limits_{\|\zeta - z\| = \varepsilon} [f_k(\zeta) - f_k(z)] \wedge B^k - \int\limits_{\|\zeta - z\| = \varepsilon} f_k(z) \wedge B^k(\zeta, z)$$

$$= O(\varepsilon) - \int\limits_{\|\zeta - z\| = \varepsilon} f_k(z) \wedge B^k(\zeta, z).$$

Call the last integral $J_1(\varepsilon)$. Then as in (6.10)

$$J_1(\varepsilon) = \frac{1}{\varepsilon^{2n}} \int\limits_{\|\zeta - z\| = \varepsilon} f_k(z) \wedge \widetilde{B}^k(\zeta, z) \tag{6.23}$$

$$= \frac{1}{\varepsilon^{2n}} \int\limits_{\|\zeta - z\| \leq \varepsilon} d(f_k(z) \wedge \widetilde{B}^k(\zeta, z))$$

$$= \frac{1}{\varepsilon^{2n}} \int\limits_{\|\zeta - z\| \leq \varepsilon} d\bar{\zeta}^k \wedge f_k(z) \wedge \frac{\partial \widetilde{B}^k}{\partial \bar{\zeta}^k}(\zeta, z).$$

$\widetilde{B}^k$ can be determined from (6.11) and the definition of B and $\widetilde{B}$. The result:

$$\widetilde{B}^k = \gamma \sideset{}{'}\sum_{\substack{l=1\ldots n \\ |K|=n-q \\ |L|=q-1 \\ k \notin K}} \left(\bar{\zeta}^l - \bar{z}^l \right) d\zeta^k \wedge (d\bar{\zeta} \wedge d\zeta)^K \wedge (d\bar{z} \wedge d\zeta)^L \tag{6.24}$$

with

$$\gamma = (-1)^{(q-1)(q-2)/2 + q - 1} \binom{n-1}{q-1} (n-q)!(q-1)!(2\pi i)^{-n}. \tag{6.25}$$

This means that we have

$$\frac{\partial \widetilde{B}^k}{\partial \bar{\zeta}^k} = \gamma d\zeta^k \wedge \sideset{}{'}\sum_{|K|=n-q} (d\bar{\zeta} \wedge d\zeta)^K \wedge \sum_{|L|=q-1} (d\bar{z} \wedge d\zeta)^L. \tag{6.26}$$

If we introduce

$$f_k(z) = \sideset{}{'}\sum_{|J|=q-1} f_{k,J}(z) \, d\bar{\zeta}^J,$$

we have

$$J_1(\varepsilon) = \frac{1}{\varepsilon^{2n}}\gamma(-1)^{q-1} \int\limits_{\|\zeta-z\|\leq\varepsilon} \sum f_{k,J}\, d\bar{\zeta}^J \wedge$$

$$\wedge d\bar{\zeta}^k \wedge d\zeta^k \wedge (d\bar{\zeta}\wedge d\zeta)^K \wedge (d\bar{z}\wedge d\zeta)^L, \quad (6.27)$$

where the summation is extended over strictly increasing multiindices J, K, L of length $q-1, n-q, q-1$, respectively. The only non-zero terms satisfy

$$k\cup K\cup L = N = k\cup K\cup J, \tag{6.28}$$

thus $L = J$, and (6.27) simplifies to

$$J_1(\varepsilon) = \frac{1}{\varepsilon^{2n}}\gamma(-1)^{q-1} \sum_{\substack{|L|=q-1\\|K|=n-q\\k\cup K\cup L=N}} \int\limits_{\|\zeta-z\|\leq\varepsilon} f_{k,L}\, d\bar{\zeta}^L \wedge d\bar{\zeta}^k \wedge$$

$$\wedge d\zeta^k \wedge (d\bar{\zeta}\wedge d\zeta)^K \wedge d\zeta^L \wedge d\bar{z}^L \quad (6.29)$$

$$= \frac{1}{\varepsilon^{2n}}\gamma(-1)^{q-1+(q-1)(q-2)/2} f_k(z)\cdot J_0,$$

$$J_0 = \int\limits_{\|\zeta-z\|\leq\varepsilon} d\bar{\zeta}^k \wedge d\zeta^k \wedge (d\bar{\zeta}\wedge d\zeta)^K \wedge (d\bar{\zeta}\wedge d\zeta)^L, \tag{6.30}$$

where now

$$f_k(z) = \sum_{|L|=q-1} f_{k,L}(z)\, d\bar{z}^L.$$

But

$$J_0 = (2i)^n \varepsilon^{2n}\frac{\pi^n}{n!}; \tag{6.31}$$

plugging this into (6.29) we get

$$J_1(\varepsilon) = -\frac{1}{n}f_k(z) \tag{6.32}$$

and finally from (6.21)

$$\lim_{\varepsilon\to 0} J(\varepsilon) = -\frac{1}{n}f_k(z)$$

$$J = -\frac{1}{n}f_k(z) + \text{p.v.} \int f\wedge\frac{\partial B}{\partial\bar{z}^k}.$$

8.   The theorem has now been proved for forms with compact support. The passage to arbitrary support is easy: decompose

$$f(\zeta) = \chi(\zeta)f(\zeta) + (1 - \chi(\zeta))f(\zeta), \tag{6.33}$$

where χ is 1 in $\Omega_0 \subset\subset \Omega$, has compact support in Ω, with $0 \le \chi \le 1$, and is smooth. The BM transform of the second term in (6.33) can be differentiated under the integral (for $z \in \Omega_0$), and the first term has compact support. Adding up gives the required formulae on Ω_0; since Ω_0 was arbitrary the claim is true for all $z \in \Omega$. (cf. the corresponding argument in §5)

9.   Theorem 6.1 can be combined with a second limit theorem to give an alternative proof of the BMK formula, independent of potential theory. We first need

Proposition 6.34 *Let f be a differentiable $(0, q)$-form and B_{nq} the BMK kernel for $(0, q)$-forms. Then*

$$\lim_{\varepsilon \to 0} \int_{\|\zeta - z\| = \varepsilon} f(\zeta) \wedge B_{nq}(\zeta, z) = \left(1 - \frac{q}{n}\right) f(z).$$

Proof   Because of the type of singularity of B_{nq} it is clear that $f(\zeta)$ can be replaced by a constant form; so let

$$f(\zeta) = d\bar{\zeta}^J$$

$$B_{nq} = \frac{1}{\|\zeta - z\|^{2n}} \widetilde{B}_{nq} \tag{6.35}$$

$$\widetilde{B}_{nq} = \gamma \sum_{\substack{j=1\ldots n \\ |K|=n-q-1 \\ |L|=q}}{}' (\bar{\zeta}^j - \bar{z}^j)\, d\zeta^j \wedge (d\bar{\zeta} \wedge d\zeta)^K \wedge d\zeta^L \wedge d\bar{z}^L \tag{6.36}$$

$$\gamma = (-1)^{q(q+1)/2} \binom{n-1}{q} q!(n-q-1)!(2\pi i)^{-n} \tag{6.37}$$

we substitute (6.35) – (6.36) into 6.34 and use Stokes' formula to obtain

$$\lim_{\varepsilon \to 0} \int_{\|\zeta - z\| = \varepsilon} f(\zeta) \wedge B_{nq}(\zeta, z) \tag{6.38}$$

$$= (-1)^q \gamma \frac{1}{\varepsilon^{2n}} \int_{\|\zeta - z\| \le \varepsilon} \sum_{\substack{j=1\ldots n \\ |K|=n-q-1 \\ |L|=q}}{}' d\bar{\zeta}^j \wedge d\zeta^j \wedge d\bar{\zeta}^J \wedge (d\bar{\zeta} \wedge d\zeta)^K \wedge d\zeta^L \wedge d\bar{z}^L$$

We have necessarily $J = L$ in the last expression, so (6.38) becomes

$$(-1)^{q+\frac{q}{2}(q-1)} \gamma \frac{1}{\varepsilon^{2n}} \sum_{\substack{j=1\ldots n \\ |K|=n-q-1}}{}' \int_{\|\zeta - z\| \le \varepsilon} d\bar{\zeta}^j \wedge d\zeta^j \wedge (d\bar{\zeta} \wedge d\zeta)^J \wedge (d\bar{\zeta} \wedge d\zeta)^K \wedge d\bar{z}^J \tag{6.39}$$

There are exactly $n - q$ possibilities for j and K to form the complement of J. Then (6.39) can be written as

$$(-1)^{\frac{q}{2}(q+1)}\gamma(n-q)\frac{1}{\varepsilon^{2n}}\int\limits_{\|\zeta-z\|\leq\varepsilon}(d\bar{\zeta}\wedge d\zeta)^N\cdot f(z),$$

which is in view of (6.31) and (6.37):

$$(-1)^{\frac{q}{2}(q+1)}(-1)^{\frac{q}{2}(q+1)}\binom{n-1}{q}q!(n-q-1)!(2\pi i)^{-n}(n-q)\frac{1}{\varepsilon^{2n}}(2i)^n\varepsilon^{2n}\frac{\pi^n}{n!}d\bar{z}^J$$

$$=\frac{n-q}{n}d\bar{z}^J.$$

$\square$

10. Now suppose f is a form in $C^1_{0,q}(\overline{\Omega})$. For $z\in\Omega$ we consider

$$\int\limits_{b\Omega}f(\zeta)\wedge B_{nq}(\zeta,z)-\int\limits_{\|\zeta-z\|=\varepsilon}f(\zeta)\wedge B_{nq}(\zeta,z). \tag{6.40}$$

This becomes, for $\varepsilon\to 0$,

$$\int\limits_{b\Omega}f(\zeta)\wedge B_{nq}(\zeta,z)-\left(1-\frac{q}{n}\right)f(z) \tag{6.41}$$

in view of 6.34. On the other hand, Stokes' theorem transforms (6.40) into

$$\int\limits_{\Omega\cap\{\|\zeta-z\|\leq\varepsilon\}}\bar{\partial}[\,f(\zeta)\wedge B_{nq}(\zeta,z)]$$

$$=\int\limits_{\Omega\cap\{\|\zeta-z\|\leq\varepsilon\}}\bar{\partial}f(\zeta)\wedge B_{nq}(\zeta,q)+(-1)^q\int\limits_{\Omega\cap\{\|\zeta-z\|\leq\varepsilon\}}f(\zeta)\wedge\bar{\partial}_\zeta B_{nq}(\zeta,z). \tag{6.42}$$

But – by 4.12 –

$$(-1)^q\bar{\partial}_\zeta B_{nq}=\bar{\partial}_z B_{n,q-1},$$

so (6.42) becomes

$$\int\limits_{\Omega\cap\{\|\zeta-z\|\leq\varepsilon\}}\bar{\partial}f\wedge B_{nq}(\zeta,z)+\int\limits_{\Omega\cap\{\|\zeta-z\|\leq\varepsilon\}}f(\zeta)\wedge\bar{\partial}_z B_{n,q-1}(\zeta,z). \tag{6.43}$$

Let now $\varepsilon\to 0$ and apply Corollary 6.5; this transforms (6.43) into

$$\int\limits_{\Omega}\bar{\partial}f(\zeta)\wedge B_{nq}(\zeta,z)+\bar{\partial}_z\int\limits_{\Omega}f(\zeta)\wedge B_{n,q-1}(\zeta,z)+\frac{q}{n}f(z) \tag{6.44}$$

Equating (6.41) and (6.44) yields the BMK formula

$$f(z)=\int\limits_{b\Omega}f(\zeta)\wedge B_{nq}(\zeta,z)-\int\limits_{\Omega}\bar{\partial}f(\zeta)\wedge B_{nq}(\zeta,z)-\bar{\partial}_z\int\limits_{\Omega}f(\zeta)\wedge B_{n,q-1}(\zeta,z).$$

11. We finally analyse the principal value integrals of 6.1. The BMK transform

$$B_{q-1}f = \int_\Omega f(\zeta) \wedge B_{n,q-1}(\zeta, z)$$

decomposes into a sum of integrals

$$\text{const} \cdot \int_\Omega g(\zeta) \frac{\overline{\zeta}^j - \overline{z}^j}{\|\zeta - z\|^{2n}} \, dV(\zeta);$$

so we have to consider the principal value integral transforms with kernels

$$\frac{\partial}{\partial z^k} \frac{\overline{\zeta}^j - \overline{z}^j}{\|\zeta - z\|^{2n}} \tag{6.45}$$

and

$$\frac{\partial}{\partial \overline{z}^k} \frac{\overline{\zeta}^j - \overline{z}^j}{\|\zeta - z\|^{2n}}. \tag{6.46}$$

We may set $z = 0$ and introduce real coordinates $\zeta^j = \xi^j + i\eta^j$. Then (6.45) becomes

$$n \frac{1}{\|\zeta\|^{2n}} \left[\frac{\xi^j \xi^k - \eta^j \eta^k}{\|\zeta\|^2} - i \frac{\xi^j \eta^k + \xi^k \eta^j}{\|\zeta\|^2} \right], \tag{6.47}$$

and from (6.46) we get

$$\frac{1}{\|\zeta\|^{2n}} \left[\frac{-\delta_{jk}\|\zeta\|^2 + n(\xi^j \xi^k + \eta^j \eta^k)}{\|\zeta\|^2} + i \frac{n(\xi^j \eta^k + \xi^k \eta^j)}{\|\zeta\|^2} \right]. \tag{6.48}$$

All these kernels are Riesz kernels of order 2 in the sense of E. Stein [Ste 70]. We state this as

Proposition 6.49 *The transformations*

$$\text{p.v.}(D^z B_{q-1})f(z) = \text{p.v.} \int_\Omega f(\zeta) \wedge D^z B_{n,q-1}(\zeta, z)$$

are Riesz transforms of second order.

§7 Applications of the BMK Formula

1. Traditionally the Bochner-Martinelli formula is used to solve certain extension problems in complex analysis, either directly or by first solving the Cauchy-Riemann equations

with compact support. We are going to describe these classical applications in this paragraph. We want to stress, however, that the most important application of the BMK formula is far more indirect: it serves as starting point for the construction of integral formulae which are adapted to special manifolds, e. g.convex or strictly pseudoconvex domains. The passage from the BMK formula to these new formulae requires the calculus of Cauchy-Fantappiè forms which will be developed in the next chapter; the connection between a fundamental solution of the Laplacian and the BMK formula, as developed in the preceding paragraphs, is crucial for the use of these formulae on complex manifolds.

Let us first state the most important special case of the BMK formula:

Theorem 7.1 *If f is holomorphic on the closure $\overline{\Omega}$ of a bounded domain with piecewise smooth boundary, then*

$$f(z) = \int_{b\Omega} f(\zeta) \wedge B_{n0}(\zeta, z)$$

for all $z \in \Omega$.

For the next application we need

Lemma 7.2 *Let S be a closed hypersurface in $\mathbb{C}^n$ and f a holomorphic function in a neighbourhood U of S. Then the function*

$$F(z) = \int_S f(\zeta) \wedge B_{n0}(\zeta, z)$$

is holomorphic in $z \notin S$.

In fact,

$$\overline{\partial}_z F(z) = \int_S f(\zeta) \overline{\partial}_z B_{n0}(\zeta, z)$$

$$= -\int_S f(\zeta) \overline{\partial}_\zeta B_{n1}(\zeta, z)$$

$$= -\int_S d_\zeta (f(\zeta) B_{n1}(\zeta, z))$$

$$= 0.$$

Theorem 7.3 (The Hartogs-Bochner Kugelsatz) *Let $\Omega \subset\subset \mathbb{C}^n$, $n > 1$, be a domain with connected boundary $b\Omega$. If f is a function holomorphic in a neighbourhood U of $b\Omega$, then f can be holomorphically extended to $\overline{\Omega}$.*

Proof Choose domains Ω_1, Ω_2 with connected piecewise smooth boundaries such that

$$\Omega_1 \subset\subset \Omega \subset\subset \Omega_2, \quad \Omega_2 - \overline{\Omega}_1 \subset\subset U.$$

For $z \in \Omega_2 - \overline{\Omega}_1$, the BMK formula gives

$$f(z) = \int\limits_{b\Omega_2} f(\zeta) B_{n0}(\zeta, z) - \int\limits_{b\Omega_1} f(\zeta) B_{n0}(\zeta, z)$$

$$\overset{\text{def}}{=} f_2(z) - f_1(z).$$

The functions f_1 and f_2 are holomorphic outside $b\Omega_1$ resp. $b\Omega_2$. We have, moreover,

$$\lim_{z \to \infty} f_1(z) = 0; \tag{7.4}$$

if $|c|$ is sufficiently large, the function of $n - 1$ variables

$$\widetilde{f}_1(z_1, \ldots, z_{n-1}) = f_1(z_1, \ldots, z_{n-1}, c)$$

is a bounded entire function, therefore $\equiv 0$ because of (7.4). This shows $f_1 \equiv 0$ on the unbounded component of $\mathbb{C}^n - b\Omega_1$, in particular on $\Omega_2 - \overline{\Omega}_1$. – Since, on the other hand, $f_2(z)$ is holomorphic in Ω_2, the required extension is given by f_2. $\qquad\square$

Theorem 7.5 (Cauchy-Riemann equations with compact support) *Let $n > 1$, f a C^1-smooth $\overline{\partial}$-closed $(0,1)$-form in $\mathbb{C}^n$ with compact support. Then there is a C^1-function u with compact support and*

$$\overline{\partial} u = f.$$

More precisely,

$$\text{supp}\, u \subset \text{supp}\, f \cup \Omega,$$

where Ω is the union of the bounded components of $\mathbb{C}^n - \text{supp}\, f$.

Proof The BMK formula for $(0,1)$-forms can be applied to a sufficiently large ball (containing the support of f) and gives

$$f(z) = \overline{\partial}_z \int\limits_{\mathbb{C}^n} f(\zeta) \wedge B_{n0}(\zeta, z)$$

$$\overset{\text{def}}{=} \overline{\partial} u(z).$$

u is holomorphic outside the support of f and vanishes at ∞ (as in the previous proof). From there the statement on the support of u follows as before. $\qquad\square$

The theorem does not hold for $n = 1$. It can in fact be shown that Theorem 7.5 implies Theorem 7.3 which is clearly false in the complex plane.

If one does not insist on compactly supported solutions, however, the Cauchy-Riemann system can always be solved for compactly supported data by the BMK formula.

Proposition 7.6 *Let $f \in C^1_{0q;c}(\mathbb{C}^n)$ and $\overline{\partial} f = 0$. Then*

$$f(z) = \overline{\partial} u(z) = \overline{\partial}_z \int_{\mathbb{C}^n} f(\zeta) \wedge B_{n,q-1}(\zeta, z).$$

(Moreover, if f is C^k-smooth, so is u).

2.   As we have repeatedly mentioned, the kernel $B_{n0}(\zeta, z)$ is not holomorphic in z; this means that the function

$$F(z) = \int_{b\Omega} f(\zeta) B_{n0}(\zeta, z)$$

need not be holomorphic if f is a continuous function on $b\Omega$. We can all the same characterise holomorphic functions by the BM integral:

Proposition 7.7 *Let f be a C^2-function on $\overline{\Omega}$ such that for all $z \in \Omega$:*

$$f(z) = \int_{b\Omega} f(\zeta) B_{n0}(\zeta, z).$$

Then f is holomorphic.

Proof i) Set $B_{n0} = - * \partial_\zeta \Gamma$. By the BMK formula 4.5 resp. 4.11 the assumption yields

$$(\overline{\partial} f, \overline{\partial} \Gamma) \equiv 0$$

on Ω. Moreover, the function

$$z \mapsto (\overline{\partial} f, \overline{\partial} \Gamma)(z) = - \int_\Omega f(\zeta) \wedge B_{n0}(\zeta, z)$$

is continuous in $\mathbb{C}^n$, harmonic outside $\overline{\Omega}$ and vanishes at ∞. Since its values on $b\Omega$ are 0 it vanishes everywhere:

$$(\overline{\partial} f, \overline{\partial} \Gamma) \equiv 0 \qquad \text{on } \mathbb{C}^n. \tag{7.8}$$

ii) We apply Green's formula (2.27) to (7.8):

$$(\overline{\partial} f, \overline{\partial} \Gamma) = (\vartheta \overline{\partial} f, \Gamma) + \int_{b\Omega} \overline{\Gamma} \wedge *\overline{\partial} f;$$

the parameter z can be anywhere in $\mathbb{C}^n$ since the singularity of Γ is sufficiently weak. Thus we get – since f is harmonic –

$$\int_{b\Omega} \overline{\Gamma}(\zeta, z) \wedge *\overline{\partial}f(\zeta) \equiv 0 \tag{7.9}$$

on $\mathbb{C}^n$. We now need

Lemma 7.10 *Let K be a compact hypersurface with surface measure $d\sigma$ (with or without boundary) in $\mathbb{R}^n$, $n \geq 3$. Let f be a continuous function on K such that*

$$F(y) = \int_K \frac{f(x)}{\|x - y\|^{n-2}} \, d\sigma(x)$$

vanishes on $\mathbb{R}^n - K$. Then $f \equiv 0$ on K.

Proof Choose an arbitrary C^2-function g with compact support in $\mathbb{R}^n$. Then

$$g(y) = \int_{\mathbb{R}^n} \Delta g(x)G(x, y) \, dV(x),$$

where $G(x, y)$ is a fundamental solution to the real Laplacian Δ:

$$G(x, y) = \frac{\text{const}}{\|x - y\|^{n-2}}.$$

Then by Fubini's theorem:

$$\int_K f(x)g(x)d\sigma(x) = \int_K f(x) \int_{\mathbb{R}^n} \Delta g(y)G(y, x) \, dV(y) \, d\sigma(x)$$

$$= \int_{\mathbb{R}^n} \Delta g(y) \int_K f(x)G(y, x) \, d\sigma(x) \, dV(y)$$

$$= 0.$$

This means that f vanishes $d\sigma$-almost-everywhere; since it is continuous it vanishes on all of K.   □

iii) We can now finish the proof of 7.7. From (7.9) and 7.10 we obtain

$$*\overline{\partial}f = 0 \qquad \text{on } b\Omega. \tag{7.11}$$

But

$$(\overline{\partial}f, \overline{\partial}f) = (\vartheta\overline{\partial}f, f) + \int_{b\Omega} \overline{f} \wedge *\overline{\partial}f;$$

both terms on the right side are zero since f – given by its BM projection – is harmonic, and in view of (7.11). This proves $\overline{\partial}f = 0$ as required.   □

§8   Cauchy-Riemann Functions

The Hartogs-Bochner Kugelsatz can be considerably generalised: the function to be holomorphically extended need only be defined on the boundary of Ω and satisfy a certain system of differential equations, the *tangential Cauchy-Riemann system*. We need some preparation.

1.   Let S be a C^1-hypersurface in some domain $U \subset \mathbb{C}^n$ given by a C^1-function r with non-vanishing gradient:

$$S = \{z \in U : r(z) = 0\}; \quad dr(z) \neq 0 \text{ on } U.$$

Among the holomorphic vector fields

$$L = \sum_{j=1}^{n} a_j(z) \frac{\partial}{\partial z^j}$$

with continuous coefficients a_j on S we single out the *tangential holomorphic fields* by the condition

$$\partial r[L] = Lr = \sum_{j=1}^{n} a_j \frac{\partial r}{\partial z^j} = 0. \tag{8.1}$$

They form a module over the continuous functions on S which is locally freely generated by $n - 1$ elements; a system of generators on all of S consists of

$$L_{jk} = \frac{\partial r}{\partial z^j} \frac{\partial}{\partial z^k} - \frac{\partial r}{\partial z^k} \frac{\partial}{\partial z^j}, \quad 1 \leq j < k \leq n \tag{8.2}$$

Near each point a of S we can choose $n - 1$ among these vector fields as generators, say $L_1, \ldots, L_{n-1}$; moreover, we can find a holomorphic non-tangential vector field L_n with

$$L_n r \equiv 1 \quad \text{on } S \tag{8.3}$$

such that $L_1, \ldots, L_n$ generate the module of all holomorphic vector fields on S in a neighbourhood of a.

All the above fields are naturally defined in an open neighbourhood of S.

A differentiable function f on S is the restriction of a C^1-function $\widetilde{f}$ on U to S; for each tangential (holomorphic or not holomorphic) field L the values $L\widetilde{f}$ on S only depend on f; so

$$Lf = L\widetilde{f} \tag{8.4}$$

is well defined.

Definition 8.5   *A C^1-function f on S is called a* CR-*function (i. e. Cauchy-Riemann function) if it satisfies the tangential Cauchy-Riemann equations*

$$\overline{L}f = 0 \tag{8.6}$$

for each holomorphic tangential vector field L to S.

$\overline{L}$ is the complex conjugate of L.

Clearly (8.6) is equivalent to the system

$$\overline{L}_{jk}f = 0, \tag{8.7}$$

with L_{jk} given in (8.2).

If $\widetilde{f}$ is a C^1-extension of f to U, then (8.7) is equivalent to

$$\overline{\partial}r \wedge \overline{\partial}\widetilde{f} = 0 \quad \text{on } S. \tag{8.8}$$

This condition therefore only depends on f, and we write it simply as

$$\overline{\partial}r \wedge \overline{\partial}f = 0 \quad \text{on } S. \tag{8.8'}$$

In particular, the restriction of a holomorphic function on U to S is a CR-function. Slightly more generally we note

Proposition 8.9 *Let $\Omega \subset\subset \mathbb{C}^n$ be a domain with smooth boundary $b\Omega$ of class C^1 and f a C^1-function on $\overline{\Omega}$ which is holomorphic on Ω. Then $f|_{b\Omega}$ is a CR-function.*

Let us write the CR condition in yet another way. If f (with C^1-extension $\widetilde{f}$) satisfies (8.8) and g is a C^∞-smooth form of type $(n, n-2)$ with compact support we have

$$\int_S f \wedge \overline{\partial}g = \int_S \overline{\partial}(\widetilde{f}g) - \int_S \overline{\partial}\widetilde{f} \wedge g. \tag{8.10}$$

The first integral on the right-hand side is 0 because g has compact support (and $\overline{\partial} = d$ for reasons of type). Now (8.8) means

$$\overline{\partial}\widetilde{f} = a\overline{\partial}r + b \tag{8.11}$$

where a and b are continuous and b vanishes on S. So

$$\int_S \overline{\partial}\widetilde{f} \wedge g = \int_S a\overline{\partial}r \wedge g = \int_S a dr \wedge g = 0.$$

To sum up: *if f is a CR-function of class C^1, then for any C^∞-smooth form g such that* supp $g \cap S$ *is compact,*

$$\int_S f \wedge \overline{\partial}g = 0. \tag{8.12}$$

The converse of this statement is also true. In fact, let $\widetilde{f}$ be a C^1-extension of f to U, and suppose (8.12) holds. Choose local base fields $L_1, \ldots, L_{n-1}, L_n$ as in (8.2) and (8.3) and let $\omega^1, \ldots, \omega^{n-1}, \omega^n = \partial r$ be dual base forms. Then

$$\overline{\partial}\widetilde{f} = \sum (\overline{L}_j\widetilde{f})\overline{\omega}^j. \tag{8.13}$$

Let g be a compactly supported $(n, n-2)$-form. Then

$$0 = \int_S f\,\overline{\partial}g = \int_S \overline{\partial}(\tilde{f}g) - \int_S \overline{\partial}\tilde{f} \wedge g = - \int_S \overline{\partial}\tilde{f} \wedge g. \tag{8.14}$$

Take as g the form

$$g = h\,\omega^1 \wedge \ldots \wedge \omega^n \wedge \overline{\omega}^1 \wedge \ldots \wedge \widehat{\overline{\omega}^j} \wedge \ldots \wedge \overline{\omega}^{n-1},$$

where h is a smooth function with sufficiently small support. Now (8.14) reads as

$$\int_S (\overline{L}_j\tilde{f})h\,\omega^j \wedge \omega^1 \wedge \ldots \wedge \omega^n \wedge \overline{\omega}^1 \wedge \ldots \wedge \widehat{\overline{\omega}^j} \wedge \ldots \wedge \overline{\omega}^{n-1} = 0, \tag{8.15}$$

because

$$\omega^n \wedge \overline{\omega}^n = \partial r \wedge \overline{\partial}r = 0$$

on S. The $\widehat{}$ means that the corresponding form has to be cancelled. As we can sufficiently vary h, this implies

$$\overline{L}_j\tilde{f} = \overline{L}_j f = 0 \quad \text{on } S. \tag{8.16}$$

Now condition (8.12) makes sense even for continuous functions and leads us to

Definition 8.17 *A continuous function f on S is a CR-function if for all smooth $(n, n-2)$-forms g in a neighbourhood of S, whose support intersects S in a compact set,*

$$\int_S f\overline{\partial}g = 0.$$

Proposition 8.9 easily extends to the continuous case:

Proposition 8.18 *Let f be continuous on $\overline{\Omega}$, holomorphic on Ω, where $\Omega \subset\subset \mathbb{C}^n$ has C^1-smooth boundary. Then the restriction f_0 of f to the boundary is a continuous CR-function.*

Proof   Suppose g is a smooth form of type $(n, n-2)$ in a neighbourhood of $b\Omega$. We may assume that g is even defined in a neighbourhood of $\overline{\Omega}$. Ω is given by an inequality $r(z) < 0$ with a C^1-function r on $\mathbb{C}^n$ with $dr(z) \neq 0$ for $r(z) = 0$. Set

$$\Omega_\varepsilon = \{z : r(z) < -\varepsilon\}.$$

Then

$$\int_{b\Omega} f\overline{\partial}g = \lim_{\varepsilon \to 0} \int_{b\Omega_\varepsilon} f\overline{\partial}g = \lim_{\varepsilon \to 0} \int_{\Omega_\varepsilon} \overline{\partial}f \wedge \overline{\partial}g = 0$$

$\square$

2.   We now turn to the converse of the above propositions. The most general case depends on a jump formula for the Bochner-Martinelli boundary integral which we shall not prove here – see [Ran 86]. We state the consequence which we need.

Proposition 8.19 *Let f be a continuous function on the boundary $b\Omega$ of a bounded domain; $b\Omega$ should be smooth of class C^1. Suppose that*

$$F(z) = \int_{b\Omega} f(\zeta) B_{n0}(\zeta, z) \equiv 0$$

on $\mathbb{C}^n - \overline{\Omega}$. Then the function

$$F(z) = \begin{cases} \int_{b\Omega} f(\zeta) B_{n0}(\zeta, z), & z \in \Omega \\ f(z), & z \in b\Omega \end{cases}$$

is continuous on $\overline{\Omega}$.

Using 8.19 we can prove a deep generalisation of the Kugelsatz:

Theorem 8.20 *Let $\Omega \subset\subset \mathbb{C}^n$, $n > 1$, be a domain with C^1-smooth boundary whose complement is connected. Let f be a continuous CR-function on $b\Omega$. Then there is a unique function F on $\overline{\Omega}$ which is holomorphic on Ω, continuous on $\overline{\Omega}$, with $F|_{b\Omega} \equiv f$. If $b\Omega$ and f are of class C^k, then $F \in C^k(\overline{\Omega})$.*

Proof   Set for $z \notin b\Omega$

$$F(z) = \int_{b\Omega} f(\zeta) B_{n0}(\zeta, z).$$

F is C^∞; moreover

$$\overline{\partial} F(z) = \int_{b\Omega} f(\zeta) \overline{\partial}_z B_{n0}(\zeta, z)$$

$$= - \int_{b\Omega} f(\zeta) \overline{\partial}_\zeta B_{n1}(\zeta, z)$$

$$= 0$$

by the definition of CR-functions. So F is holomorphic on Ω and $\mathbb{C}^n - \overline{\Omega}$. But for $|z_1|$ sufficiently large, $F(z_1, z_2, \ldots, z_n)$ is an entire function of $z_2, \ldots, z_n$, which goes to zero at infinity in view of the definition of B_{n0}. This means that $F \equiv 0$ on $\mathbb{C}^n - \overline{\Omega}$; the jump formula 8.19 now gives the assertion.

Before we can prove the C^k-regularity we need a nice consequence of formula (8.11) which we state as

Lemma 8.21 *Let S be of class C^m, $1 \le k \le m$, and f a C^k-CR-function on S. Then there is an extension $\widehat{f}$ of f with the following properties:*

i. $\widehat{f} \in C^{k-1}(U)$.

ii. f is k-times differentiable in all points $z \in S$, with continuous derivatives on S.

iii. $\overline{\partial}\widehat{f}(z) = 0$ for $z \in S$.

In fact, take any extension $\widetilde{f}$ of class C^k, set

$$\widehat{f} = \widetilde{f} - ar,$$

where a is taken from (8.11), and verify.

Now consider the imbedding $\iota : S \to U$. Then, on the one hand, $d_\zeta f$ is a well-defined 1-form on S, associating, with $\zeta \in S$, a cotangent vector in $T^*_\zeta S$, on the other hand, the form $d_\zeta \widehat{f}$ associates with ζ a cotangent vector in $T^*_\zeta \mathbb{C}^n$. These forms are related as follows:

$$d_\zeta \widehat{f} \circ \iota = d_\zeta f. \tag{8.22}$$

This relation explains the following calculations (8.24).

To continue the proof 8.20 we work with an extension $\widehat{f}$ as above and consider, for $z \notin b\Omega$,

$$\frac{\partial F}{\partial z^k}(z) = \frac{\partial}{\partial z^k} \int_\Omega f(\zeta) B_{n0}(\zeta, z) = -\int_{b\Omega} f(\zeta) \frac{\partial B_{n0}(\zeta, z)}{\partial \zeta^k}. \tag{8.23}$$

We can write as in §6

$$B_{n0}(\zeta, z) = B(\zeta, z) = d\zeta^k \wedge B_k(\zeta, z)$$

and we note $\overline{\partial}_\zeta B_k(\zeta, z) = 0$. Then (8.22) becomes

$$\frac{\partial F}{\partial z^k}(z) = -\int_{b\Omega} f(\zeta)\, d\zeta^k \wedge \frac{\partial B_k(\zeta, z)}{\partial \zeta^k} = -\int_{b\Omega} f(\zeta)\, \partial_\zeta B_k(\zeta, z)$$

$$= -\int_{b\Omega} f(\zeta)\, d_\zeta B_k(\zeta, z) = \int_{b\Omega} d_\zeta f(\zeta) \wedge B_k(\zeta, z)$$

$$= \int_{b\Omega} d_\zeta \widehat{f}(\zeta) \wedge B_k(\zeta, z) = \int_{b\Omega} \partial_\zeta \widehat{f}(\zeta) \wedge B_k(\zeta, z) \tag{8.24}$$

(because of $\overline{\partial}_\zeta \widehat{f} = 0$)

$$= \int_{b\Omega} \frac{\partial \widehat{f}}{\partial \zeta^k}(\zeta)\, d\zeta^k \wedge B_k(\zeta, z) = \int_{b\Omega} \frac{\partial \widehat{f}}{\partial \zeta^k}(\zeta)\, B(\zeta, z).$$

Thus,

$$\frac{\partial F}{\partial z^k}(z) = \int\limits_{b\Omega} \frac{\partial \hat{f}}{\partial \zeta^k}(\zeta)\, B(\zeta, z); \qquad (8.25)$$

since $F \equiv 0$ outside $\overline{\Omega}$, we also have $\frac{\partial F}{\partial z^k}(z) \equiv 0$ outside $\overline{\Omega}$. The jump formula now gives the claim (for first derivatives).

The case of higher derivatives is treated in the same way. Finally F is unique in view of the BM integral formula (which is valid, by approximation, for functions continuous on $\overline{\Omega}$ and holomorphic in the interior). □

3.   Theorem 8.20 can be proved without taking recourse to the jump formula if one assumes better differentiability properties from the very beginning. The key to the proof is the following extension theorem:

Proposition 8.26 *Let S be a hypersurface given by $r = 0$, where r is a C^∞-smooth function on U with $dr(z) \neq 0$. Let f_0 be a C^∞ CR-function on S. Then for each $k = 0, 1, 2, \ldots$ there is a smooth extension f of f_0 to U satisfying the estimate (coefficientwise)*

$$\overline{\partial} f = O(|r|^k).$$

Proof   For $k = 0$ there is nothing to prove. Suppose the statement true for $k - 1 \geq 0$, and let $\tilde{f}$ be the corresponding extension. This yields

$$\overline{\partial} \tilde{f} = r^{k-1} g;$$

therefore

$$0 = \overline{\partial}^2 \tilde{f} = (k-1)r^{k-2} \overline{\partial} r \wedge g + r^{k-1} \overline{\partial} g,$$
$$0 = (k-1)\overline{\partial} r \wedge g + r \overline{\partial} g.$$

This implies

$$g = a\overline{\partial} r + rh, \qquad (8.27)$$

(also in the case $k - 1 = 0$, because then $g = \overline{\partial} \tilde{f}$ and (8.22) is the CRcondition).

Setting

$$f = \tilde{f} - \frac{1}{k} ar^k$$

we obtain the desired extension:

$$\overline{\partial} f = \overline{\partial} \tilde{f} - r^{k-1} \overline{\partial} r a - \frac{1}{k} r^k \overline{\partial} a$$
$$= r^{k-1}(g - a\overline{\partial} r) - \frac{1}{k} r^k \overline{\partial} a$$
$$= r^k h - \frac{1}{k} r^k \overline{\partial} a$$
$$= O(|r|^k).$$

□

We can now easily prove the C^∞-version of Theorem 8.20:

Proposition 8.28 *Let $\Omega \subset\subset \mathbb{C}^n$, $n > 1$, be a domain with C^∞-smooth boundary and f_0 a C^∞-CR-function on $b\Omega$. Then there is a C^∞ function f on $\overline{\Omega}$, holomorphic on Ω, with $f|_{b\Omega} \equiv f_0$.*

Proof   We use proposition 8.26 with $k \geq 2$. f_0 can be extended to a smooth function f_1 on $\mathbb{C}^n$ with $\overline{\partial} f_1 = O(|r|^k)$, where r is a defining function for $b\Omega$. This means that the $(0,1)$-form

$$g = \begin{cases} \overline{\partial} f_1 & \text{on } \overline{\Omega} \\ 0 & \text{on } \mathbb{C}^n - \overline{\Omega} \end{cases}$$

is in $C_{01,c}^{k-1}(\mathbb{C}^n)$ with support in $\overline{\Omega}$. Since $n > 1$ there is a C^{k-1} function u on $\mathbb{C}^n$ satisfying

$$\overline{\partial} u = g, \quad u \equiv 0 \text{ on } \mathbb{C}^n - \overline{\Omega}.$$

Then

$$f = f_1 - u$$

is holomorphic on Ω and of class C^{k-1} on $\overline{\Omega}$, with $f|_{b\Omega} = f_0$. But f is independent of k, namely

$$f(z) = \int\limits_{b\Omega} f(\zeta) B_{n0}(\zeta, z),$$

so f has to be C^∞-smooth on $\overline{\Omega}$. $\square$

§9   The Bochner-Martinelli Transform for Currents

Our last application of the BMK formula is a proof of the local regularity of the $\overline{\partial}$-operator. This can best be handled in the framework of distributions and currents. We recall the essential notions.

1.   Test forms on Ω are compactly supported smooth forms. A *current of type (p,q) and dimension $(n-p, n-q)$* is a continuous (in the *Schwartz topology* on test forms) linear form on the space of test forms of type $(n-p, n-q)$. Each locally integrable form f of type (p,q) is a (p,q)-current if one sets

$$\langle f, t \rangle = \int\limits_{\Omega} f(z) \wedge t(z). \tag{9.1}$$

$\langle , \rangle$ stands, here and in the sequel, for the application of a current to a test form. A current T *vanishes at a point z* if there is a neighbourhood U of z such that

$$\langle T, t \rangle = 0 \tag{9.2}$$

for all test forms supported in U. This defines the *support of T*, supp T, as the complement of the points where T vanishes. T is smooth at a point z if there is a smooth form f such that $T - f$ vanishes at z (where f is considered as a current). Currents with compact support can be applied to smooth forms with arbitrary support. Important operations on currents are *exterior multiplication with smooth forms* and the *exterior derivative d* (and the $\bar{\partial}$-derivative $\bar{\partial}$):

$$\langle T \wedge f, t \rangle = \langle T, f \wedge t \rangle \tag{9.3}$$

$$f \wedge T = (-1)^{(p+q)(r+s)} T \wedge f \tag{9.4}$$

$$\langle dT, t \rangle = (-1)^{p+q-1} \langle T, dt \rangle \tag{9.5}$$

$$\langle \bar{\partial}T, t \rangle = (-1)^{p+q-1} \langle T, \bar{\partial}t \rangle; \tag{9.6}$$

in these formulas T was supposed to be of type (p, q) and f of type (r, s). The usual Leibniz rule holds in this context.

2. We can now define the Bochner-Martinelli-Koppelman transform for currents.

Definition 9.7 *Let T be a $(0, q+1)$-current on Ω with compact support. The BM transform of T is the $(0, q)$-current $\boldsymbol{B}_q T$ defined by the formula*

$$\langle \boldsymbol{B}_q T, t \wedge dz^N \rangle = (-1)^{q+1} \langle T, \boldsymbol{B}_{n-q-1} t \wedge dz^N \rangle$$

In (9.7) t is a test form of type $(0, n - q)$, and $dz^N = dz^1 \wedge \ldots \wedge dz^n$. – We first note that for continuous forms f this definition coincides with the previous definition (5.3). In other words we have to show

Lemma 9.8

$$\int_\Omega \boldsymbol{B}_q f(z) \wedge t(z) \wedge dz^N = (-1)^{q+1} \int_\Omega f(z) \wedge \boldsymbol{B}_{n-q-1} t(z) \wedge dz^N,$$

where $\boldsymbol{B}_q$ and $\boldsymbol{B}_{n-q-1}$ are defined in 5.3.

Inserting the definitions we get on the left-hand side

$$\int_\Omega \int_\Omega f(\zeta) \wedge B_{nq}(\zeta, z) \wedge t(z) \wedge dz^N,$$

whereas the right-hand side yields

$$(-1)^{q+1} \int_\Omega f(z) \wedge \int_\Omega t(\zeta) \wedge B_{n,n-q-1}(\zeta, z) \wedge dz^N$$

$$= (-1)^{q+1} \int_\Omega \int_\Omega f(\zeta) \wedge t(z) \wedge B_{n,n-q-1}(z, \zeta) \wedge d\zeta^N.$$

So 9.8 results from

Lemma 9.9

$$(-1)^{q(n-q)} B_{nq}(\zeta, z) \wedge dz^N = (-1)^{q+1} B_{n,n-q-1}(z, \zeta) \wedge d\zeta^N,$$

which is proved by checking the definitions and carefully keeping track of all the signs involved.

Lemma 9.10 *If $z \notin \operatorname{supp} T$ then $B_q T$ is smooth in z.*

Proof Choose a smooth function χ with compact support, $0 \leq \chi \leq 1$, which is $\equiv 1$ on $\operatorname{supp} T$ and 0 in a neighbourhood V of z. Let t be a test form with support in V. Then

$$\langle B_q T, t \wedge dz^N \rangle \tag{9.11}$$

$$= \langle B_q \chi T, (1 - \chi) t \wedge dz^N \rangle$$

$$= (-1)^{q+1} \langle \chi T, B_{n-q-1}(1 - \chi) t \wedge dz^N \rangle$$

$$= (-1)^{q+1} \langle T, \chi B_{n-q-1}(1 - \chi) t \wedge dz^N \rangle$$

$$= (-1)^{q+1} \left\langle T, \int_\Omega t(\zeta) \wedge \chi(z)(1 - \chi(\zeta)) B_{n,n-q-1}(\zeta, z) \wedge dz^N \right\rangle.$$

Now the form

$$H(\zeta, z) = \chi(z)(1 - \chi(\zeta)) B_{n,n-q-1}(\zeta, z) \wedge dz^N$$

is a smooth family of forms in z with compact support; ζ is in V. So we can apply, for fixed ζ, T to this form and obtain a smooth form in ζ, say $h(\zeta)$. Thus we obtain from (9.11)

$$\langle B_q T, t \wedge dz^N \rangle \tag{9.12}$$

$$= (-1)^{q+1} \left\langle T, \int_\Omega t(\zeta) \wedge H(\zeta, z) \right\rangle$$

$$= (-1)^{q+1} \int_\Omega t(\zeta) \wedge \langle T, H(\zeta, \cdot) \rangle$$

$$= (-1)^{q+1} \int_\Omega t(\zeta) \wedge h(\zeta).$$

$\square$

The Bochner-Martinelli-Koppelman formula carries over to currents. In fact, let T be a $(0, q)$-current with compact support, and consider, for a test form t of type $(0, n - q)$,

$$\langle B_q \bar{\partial} T + \bar{\partial} B_{q-1} T, t \wedge dz^N \rangle$$

$$= (-1)^{q+1} \langle \bar{\partial} T, B_{n-q-1} t \wedge dz^N \rangle + (-1)^q \langle B_{q-1} T, \bar{\partial} t \wedge dz^N \rangle$$

$$= \langle T, (\bar{\partial} B_{n-q-1} t + B_{n-q} \bar{\partial} t) \wedge dz^N \rangle$$

$$= -\langle T, t \wedge dz^N \rangle.$$

Thus we have

Proposition 9.13 (BMK formula for currents) *If T is a $(0,q)$-current with compact support, then*

$$T = -\bar{\partial}B_{q-1}T - B_q\bar{\partial}T.$$

3.   Turning to the Cauchy-Riemann equations we easily arrive at an important regularity result:

Theorem 9.14 *Let T be a distribution (i.e. current of type $(0,0)$). If $\bar{\partial}T$ is (C^k-) smooth in an open set Ω then T is smooth on Ω.*

Proof   Choose $\Omega_0 \subset\subset \Omega$ and a smooth function χ, $0 \leq \chi \leq 1$, $\mathrm{supp}\,\chi \subset \Omega$, $\chi \equiv 1$ on Ω_0. It suffices to show that χT is smooth on Ω_0. By the BMK formula 9.13 we have

$$\chi T = -B_0\bar{\partial}\chi T = -B_0(\bar{\partial}\chi \cdot T + \chi\bar{\partial}T)$$
$$= -B_0(\bar{\partial}\chi \cdot T) - B_0(\chi\bar{\partial}T) \tag{9.15}$$

The second term on the right-hand side of 9.15 is smooth because $\chi\bar{\partial}T$ is everywhere C^k, the first term is smooth on Ω_0 because $\mathrm{supp}\,\bar{\partial}\chi \cdot T \cap \Omega_0 = \emptyset$: this allows to apply Lemma 9.10.   □

§10   Regularity Properties of Isotropic Operators

We study isotropic operators in some detail.

Proposition 10.1 *Suppose the kernel $\mathcal{E}_m(\zeta, z)$ on $\Omega \times \Theta \subset\subset \mathbb{C}^n \times \mathbb{C}^n$ is uniformly of order m, i.e.*

$$|\mathcal{E}_m(\zeta, z)| \leq C_0\|\zeta - z\|^m$$

for all $\zeta, z \in \Omega \times \Theta$. Assume $-2n < m \leq 0$. Then there are constants C_s, for $1 \leq s < \frac{2n}{|m|}$, such that

$$\int\limits_\Omega |\mathcal{E}_m(\zeta, z)|^s\, dV(\zeta) \leq C_s, \quad z \in \Theta,$$

$$\int\limits_\Theta |\mathcal{E}_m(\zeta, z)|^s\, dV(z) \leq C_s, \quad \zeta \in \Omega.$$

(The integrands denote the s-th power of the absolute values of arbitrary coefficients of $\mathcal{E}_m$.) The above statement is obvious but very important: it implies, in view of a generalised Young inequality, the following boundedness theorem:

Theorem 10.2 *Let $\mathcal{E}_m$ as above. Then the operator E_m with kernel $\mathcal{E}_m$:*

$$E_m f = (f, \mathcal{E}_m) = \int_\Omega f(\zeta) \wedge *\overline{\mathcal{E}_m(\zeta, z)}$$

is bounded from L^p into L^s:

$$E_m : L^p(\Omega) \longrightarrow L^s(\Theta),$$

for $1 \leq p \leq \infty$ and $\frac{1}{s} > \frac{1}{p} - \frac{2n+m}{2n}$.

This statement contains 5.13 as a special case. More is true in the L^∞ case:

Theorem 10.3 *Let the kernel $\mathcal{E}_{1-2n}(\zeta, z)$ be uniformly (see 10.1) of order $1-2n$ on $\Omega \times \Theta$. Then the corresponding operator E_{1-2n} defines a continuous map from $L^\infty \to C^\alpha$,*

$$E_{1-2n} : L^\infty(\Omega) \longrightarrow C^\alpha(\Theta),$$

for any $\alpha < 1$.

C^α, for $0 < \alpha < 1$, is the *space of Hölder continuous functions of exponent* α, provided with the *norm*

$$\|f\|_{C^\alpha} = \|f\|_{L^\infty} + \sup_{z_1 \neq z_2} \frac{|f(z_1) - f(z_2)|}{\|z_1 - z_2\|^\alpha}.$$

We will later introduce spaces C^α for $\alpha > 1$; in case $0 < \alpha < 1$ we shall also use the notation Λ_α for C^α.

Proof   We may assume $\Omega = \Theta$ and even choose for Ω a ball $B_R(0) \subset\subset \mathbb{C}^n$. Evidently,

$$\left| \int_\Omega f(\zeta) \mathcal{E}(\zeta, z) - \int_\Omega f(\zeta) \mathcal{E}(\zeta, z_0) \right| \leq J \|f\|_{L^\infty}, \tag{10.4}$$

with

$$J = \int_\Omega |\mathcal{E}(\zeta, z) - \mathcal{E}(\zeta, z_0)| dV(\zeta)$$

(we have dropped the index $1 - 2n$). So we have to estimate the integral J, for z and z_0 in Ω.

Set

$$\delta = \|z - z_0\|, \quad w = \frac{1}{2}(z_0 + z),$$
$$B = B_{3\delta}(w) = \{\zeta : \|\zeta - w\| \leq 3\delta\}. \tag{10.5}$$

$$J = \int_B |\mathcal{E}(\zeta, z) - \mathcal{E}(\zeta, z_0)| dV(\zeta) + \int_{\Omega - B} |\mathcal{E}(\zeta, z) - \mathcal{E}(\zeta, z_0)| dV(\zeta)$$

$$= J_1 + J_2. \tag{10.6}$$

J_1 is easy to deal with:

$$J_1 \leq \int_B |\mathcal{E}(\zeta, z)| dV(\zeta) + \int_B |\mathcal{E}(\zeta, z_0)| dV(\zeta) \tag{10.7}$$

$$\leq \text{const} \int_{B_{3\delta}(w)} \left(\frac{1}{\|\zeta - z\|^{2n-1}} + \frac{1}{\|\zeta - z_0\|^{2n-1}} \right) dV(\zeta)$$

$$\leq \text{const} \int_{B_{3\delta}(w)} \frac{dV(\zeta)}{\|\zeta - w\|^{2n-1}} \leq \text{const} \cdot \delta.$$

For J_2, we have $\|\zeta - w\| \geq 3\delta$, which means that the integrand is smooth. Moreover,

$$\nabla^z \mathcal{E}(\zeta, z) = \mathcal{E}_{-2n}(\zeta, z). \tag{10.8}$$

By the mean value theorem,

$$|\mathcal{E}(\zeta, z) - \mathcal{E}(\zeta, z_0)| \leq \|z - z_0\| \max_{v \in [z_0, z]} |\mathcal{E}_{-2n}(\zeta, v)|$$

$$\leq \text{const} \max_{[z_0, z]} \frac{1}{\|\zeta - v\|^{2n}} \|z - z_0\|. \tag{10.9}$$

But for $\zeta \in \Omega - B$,

$$\|\zeta - v\| \geq \text{const}\|\zeta - w\|. \tag{10.10}$$

(10.9) and (10.10) combined give

$$J_2 \leq \text{const} \int_{\substack{\|\zeta - w\| \geq 3\delta \\ \zeta \in \Omega}} \frac{dV(\zeta)}{\|\zeta - w\|^{2n}} \|z - z_0\| \tag{10.11}$$

$$\leq \text{const} \int_{3\delta}^R \frac{dr}{r} \|z - z_0\|$$

$$\leq \text{const}(\delta + \delta |\log \delta|) \leq \text{const} \, \delta^\alpha$$

for any $\alpha < 1$. – (10.7) and (10.11) establish the claim. $\qquad\square$

As an immediate corollary we have

Proposition 10.12 *If $f \in L^\infty_{\text{loc}}(\Omega) \cap L^1(\Omega)$, then $E_{1-2n}f$ is continuous.*

A more delicate regularity statement is

Theorem 10.13 *Let*

$$\mathcal{E}_{1-2n} = \frac{\mathcal{E}_m}{\rho^{2t}}, \quad \text{that is } m - 2t = 1 - 2n,$$

be an isotropic kernel uniformly of order $1 - 2n$. *If* f *is in* $C^k(\Omega) \cap L^1(\Omega)$, *then*

$$E_{1-2n}f \in C^k(\Theta) \cap L^1(\Theta).$$

This contains Proposition 5.14 as a special case.

Proof   We first need

Lemma 10.14 *Let* D^z *be a differentiation with respect to* z *or* $\bar{z}$, D^ς *the corresponding differentiation with respect to* ς, $\bar{\varsigma}$. *Then for* $m \geq 0$,

$$D^\varsigma \mathcal{E}_m(\varsigma, z) + D^z \mathcal{E}_m(\varsigma, z) = \mathcal{E}_m(\varsigma, z)$$

(where $\mathcal{E}_m$ *stands for different forms on the two sides of the equation).*

This follows easily from Taylor's formula. – Turning now to 10.13 we may again assume $\Omega = \Theta$; it suffices to consider forms with compact support since differentiability of $E_{1-2n}f$ at points off the support of f is clear. Moreover, let us stick to the case $k = 1$ – higher order derivatives are treated by iterating the argument for $k = 1$. Finally, without loss of generality we may assume that f is a function.

Set

$$F(z) = \int\limits_{\mathbb{C}^n} f(\varsigma) \wedge *\overline{\mathcal{E}_{1-2n}(\varsigma, z)}$$

$$= \int\limits_{\mathbb{C}^n} f(\varsigma) \cdot \frac{\mathcal{E}_m(\varsigma, z)}{\rho^{2t}} \, dV(\varsigma) \qquad (10.15)$$

where the meaning of $\mathcal{E}_m$ has changed but the new function $\mathcal{E}_m$ still satisfies (10.14). We now introduce, for each $\varepsilon > 0$, the functions

$$F_\varepsilon(z) = \int\limits_{\mathbb{C}^n} f(\varsigma) \cdot \frac{\mathcal{E}_m(\varsigma, z)}{\rho^{2t}(\varsigma, z) + \varepsilon} \, dV(\varsigma) \qquad (10.16)$$

and note that the F_ε are C^∞ smooth.

Lemma 10.17 *The convergence relation*

$$\lim_{\varepsilon \to 0} F_\varepsilon(z) = F(z)$$

holds pointwise and in the sense of distributions.

Proof   For z fixed the integrands converge pointwise. The inequality

$$|f(\zeta)|\frac{\mathcal{E}_m(\zeta, z)}{\rho^{2t}(\zeta, z) + \varepsilon} \leq |f(\zeta)|\frac{\mathcal{E}_m(\zeta, z)}{\rho^{2t}(\zeta, z)} \tag{10.18}$$

proves the convergence of the integrals, i. e.the pointwise convergence

$$F_\varepsilon(z) \to F(z).$$

If $\varphi(z)$ is a test function, the integral

$$\int_{\mathbb{C}^n} |F(z)||\varphi(z)|\, dV(z)$$

exists because F is continuous, consequently the function

$$f(\zeta)\frac{\mathcal{E}_m(\zeta, z)}{\rho^{2t}(\zeta, z)}\varphi(z)$$

is integrable in $\mathbb{C}^n \times \mathbb{C}^n$. Multiplying (10.18) by $|\varphi(z)|$ we see, that Lebesgue's theorem applies and yields the convergence

$$\int F_\varepsilon(z)\varphi(z)\, dV(z) \to \int F(z)\varphi(z)\, dV(z),$$

that is the distribution convergence $F_\varepsilon \to F$. This proves lemma 10.17. $\square$

To continue the proof of 10.13 we consider a first order differentiation D^z with respect to z or $\bar{z}$:

$$D^z F_\varepsilon(z) = D^z \int f(\zeta)\frac{\mathcal{E}_m(\zeta, z)}{\rho^{2t}(\zeta, z) + \varepsilon}\, dV(\zeta) \tag{10.19}$$

$$= \int f(\zeta)D^z\frac{\mathcal{E}_m(\zeta, z)}{\rho^{2t}(\zeta, z) + \varepsilon}\, dV(\zeta)$$

$$= -\int f(\zeta)D^\zeta\frac{\mathcal{E}_m(\zeta, z)}{\rho^{2t}(\zeta, z) + \varepsilon}\, dV(\zeta) + \int f(\zeta)\frac{\mathcal{E}_m(\zeta, z)}{\rho^{2t}(\zeta, z) + \varepsilon}\, dV(\zeta)$$

(because of (10.14) – the meaning of $\mathcal{E}_m$ in the second integral has changed, but is still uniformly isotropic of order m)

$$= -\int D^\zeta f(\zeta)\frac{\mathcal{E}_m(\zeta, z)}{\rho^{2t}(\zeta, z) + \varepsilon}\, dV(\zeta) + \int f(\zeta)\frac{\mathcal{E}_m(\zeta, z)}{\rho^{2t}(\zeta, z) + \varepsilon}\, dV(\zeta)$$

$$\stackrel{\text{def}}{=} G_\varepsilon(z) + H_\varepsilon(z).$$

Let us introduce the functions

$$G(z) = -\int D^\zeta f(\zeta) \cdot \frac{\mathcal{E}_m(\zeta, z)}{\rho^{2t}(\zeta, z)}\, dV(\zeta) \tag{10.20}$$

$$H(z) = \int f(\zeta)\frac{\mathcal{E}_m(\zeta, z)}{\rho^{2t}(\zeta, z)}\, dV(\zeta) \tag{10.21}$$

where the sign in (10.20) is the sign of $G_\varepsilon(z)$ in (10.19) and the kernels $\mathcal{E}_m(\zeta, z)$ are the kernels arising in $G_\varepsilon(z)$ resp. $H_\varepsilon(z)$.

Then as in Lemma 10.17:

$$\lim_{\varepsilon \to 0} G_\varepsilon(z) = G(z)$$

$$\lim_{\varepsilon \to 0} H_\varepsilon(z) = H(z)$$

both pointwise and in the sense of distributions.

So we know

$$F_\varepsilon \to F, \quad D^z F_\varepsilon \to G + H$$

as distributions. Moreover, all the above functions are continuous. But then $G + H$ is the distribution derivative of F; F is differentiable and the usual derivative coincides with its distribution derivative:

$$D^z F = G + H.$$

This proves the theorem.   □

We remark, that we have also proved the commutator relation

$$DE_{1-2n}f = E_{1-2n}Df + E_{1-2n}f \qquad (10.22)$$

valid for forms with compact support (The last E_{1-2n}-operator is of course different from the given operator).

§11 Notes

The main result of the chapter is Theorem 4.11, the Bochner-Martinelli-Koppelman formula. It was proved, for $q = 0$ by Martinelli [Mar 38] and Bochner [Boc 43] between 1938 and 1943. The complicated history is carefully presented in [Ran 86]. The case $q > 0$ is due to Koppelman 1967 [Kop 67₁, Kop 67₂], the case of one complex dimension goes back to Pompeiu 1912 [Pom 58], it seems to have been forgotten for many years until resurfacing around 1950 in the modern development of multidimensional complex analysis – see [Car 51] and [Dol 56, Dol 57]. The first published proof of Koppelman's formula is in [Lie 70]; our proof closely follows [Ran 86] whose proof is again based on [LiR 83]. The relation of the BM formula to a fundamental solution of the Laplacian is already basic for Bochner's original proof. The surprisingly useful proposition 3.22 was established by Range [Ran 86]. The results on the Laplacian §§2 and 3 are classical and by now difficult to correctly attribute. For the concepts of §1 compare [Rha 60] and [Wei 58]; the *-operator was introduced and extensively used in [Hod 41] (1941).

Derivatives of the Cauchy transform have been studied in the theory of pseudoanalytic functions [Vek 62]; the corresponding results for the Bochner-Martinelli-Koppelman transform (Theorem 6.1) are also presented in [Kyt 95] from where we have taken the application to the BMK formula; Kytmanov ascribes this proof to Tarkhanov. For properties of the Riesz transform compare [Ste 70].

The applications of the Bochner-Martinelli-Koppelman formula (§7) to several complex variables are by now classical; we follow [Ran 86] and [LTh 97]. More precisely: the Kugelsatz (Theorem 7.3) is due to Hartogs [Har 06] (1906) – it is one of the starting points of the development of multidimensional complex analysis. Hartogs used Cauchy's integral; our proof essentially coincides with Bochner's. The connection between the Kugelsatz and compactly supported solutions to the Cauchy-Riemann system is due to Ehrenpreis [Ehr 61], the characterization 7.7 of holomorphic functions by the Bochner-Martinelli formula is taken from [Hen 90] where it is attributed to Aronov, Aizenberg and Kytmanov.

The theory of Cauchy-Riemann functions (§8) declenched by the Kugelsatz is in rapid development, we have only given the bare minimum, following [Ran 86] and [Hör 66]. The – once again fairly involved – early history is described in [Ran 98] where, in particular, the somewhat overlooked contribution of the Italian analytic school of the '30s is exhibited. For the jump formula 8.19 we refer to [Ran 86]. An extensive exposition of the theory of Cauchy-Riemann functions including many recent results is given in [LTh 97] and [ChS 2001].

Theorem 9.14 is a special – and particularly easy – case of interior regularity of elliptic operators; an early proof can be found in [Dol 56]; the proof we present is also in [LTh 97]. The proofs of the regularity theorems in §§5 and 10 follow classical methods; L^p-estimates are based on Young's inequality which we state later on (Ch. III.5.35) and whose proof can be found in [Ran 86]. The proofs of interior C^k-regularity of isotropic operators are slightly more difficult than expected because we cannot use translation invariance of the kernels – cf. the proofs of Prop. 5.14 and 10.13.

Chapter II

The Calculus of Cauchy-Fantappiè Forms

The calculus of Cauchy-Fantappiè forms (CF forms) provides the passage from integral formulae of Bochner-Martinelli type as developed in the previous chapter, to more flexible integral representations which can, in particular, be adapted to the geometry of the domains we shall consider. We give a self-contained exposition of this calculus based on ideas of Koppelman [Kop 67$_2$] and Berndtsson and Andersson [BeA 83]; different (but – of course – equivalent) expositions can be found in [Ran 86], [HeL 84] resp. [Lie 70] and [FiL 74].

§1 The Koppelman Formula

We shall introduce a large class of kernels (i.e. double differential forms) which contains the Bochner-Martinelli kernel as a special case, and investigate the relationship between different members of that class.

Let us denote points in $\mathbb{C}^n \times \mathbb{C}^n$ by $(\xi, \eta) = (\xi_1 \ldots \xi_n, \eta_1, \ldots, \eta_n)$ and consider the exterior form

$$M = e^{\langle \xi, \eta \rangle} \omega(\xi) \wedge \omega(\eta)$$

with

$$\langle \xi, \eta \rangle = \sum_{i=1}^{n} \xi_i \eta_i$$
$$\omega(\xi) = d\xi_1 \wedge \ldots \wedge d\xi_n$$
$$\omega(\eta) = d\eta_1 \wedge \ldots \wedge d\eta_n$$

M is of type $(2n, 0)$ in $\mathbb{C}^n \times \mathbb{C}^n$; consequently:

$$dM = 0 \qquad (1.1)$$

Now consider C^1-maps

$$\tilde{a} = (a_1, \ldots, a_n) : W \longrightarrow \mathbb{C}^n, \qquad (1.2)$$

where $W \subset \mathbb{C}^n \times \mathbb{C}^n$ is an open set; points of W will be denoted by

$$(\zeta, z) = (\zeta_1, \ldots, \zeta_n, z_1, \ldots, z_n).$$

Definition 1.3 *The map* (1.2) *is called a Leray map (or a generating map) on* W, *if*

$$\sum_{j=1}^{n} a_j(\zeta, z)(\zeta_j - z_j) \equiv 1 \qquad (1.4)$$

on W.

Note that the map

$$\frac{\overline{\zeta} - \overline{z}}{\|\zeta - z\|^2} : \mathbb{C}^n \times \mathbb{C}^n - \Delta \longrightarrow \mathbb{C}^n \qquad (1.5)$$

is an example of a Leray map. $\Delta = \{(\zeta, z) : \zeta = z\}$ is the diagonal in $\mathbb{C}^n \times \mathbb{C}^n$.

Suppose we are given two Leray maps $\widetilde{a} = (a_1, \ldots, a_n)$ and $\widetilde{b} = (b_1, \ldots, b_n)$ on W. We deform them into each other via the map

$$\widetilde{c} = t\widetilde{a} + (1 - t)\widetilde{b}, \quad 0 \le t \le 1. \qquad (1.6)$$

with components

$$c_j = ta_j + (1 - t)b_j;$$

So $\widetilde{c}$ is a map

$$\widetilde{c} : W \times I \longrightarrow \mathbb{C}^n,$$

where I is the unit interval. (1.4) and (1.6) immediately imply

$$\sum c_j(\zeta, z, t)(\zeta_j - z_j) = 1. \qquad (1.7)$$

We use $\widetilde{c}$ to define a map

$$\psi : W \times I \times \mathbb{R} \longrightarrow \mathbb{C}^n \times \mathbb{C}^n$$

by the formula

$$\psi(\zeta, z, t, s) = (s\widetilde{c}(\zeta, z, t), \zeta - z), \qquad (1.8)$$

in other words

$$\xi = s\widetilde{c}(\zeta, z, t)$$
$$\eta = \zeta - z.$$

Let

$$N = M \circ \psi.$$

be the pull-back of the form M. Since M is closed we also have

$$dN = 0. \qquad (1.9)$$

The form N can be decomposed into a sum

$$N = N' + N'' \wedge ds, \qquad (1.10)$$

where N' and N'' are forms which do not contain ds. Similarly, the total differential d which occurs in (1.9) is the sum

$$d = d_{\zeta,z,t} + d_s, \qquad (1.11)$$

where the indices denote the variables of differentiation. Substitution of (1.10) and (1.11) into (1.9) yields

$$d_s N' = -d_{\zeta,z,t} N'' \wedge ds. \qquad (1.12)$$

We define integration of N with respect to s over an interval J as

$$\int_J N \, ds = \int_J N'' \, ds = \int_J N'' \wedge ds \qquad (1.13)$$

and introduce the exterior form

$$H = \int_{-\infty}^{0} N \, ds \qquad (1.14)$$

on $W \times I$. (Existence of the integral will be shown a few lines later.) Then, in view of (1.12),

$$\begin{aligned} dH &= \int_{-\infty}^{0} d_{\zeta,z,t} N'' \wedge ds \\ &= -\int_{-\infty}^{0} d_s N' \\ &= (N'(0) - N'(-\infty)) \\ &= 0, \end{aligned}$$

cf. (1.24). We state this result as

Lemma 1.15 $dH = d \int_{-\infty}^{0} N ds = 0.$

This information will now be used as the corresponding result (1.9). So we decompose

$$H = H' + H'' \wedge dt, \qquad (1.16)$$

where H', H'' are forms in the ζ- and z-differentials alone, decompose correspondingly

$$d_{\zeta,z,t} = d_{\zeta,z} + d_t, \qquad (1.17)$$

and insert (1.16) and (1.17) into Lemma (1.15). This gives the analogue of (1.12) for H:

$$d_t H' = -d_{\zeta,z} H'' \wedge dt. \qquad (1.18)$$

Let us integrate once more, now with respect to t:

$$R = \int_0^1 H \, dt = \int_0^1 H'' \, dt = \int_0^1 H'' \wedge dt. \tag{1.19}$$

Then

$$dR = d \int_0^1 H'' \wedge dt = \int_0^1 d_{\zeta,z} H'' \wedge dt$$

$$= - \int_0^1 d_t H' = \int_0^1 H'_t \wedge dt$$

$$= H'(1) - H'(0)$$

$$= H(1) - H(0).$$

We state this result as

Theorem 1.20 *If H and R are defined by (1.14) and (1.19), then*

$$H(1) - H(0) = dR.$$

This tells us that certain exterior forms on W, defined in terms of two Leray maps $\widetilde{a}$ and $\widetilde{b}$, are d-homologous. It contains the complete information that we need in the study of Cauchy-Fantappiè forms. Before we show this it is time to define the differential forms associated with $\widetilde{a}$ and $\widetilde{b}$.

Definition 1.21 *Let $\widetilde{a} = (a_j)_{j=1...n}$ be a Leray map on W. The differential form*

$$a = \sum_{i=1}^n a_j \, d\zeta_j \tag{1.22}$$

is called the Leray form associated with $\widetilde{a}$, the form

$$A = \sum a_j (d\zeta_j - dz_j) \tag{1.23}$$

the complete Leray form.

Both (1.22) and (1.23) will be considered, for the time being, as exterior forms on W. – Similarly we associate with $\widetilde{b}$ and $\widetilde{c}$ the forms b, B resp. c, C defined by the formula (1.22) and (1.23).

Let us return to the definition of N and H. We have, because of (1.7),

$$N = \frac{1}{n!} e^s [d(sC)]^n \tag{1.24}$$
$$= \frac{1}{n!} e^s [s\, dC - C \wedge ds]^n$$
$$= -\frac{1}{(n-1)!} e^s s^{n-1} C \wedge (dC)^{n-1} \wedge ds + \frac{1}{n!} e^s s^n (dC)^n.$$

This shows, in particular, the existence of H, and gives $N(0) = N(-\infty) = 0$ as well as

$$H = \int_{-\infty}^{0} N\, ds = -\frac{1}{(n-1)!} \left[\int_{-\infty}^{0} e^s s^{n-1}\, ds \right] C \wedge (dC)^{n-1} \tag{1.25}$$
$$= (-1)^n C \wedge (dC)^{n-1}.$$

Now insert the definition
$$C = tA + (1-t)B \tag{1.26}$$

into (1.25) and note
$$C \wedge (B - A) = A \wedge B, \tag{1.27}$$

to obtain

$$H = (-1)^n C \wedge [d(tA + (1-t)B)]^{n-1} \tag{1.28}$$
$$= (-1)^n (n-1) A \wedge B \wedge [t\, dA + (1-t)\, dB]^{n-2} \wedge dt$$
$$+ (-1)^n C \wedge [t\, dA + (1-t)\, dB]^{n-1}.$$

From (1.28) we get

$$R = (-1)^n (n-1) A \wedge B \wedge \int_0^1 (t\, dA + (1-t)\, dB)^{n-2}\, dt; \tag{1.29}$$

since
$$H(0) = (-1)^n B \wedge (dB)^{n-1}, \quad H(1) = (-1)^n A \wedge (dA)^{n-1}, \tag{1.30}$$

so a second, more explicit, version of Theorem 1.20 reads as follows:

Theorem 1.20$'$

$$A \wedge (dA)^{n-1} - B \wedge (dB)^{n-1} = d \left[(n-1) A \wedge B \wedge \int_0^1 (t\, dA + (1-t)\, dB)^{n-2}\, dt \right].$$

The forms involved are exterior forms on $W \subset \mathbb{C}^n \times \mathbb{C}^n$; they decompose into a sum of terms which have a well defined type (p, q) in ζ and a well defined type (r, s) in z. We recall that the quadruple $(p, q; r, s)$ is the double type and $(p+q; r+s)$ is the double degree of the corresponding term.

We are interested, for fixed q, in the terms of double type

$$(n, n - q - 1; 0, q)$$

in Theorem 1.20'. In order to single them out we can replace, in Theorem 1.20', A and B by a and b and the differential

$$d = \partial_\zeta + \overline{\partial}_\zeta + \partial_z + \overline{\partial}_z$$

by its antiholomorphic part $\overline{\partial}_\zeta + \overline{\partial}_z$. Using

$$a \wedge \left[(\overline{\partial}_\zeta + \overline{\partial}_z)a \right]^{n-1} = a \wedge \sum_{k=0}^{n-1} \binom{n-1}{k} (\overline{\partial}_\zeta a)^{n-k-1} \wedge (\overline{\partial}_z a)^k$$

we get on the left hand side of the equation

$$\binom{n-1}{q} a \wedge (\overline{\partial}_\zeta a)^{n-q-1} \wedge (\overline{\partial}_z a)^q - \binom{n-1}{q} b \wedge (\overline{\partial}_\zeta b)^{n-q-1} \wedge (\overline{\partial}_z b)^q \qquad (1.31)$$

as the required component. – The right hand side causes more work. First we have, with $\overline{\partial} = \overline{\partial}_\zeta + \overline{\partial}_z$:

$$a \wedge b \wedge \int_0^1 [t\,\overline{\partial}a + (1-t)\,\overline{\partial}b]^{n-2}\, dt \qquad (1.32)$$

$$= a \wedge b \wedge \sum_{k=0}^{n-2} \binom{n-2}{k} \left[\int_0^1 t^{n-2-k}(1-t)^k\, dt \right] (\overline{\partial}a)^{n-2-k} \wedge (\overline{\partial}b)^k$$

$$\overset{\text{def}}{=} a \wedge b \wedge \sum_{k=0}^{n-2} \binom{n-2}{k} T(n,k)(\overline{\partial}a)^{n-2-k} \wedge (\overline{\partial}b)^k$$

$$= a \wedge b \wedge \sum_{k,l,m} \binom{n-2}{k} T(n,k) \binom{n-2-k}{l} \binom{k}{m} (\overline{\partial}_\zeta a)^{n-2-k-l}$$

$$\wedge (\overline{\partial}_z a)^l \wedge (\overline{\partial}_\zeta b)^{k-m} \wedge (\overline{\partial}_z b)^m$$

$$\overset{\text{def}}{=} a \wedge b \wedge \sum_{k,l,m} S(n,k,l,m).$$

Comparing double types one sees that the only terms of double type $(n, n - q - 1; 0, q)$ on the right hand side of Theorem 1.20' are

$$\overline{\partial}_\zeta \left[(n-1)a \wedge b \wedge \sum_{\substack{k,l,m \\ l+m=q}} S(n,k,l,m) \right] \qquad (1.33)$$

and

$$\overline{\partial}_z \left[(n-1) a \wedge b \wedge \sum_{\substack{k,l,m \\ l+m=q-1}} S(n,k,l,m) \right].$$ (1.34)

Now

$$T(n,k) = \frac{k!(n-2-k)!}{(n-1)!}.$$ (1.35)

We set

$$h = n - 2 - k - l$$

in (1.33), (1.34) and (1.35) and extend the summation over l and h. Putting everything together yields

Theorem 1.20″

$$\binom{n-1}{q} a \wedge (\overline{\partial}_\zeta a)^{n-1-q} \wedge (\overline{\partial}_z a)^q - \binom{n-1}{q} b \wedge (\overline{\partial}_\zeta b)^{n-1-q} \wedge (\overline{\partial}_z b)^q$$

$$= \overline{\partial}_\zeta \sum_{l,h} \binom{l+h}{l} \binom{n-2-l-h}{q-l} a \wedge b$$

$$\wedge (\overline{\partial}_\zeta a)^h \wedge (\overline{\partial}_\zeta b)^{n-2-q-h} \wedge (\overline{\partial}_z a)^l \wedge (\overline{\partial}_z b)^{q-l}$$

$$+ \overline{\partial}_z \sum_{l,h} \binom{l+h}{l} \binom{n-2-l-h}{q-1-l} a \wedge b$$

$$\wedge (\overline{\partial}_\zeta a)^h \wedge (\overline{\partial}_\zeta b)^{n-1-q-h} \wedge (\overline{\partial}_z a)^l \wedge (\overline{\partial}_z b)^{q-1-l}.$$

In our final result we want to use the language of double forms.

Definition 1.36 *Let a be a double form of double type* $(1,0;0,0)$. *The double differential forms of double type* $(n, n-q-1; 0, q)$

$$\Omega_q(a) = (-1)^{q(q-1)/2} \binom{n-1}{q} \left(\frac{1}{2\pi i} \right)^n a \wedge (\overline{\partial}_\zeta a)^{n-q-1} \wedge (\overline{\partial}_z a)^q$$

are called Cauchy-Fantappiè kernels for $(0,q)$-*forms, associated with* a.

If

$$a = \sum_{j=1}^n \frac{\overline{\zeta}_j - \overline{z}_j}{\|\zeta - z\|^2} d\zeta_j,$$

then

$$\Omega_q(a) = B_{nq},$$

the Bochner-Martinelli kernel for $(0,q)$-forms.

Definition 1.37 *Let a and b be two double forms of double type $(1, 0; 0, 0)$. The double forms*

$$A_q(a, b) = (-1)^{q(q+1)/2} \left(\frac{1}{2\pi i} \right)^n \cdot$$

$$\cdot \sum_{l,h} \binom{l + h}{l} \binom{n - 2 - l - h}{q - l} a \wedge b \wedge (\bar{\partial}_\zeta a)^h \wedge (\bar{\partial}_\zeta b)^{n-2-q-h}$$

$$\wedge (\bar{\partial}_z a)^l \wedge (\bar{\partial}_z b)^{q-l}$$

are called transition kernels between the $\Omega_q(b)$ and $\Omega_q(a)$. – Summation is over $0 \leq l \leq q$ and $0 \leq h \leq n - 2 - q$.

In either case, $\wedge$ denotes the exterior product of double forms. We have $0 \leq q \leq n - 1$ in Definition 1.36 and $0 \leq q \leq n - 2$ in Definition 1.37. It is consistent to set Ω_{-1} or $\Omega_n = 0$, and also A_{-1} or $A_{n-1} = 0$.

We can now use the translation homomorphism τ of I.§1 to pass from exterior forms to double forms. On carefully keeping track of all the sign rules of I.§1 ((1.7), (1.9), (1.10), (1.12)) we arrive, starting from 1.20′, first at

$$(-1)^{q(q+1)/2} \binom{n - 1}{q} \left[a \wedge (\bar{\partial}_\zeta a)^{n-1-q} \wedge (\bar{\partial}_z a)^q \right.$$

$$\left. - b \wedge (\bar{\partial}_\zeta b)^{n-1-q} \wedge (\bar{\partial}_z b)^q \right]$$

$$= (-1)^{q(q+1)/2} \bar{\partial}_\zeta \sum_{l,h} \binom{l + h}{l} \binom{n - 2 - l - h}{q - l} a \wedge b$$

$$\wedge (\bar{\partial}_\zeta a)^h \wedge (\bar{\partial}_\zeta b)^{n-2-q-h} \wedge (\bar{\partial}_z a)^l \wedge (\bar{\partial}_z b)^{q-l}$$

$$+ (-1)^{q-1} (-1)^{q(q-1)/2} \bar{\partial}_z \sum_{l,h} \binom{l + h}{l} \binom{n - 2 - l - h}{q - 1 - l} a \wedge b$$

$$\wedge (\bar{\partial}_\zeta a)^h \wedge (\bar{\partial}_\zeta b)^{n-1-q-h} \wedge (\bar{\partial}_z a)^l \wedge (\bar{\partial}_z b)^{q-1-l}.$$

finally, using the two previous definitions, at Koppelman's homotopy formula

Theorem 1.38 *Let a and b be Leray forms on W. Then*

$$\Omega_q(b) - \Omega_q(a) = (-1)^{q+1} \bar{\partial}_\zeta A_q(a, b) + \bar{\partial}_z A_{q-1}(a, b).$$

In most cases the detailed information on the coefficients is not required; it is enough to know the fact that two CF kernels differ by a sum of $\bar{\partial}_\zeta$- and $\bar{\partial}_z$-exact double forms which can be explicitly constructed from the two Leray forms involved.

As a corollary we have

Proposition 1.39 *Let a be a Leray form. Then*

$$\overline{\partial}_z\Omega_{q-1}(a) = (-1)^q\overline{\partial}_\zeta\Omega_q(a).$$

Proof   This is true if $a = b$, the generating form of the BM kernel – see I.4.12. Theorem 1.38 now gives

$$\overline{\partial}_z\Omega_{q-1}(b) - \overline{\partial}_z\Omega_{q-1}(a) = (-1)^q\overline{\partial}_z\overline{\partial}_\zeta A_{q-1}(a,b)$$
$$\overline{\partial}_\zeta\Omega_q(b) - \overline{\partial}_\zeta\Omega_q(a) = \qquad \overline{\partial}_\zeta\overline{\partial}_z A_{q-1}(a,b);$$

multiplying the second equation with $(-1)^q$ and adding up gives

$$\overline{\partial}_z\Omega_{q-1}(a) - (-1)^q\overline{\partial}_\zeta\Omega_q(a) = \overline{\partial}_z\Omega_{q-1}(b) - (-1)^q\overline{\partial}_\zeta\Omega_q(b) = 0.$$

□

§2   A Generalisation of the Bochner-Martinelli-Koppelman Formula

As a first application of Koppelman's homotopy formula we shall establish a BMK type formula adapted to an arbitrary hermitian metric in $\mathbb{C}^n$. The case of the Euclidean metric yields the BMK formula.

Let

$$ds^2 = \sum g_{jk}(\zeta)\, d\zeta^j\, d\overline{\zeta}^k \tag{2.1}$$

be a hermitian metric and introduce the Leray form

$$h(\zeta,z) = \frac{1}{R^2}\sum_j\left(\sum_k g_{jk}(\zeta)(\overline{\zeta}^k - \overline{z}^k)\right)d\zeta^j \tag{2.2}$$

with

$$R^2 = \sum_{j,k}g_{jk}(\zeta)(\zeta^j - z^j)(\overline{\zeta}^k - \overline{z}^k). \tag{2.3}$$

The form h is defined on $\mathbb{C}^n \times \mathbb{C}^n - \Delta$. Note that R^2 is a good approximation to the square of the geodesic distance – cf. [Rha 60]. – The associated Cauchy- Fantappiè kernels are

$$H_q(\zeta,z) = \Omega_q(h); \tag{2.4}$$

we call them generalised Bochner-Martinelli-Koppelman kernels .

$$B_q(\zeta,z) = \Omega_q(b) \tag{2.5}$$

denotes the Bochner-Martinelli-Koppelman kernel (which corresponds to the Euclidean metric). Theorem 1.38 (the homotopy formula) connects K_q with B_q:

$$B_q = H_q + (-1)^{q+1}\overline{\partial}_\zeta A_q(h,b) + \overline{\partial}_z A_{q-1}(h,b). \tag{2.6}$$

All the terms of (2.6) are regular outside the diagonal Δ; their singularity on Δ is integrable.

Let now $\Omega \subset\subset \mathbb{C}^n$ be a bounded domain with piecewise smooth boundary $b\Omega$ and f a C^1 form of type $(0, q)$ on $\overline{\Omega}$. By the BMK formula we have for $z \in \Omega$

$$f(z) = \int_{b\Omega} f(\zeta) \wedge B_q(\zeta, z) - \int_{\Omega} \overline{\partial} f(\zeta) \wedge B_q(\zeta, z) - \overline{\partial}_z \int_{\Omega} f(\zeta) \wedge B_{q-1}(\zeta, z) \quad (2.7)$$

$$\overset{\text{def}}{=} B[f].$$

We define $H[f]$ in analogy to $B[f]$ and insert (2.6) into (2.7):

$$f = B[f] \tag{2.8}$$

$$= H[f] + (-1)^{q+1} \int_{b\Omega} f \wedge \overline{\partial} A_q + \overline{\partial}_z \int_{b\Omega} f \wedge A_{q-1}$$

$$- (-1)^{q+1} \int_{\Omega} \overline{\partial} f \wedge \overline{\partial} A_q - \overline{\partial}_z \int_{\Omega} \overline{\partial} f \wedge A_{q-1}$$

$$- \overline{\partial}_z \left[(-1)^q \int_{\Omega} f \wedge \overline{\partial} A_{q-1} + \overline{\partial}_z \int_{\Omega} f \wedge A_{q-2} \right]$$

$$= H[f] + \mathrm{I} + \cdots + \mathrm{VI}.$$

Now

$$\mathrm{I} = (-1)^{q+1} \int_{b\Omega} f \wedge \overline{\partial} A_q$$

$$= -(-1)^q (-1)^{q+1} \int_{b\Omega} \overline{\partial} f \wedge A_q$$

$$= (-1)^{q+1} \int_{\Omega} \overline{\partial} f \wedge \overline{\partial} A_q$$

$$= -\mathrm{III};$$

we had to apply Stokes' theorem in two steps because of the non-integrability of the differentiated kernel. Similarly

$$\mathrm{II} + \mathrm{IV} + \mathrm{V}$$

$$= \overline{\partial}_z \left[\int_{b\Omega} f \wedge A_{q-1} - \int_{\Omega} \overline{\partial} f \wedge A_{q-1} - (-1)^q \int_{\Omega} f \wedge \overline{\partial} A_{q-1} \right]$$

$$= 0,$$

again by Stokes' theorem. Finally $\mathrm{VI} = 0$. So (2.8) simply reads as

Theorem 2.9 (Generalised BMK formula) *Let the kernels H_q be defined as in (2.4) in terms of the metric (2.1). Then if f is a C^1-form of type $(0, q)$ on $\overline{\Omega}$, $\Omega \subset\subset \mathbb{C}^n$ a domain with piecewise smooth boundary, we have for $z \in \Omega$:*

$$f(z) = \int_{b\Omega} f(\zeta) \wedge H_q(\zeta, z) - \int_{\Omega} \overline{\partial} f(\zeta) \wedge H_q(\zeta, z) - \overline{\partial}_z \int_{\Omega} f(\zeta) \wedge H_{q-1}(\zeta, z)$$

We still want to study the derivatives of the generalised BM transform

$$Hf = \int_{\Omega} f(\zeta) \wedge H_{q-1}(\zeta, z),$$

where f is a $(0, q)$-form which we take in $C^1(\Omega) \cap L^1(\Omega)$. The result is

Theorem 2.10

$$\frac{\partial}{\partial z^k} \int_{\Omega} f(\zeta) \wedge H_{q-1}(\zeta, z) = \text{p.v.} \int_{\Omega} f(\zeta) \wedge \frac{\partial}{\partial z^k} H_{q-1}(\zeta, z) + E_{1-2n} f$$

$$\frac{\partial}{\partial \overline{z}^k} \int_{\Omega} f(\zeta) \wedge H_{q-1}(\zeta, z) = -\frac{g(z)}{n} f_k(z) + \text{p.v.} \int_{\Omega} f(\zeta) \wedge \frac{\partial}{\partial \overline{z}^k} H_{q-1}(\zeta, z) + E_{1-2n} f$$

Some comments are in order. The principal value integrals are defined as

$$\lim_{\varepsilon \to 0} \int_{U - B_\varepsilon(z)} f(\zeta) \wedge \ldots$$

with

$$B_\varepsilon(z) = \left\{ \zeta : R_z^2 = \sum g_{ij}(z)(\zeta^2 - z^2)(\overline{\zeta}^j - \overline{z}^j) \leq \varepsilon^2 \right\}. \tag{2.11}$$

The form f_k is defined as before by

$$f = d\overline{\zeta}^k \wedge f_k + f^k,$$

where f_k and f^k do not contain $d\overline{\zeta}^k$. Finally,

$$g(z) = \det g_{ij}(z).$$

E_{1-2n} is an integral operator with isotropic kernel $\mathcal{E}_{1-2n}$.

The proof of 2.10 runs completely parallel to the corresponding result for the BM kernel; so we only point out the essential steps.

The essential work has to be done for compactly supported forms f. If D^z is a derivative with respect to z^k or $\overline{z}^k$, we have – setting $H = H_{q-1}$ –

$$D^z \int_{\Omega} f(\zeta) \wedge H(\zeta, z) = D^z \int_{\mathbb{C}^n} f(\zeta) \wedge H(\zeta, z) \tag{2.12}$$

$$= D^z \int_{\mathbb{C}^n} f(z + u) \wedge H(z + u, z).$$

Now set

$$H = \frac{1}{R^{2n}} \widetilde{H}. \tag{2.13}$$

Then

$$D^z H(z + u, z) = D^z \frac{\widetilde{H}(z + u, z)}{(\sum g_{ij}(z + u) u_i \overline{u}_j)^n}$$

$$= O\left(\frac{1}{\|u\|^{2n-1}}\right);$$

hence, by deriving under the integral sign, (2.12) becomes

$$\int\limits_{\mathbb{C}^n} D^z f(z + u) \wedge H(z + u, z) + \int\limits_{\mathbb{C}^n} f(z + u) \wedge D^z H(z + u, z)$$

$$= \int\limits_{\Omega} D^\zeta f(\zeta) \wedge H(\zeta, z) + E_{1-2n} f.$$

So we have to compute

$$J = \int\limits_{\Omega} D^\zeta f(\zeta) \wedge H(\zeta, z). \tag{2.14}$$

One has to distinguish the cases $D^z = \frac{\partial}{\partial z^k}$ and $D^z = \frac{\partial}{\partial \overline{z}^k}$. As before, the first case is easier; so let us immediately turn to the second case.

We introduce the decomposition

$$\begin{aligned} f &= d\overline{\zeta}^k \wedge f_k + f^k, \\ \widetilde{H} &= d\overline{\zeta}^k \wedge \widetilde{H}_k + \widetilde{H}^k, \\ H &= d\overline{\zeta}^k \wedge H_k + H^k, \end{aligned} \tag{2.15}$$

by extracting $d\overline{\zeta}^k$, and obtain

$$J = \int\limits_{\Omega} d\overline{\zeta}^k \wedge \frac{\partial f_k}{\partial \overline{\zeta}^k} \wedge H^k + \int\limits_{\Omega} \frac{\partial f^k}{\partial \overline{\zeta}^k} \wedge d\overline{\zeta}^k \wedge H_k, \tag{2.16}$$

$$\overset{\text{def}}{=} J_1 + J_2.$$

(Note, that k does not indicate the type of the forms in question; H_k is of double type $(n, n - q - 1; 0, q - 1)$).

To treat J_2 we use

$$d(f^k \wedge H_k) = d\overline{\zeta}^k \wedge \frac{\partial f^k}{\partial \overline{\zeta}^k} \wedge H_k + d\overline{\zeta}^k \wedge f^k \wedge \frac{\partial H_k}{\partial \overline{\zeta}^k}$$

and obtain, setting

$$U_\varepsilon(z) = \{\zeta : R^2(\zeta, z) \le \varepsilon^2\}$$
$$\Gamma_\varepsilon(z) = \{\zeta : R^2(\zeta, z) = \varepsilon^2\}, \tag{2.17}$$

$$J_2 = \lim_{\varepsilon \to 0} \int_{\Omega - U_\varepsilon(z)} \frac{\partial f^k}{\partial \bar\zeta^k} \wedge d\bar\zeta^k \wedge H_k$$

$$= (-1)^q \lim_{\varepsilon \to 0} \int_{\Omega - U_\varepsilon(z)} d(f^k \wedge H_k) \tag{2.18}$$

$$- \lim_{\varepsilon \to 0} \int_{\Omega - U_\varepsilon(z)} f^k \wedge d\bar\zeta^k \wedge \frac{\partial H_k}{\partial \bar\zeta^k}$$

If the first limit exists, then so does the second limit, with value

$$\text{p.v.} \int_\Omega f^k \wedge d\bar\zeta^k \wedge \frac{\partial H_k}{\partial \bar\zeta^k} \tag{2.19}$$

Now

$$\int_{\Omega - U_\varepsilon(z)} d(f^k \wedge H_k) = - \int_{\Gamma_\varepsilon(z)} f^k \wedge H_k$$

$$= O(\varepsilon) - \int_{\Gamma_\varepsilon(z)} f^k(z) \wedge H_k(\zeta, z), \tag{2.20}$$

where $f^k(z)$, in (2.20), is the form $f^k(\zeta)$ with coefficients evaluated at $\zeta = z$.
The last integral in (2.20) yields

$$\frac{1}{\varepsilon^{2n}} \int_{\Gamma_\varepsilon(z)} f^k(z) \wedge \tilde H_k(\zeta, z) = \frac{1}{\varepsilon^{2n}} \int_{U_\varepsilon(z)} (-1)^q f^k(z) \wedge d\tilde H_k(\zeta, z)$$

$$= (-1)^q \int_{U_\varepsilon(z)} f^k(z) \wedge d\bar\zeta^k \wedge \frac{\partial \tilde H_k}{\partial \bar\zeta^k} \tag{2.21}$$

But by the very definition of H_k, we have

$$\frac{\partial \tilde H_k}{\partial \bar\zeta^k} = \mathcal{E}_1(\zeta, z). \tag{2.22}$$

Combining (2.20), (2.21) and (2.22) we obtain

$$\lim_{\varepsilon \to 0} \int_{\Omega - U_\varepsilon(z)} d(f^k \wedge H_k) = 0,$$

and so

$$J_2 = - \text{p.v.} \int_\Omega f^k \wedge d\bar{\zeta}^k \wedge \frac{\partial H_k}{\partial \bar{\zeta}^k}(\zeta, z)$$

$$= - \text{p.v.} \int_\Omega f^k(\zeta) \wedge \frac{\partial H}{\partial \bar{\zeta}^k}(\zeta, z),$$

(2.23)

because of

$$f^k \wedge \frac{\partial H^k}{\partial \bar{\zeta}^k} = 0.$$

To compute J_1 we use the identity

$$d(f_k \wedge H^k) = d\bar{\zeta}^k \wedge \frac{\partial f_k}{\partial \bar{\zeta}^k} \wedge H^k + d\bar{\zeta}^k \wedge f_k \wedge \frac{\partial H^k}{\partial \bar{\zeta}^k}$$

(2.24)

and proceed as before to arrive at

$$J = \lim_{\varepsilon \to 0} \int_{\Omega - U_\varepsilon(z)} d(f_k \wedge H^k) - \text{p.v.} \int_\Omega f \wedge \frac{\partial H}{\partial \bar{\zeta}^k}$$

(2.25)

The limit can be computed in the same way as the corresponding limit for B^k; but now the volume of the unit ball which appears in the BM case has to be replaced by the corresponding volume in the metric $g_{ij}(z)$, that is the Euclidean volume must be multiplied by $\det g_{ij}(z)$.

Finally for the principal values $U_\varepsilon(z)$ can be replaced by $B_\varepsilon(z)$ from (2.11), and

$$\frac{\partial}{\partial \bar{\zeta}^k} H + \frac{\partial}{\partial \bar{z}^k} H = \mathcal{E}_1.$$

(2.26)

This gives the required result.

§3 Notes

The first published proof of Koppelman's homotopy formula 1.38 (which was stated in [Kop 67$_2$]) is in [Lie 70]. Our proof seems to be new; it is an adaption of the Berndtsson-Andersson theory [BeA 83]. The original calculus of Cauchy-Fantappiè forms is due to Leray [Ler 59] and Norguet [Nor 60]; the history of those ideas is described in [Ran 98]. The generalised BMK formula was discovered by Lieb and Range (unpublished); the content of §2 is new.

A different treatment of the Cauchy-Fantappiè calculus is given by many authors [HeL 84], [HaP 79], [Øvr 71$_1$], [Ran 86]. Sign differences occur all the time depending on conventions on the orientation of $\mathbb{C}^n$ and whether exterior forms or double forms on product manifolds are considered. Chapter I, §1 should allow to clean up all those differences.

Chapter III

Strictly Pseudoconvex Domains in $\mathbb{C}^n$

The theory which we have developed so far will be expanded in two ways: 1° In this chapter we shall use the Cauchy-Fantappiè calculus to transform the Bochner-Martinelli kernel into a kernel adapted to strictly pseudoconvex domains in $\mathbb{C}^n$ and prove a number of finiteness and vanishing theorems for these domains; 2° in the next chapter we shall pass from $\mathbb{C}^n$ to arbitrary Hermitian manifolds and establish the analogue of the BMK formula in that context, obtaining, along the way, the Hodge decomposition theorem on compact manifolds. Both lines of thought will merge in the subsequent chapters and lead to a solution of the Levi problem and finiteness theorems for strictly pseudoconvex manifolds, as well as to a solution of the $\overline{\partial}$-Neumann problem (to be formulated later).

The main result of this chapter is the homotopy formula of §4

$$f = P_q f + T_q \overline{\partial} f + \overline{\partial} S_{q-1} f$$

with explicit integral operators P_q, T_q, S_{q-1}, valid for a form f in L^2_{0q} with $\overline{\partial} f \in L^2_{0q+1}$ on a strictly pseudoconvex domain. This formula leads to linear L^2-bounded solution operators for the Cauchy-Riemann system and to a solution of Levi's problem in $\mathbb{C}^n$ (see §6).

§1   Strict Pseudoconvexity

1.   The *Levi form* of a real-valued C^2-function φ on a domain $U \subset \mathbb{C}^n$ at a point $z \in U$ is the hermitian form

$$L_\varphi(z; t) = \sum \frac{\partial^2 \varphi}{\partial z^j \partial \overline{z}^k}(z) t_j \overline{t}_k; \tag{1.1}$$

the corresponding hermitian matrix

$$L_\varphi(z) = \left(\frac{\partial^2 \varphi}{\partial z^j \partial \overline{z}^k}(z) \right)_{j,k=1\ldots n} \tag{1.2}$$

is the *Levi matrix*. It depends on the coordinates. More precisely, if $z = f(w)$ is a holomorphic map with Jacobian

$$J_f(w) = \left(\frac{\partial z^i}{\partial w^j} \right)_{i,j=1\ldots n}, \tag{1.3}$$

then

$$L_{\varphi \circ f}(w) = {}^t J_f(w) L_\varphi(f(w)) \overline{J}_f(w). \tag{1.4}$$

Consequently, for each point z we can find a linear (unitary) coordinate change f keeping z fixed such that $L_{\varphi \circ f}(z)$ is a diagonal matrix.

Definition 1.5 φ *is strictly plurisubharmonic (at z) if L_φ is positive definite (everywhere resp. at z).*

If φ is strictly plurisubharmonic the Levi matrix

$$L_\varphi = (\varphi_{i\bar{j}})_{i,j=1\ldots n}$$

defines a Hermitian metric

$$ds_\varphi^2 = \sum_{i,j} \varphi_{i\bar{j}} \, dz^i d\overline{z}^j. \tag{1.6}$$

(Here and frequently in the sequel we use the abbreviations

$$\varphi_i = \frac{\partial \varphi}{\partial z^i}, \quad \varphi_{i\bar{j}} = \frac{\partial^2 \varphi}{\partial z^i \partial \overline{z}^j}, \quad \text{etc.}) \tag{1.7}$$

From (1.4) we deduce that the notion of strict plurisubharmonicity is invariant under biholomorphic maps, and the same is true for the metric (1.6). Both concepts therefore carry over to complex manifolds.

Definition 1.8   *The metric (1.6) is called the Levi metric associated with (the strictly plurisubharmonic function) φ.*

It is a Kähler metric with fundamental form $\partial\overline{\partial}\varphi$; this latter form is also occasionally called the Levi form of φ.

The easiest example of a strictly plurisubharmonic function is the Euclidean (squared) norm

$$\|z\|^2 = \sum_{j=1}^n z^j \overline{z}^j;$$

whose Levi matrix is the identity matrix E (and the Levi metric is the Euclidean metric in $\mathbb{C}^n$).

2.   **Definition 1.9**   *A domain $\Omega \subset\subset \mathbb{C}^n$ is strictly pseudoconvex if there is a strictly plurisubharmonic function r with non-vanishing differential dr defined on a neighbourhood U of the boundary $b\Omega$ such that*

$$\Omega \cap U = \{z \in U : r(z) < 0\}. \tag{1.10}$$

The function r is called a strictly plurisubharmonic *defining function* for Ω. It is not uniquely determined, and not every defining function for Ω (i.e. a function r satisfying (1.10)) is strictly plurisubharmonic. The following facts, however, are easy to show (and well known – see for instance [Ran 86]): *If r is a defining function for a strictly pseudoconvex domain Ω, then*

$$\sum_{i,j} r_{i\bar{j}} t_i \bar{t}_j > 0 \tag{1.11}$$

for all $z \in b\Omega$ and $t \in \mathbb{C}^n - \{0\}$ satisfying

$$\sum_i r_i(z) t_i = 0.$$

Moreover, *there is a positive constant A such that*

$$r_A(z) = r(z) \exp Ar(z) \tag{1.12}$$

is a strictly plurisubharmonic defining function for Ω.

3.   Let us now consider a strictly pseudoconvex domain Ω with a strictly plurisubharmonic defining function r on $U(b\Omega)$. The *length* of the $(1,0)$-form ∂r in the Levi metric associated with r is given by the formula

$$|\partial r|_r^2 = 2 \sum_{j,k} r_j s^{j\bar{k}} r_{\bar{k}}, \tag{1.13}$$

where the matrix S is ${}^t L_r^{-1}$. It will be convenient to have

Proposition 1.14 *Let $b\Omega$ be C^3-smooth. There is a strictly plurisubharmonic defining function (of class C^2) for Ω such that*

$$|\partial r(z)|_r = 1$$

for $z \in b\Omega$.

Proof   We slightly generalise (1.12) and introduce

$$r_a(z) = r(z) \exp \left(\frac{a(z)}{2} r(z) \right)$$

with a real valued function $a(z)$ which has to be determined. Here r is a strictly plurisubharmonic defining function of class C^3. Let us introduce the following notations:

$$L_a = L_{r_a} = (r_{a,i\bar{j}})_{i,j=1\dots n}$$
$$S_a = {}^t L_a^{-1} = (s_a^{i\bar{j}})_{i,j=1\dots n}$$
$$L_0 = L, \quad S_0 = S.$$

The column vector of the r_i will be simply denoted by ∂r. Then (1.13) reads as a matrix equation:

$$\frac{1}{2}|\partial r|^2 = {}^t\partial r \cdot S \cdot \bar{\partial} r, \tag{1.13'}$$

where the norm is taken with respect to the Levi metric induced by $r = r_0$. Similarly,

$$\frac{1}{2}|\partial r_a|_a^2 = {}^t\partial r_a \cdot S_a \cdot \bar{\partial} r_a. \tag{1.13''}$$

An easy computation gives, on the boundary of Ω,

$$\begin{aligned} r_{a,i} &= r_i \\ r_{a,i\bar{j}} &= r_{i\bar{j}} + ar_ir_{\bar{j}}. \end{aligned} \tag{1.15}$$

We continue working on $b\Omega$. Then $\partial r = \partial r_a$, and in view of (1.15) we have

$$L_a - L = a\partial r \, {}^t\bar{\partial} r, \tag{1.16}$$

which implies by multiplying with L_a^{-1} and L^{-1} and transposing:

$$S - S_a = aS\bar{\partial} r \, {}^t\partial r S_a. \tag{1.17}$$

We can now compare $|\partial r|^2$ and $|\partial r|_a^2$:

$$\begin{aligned} \frac{1}{2}\left(|\partial r|^2 - |\partial r|_a^2\right) &= {}^t\partial r(S - S_a)\bar{\partial} r \\ &= a\,{}^t\partial r S\bar{\partial} r\,{}^t\partial r S_a\bar{\partial} r \\ &= \frac{a}{4}|\partial r|^2|\partial r|_a^2. \end{aligned}$$

Solving for $|\partial r|_a^2$ yields

$$|\partial r|_a^2 = \frac{|\partial r|^2}{1 + \frac{a}{2}|\partial r|^2}. \tag{1.18}$$

We first determine $a(z)$ so that $|\partial r|_a^2 \equiv c^2$, where $c > 0$ is an arbitrary constant. This leads to

$$a(z) = 2\frac{|\partial r|^2 - c^2}{c^2|\partial r|^2}. \tag{1.19}$$

If c is very small then a becomes arbitrarily large – and this guarantees the strict plurisubharmonicity of r_a. To sum up: there is a defining strictly plurisubharmonic function – which we now denote by r – such that

$$|\partial r|_r^2 \equiv c^2 \tag{1.20}$$

at $b\Omega$.

Now, if $\varkappa > 0$ is constant, we have by (1.13)

$$|\partial(\varkappa r)|_{\varkappa r}^2 \equiv \varkappa|\partial r|_r^2. \tag{1.21}$$

So setting finally

$$\tilde{r} = \frac{1}{c^2}r, \tag{1.22}$$

we get the required defining function.   □

4.   Complex analysis also deals with a more general class of domains.

Definition 1.23 *A domain $\Omega \subseteq \mathbb{C}^n$ which is the union of a sequence $\Omega_1 \subseteq \Omega_2 \cdots \subseteq \Omega$ of strictly pseudoconvex domains Ω_ν is called pseudoconvex.*

The quantitative function-theoretic results which we will establish in this book for strictly pseudoconvex domains carry over to the more general pseudoconvex domains only to a very limited extent. The precise extension is not yet known, and has been a topic of active and fruitful research. We will report on these results at various places in our book, without aiming at completeness and without developing the necessary methods which are not our concern here. At this point we simply state two elementary results for the proof of which we refer to [Hör 66] or [Ran 86].

Proposition 1.24 *A holomorphically convex domain in $\mathbb{C}^n$ is pseudoconvex.*

(Recall Definition 0.13.) The converse question, namely whether pseudoconvex domains are holomorphically convex, is the original version of Levi's problem; we shall solve this problem in the sequel.

Proposition 1.25 *Let $\Omega \subset \mathbb{C}^n$ be smoothly (C^2-) bounded, given by a defining function r. Then Ω is pseudoconvex if and only if*

$$L_r(z;t) \geq 0 \tag{1.26}$$

for all $z \in b\Omega$ and all $t \in \mathbb{C}^n$ satisfying

$$\sum_{j=1}^{n} \frac{\partial r}{\partial z^j}(z)t_j = 0. \tag{1.27}$$

More explicitly: if (1.27) holds for t and z, then

$$\sum_{j,k} \frac{\partial^2 r}{\partial z^j \overline{z}^k}(z)t_j\overline{t_k} \geq 0. \tag{1.28}$$

Condition (1.27) defines the holomorphic tangent space to $b\Omega$ at z, that is the maximal complex linear subspace in the real tangent space; on this space the Levi form of r is required to be positive semidefinite. — Note that the proposition implies that these conditions do not depend on the choice of r — this can be easily verified by a direct computation.

§2   The Levi Polynomial and Holomorphic Support Functions

1.   Let Ω be a strictly pseudoconvex domain given by a strictly plurisubharmonic defining C^k function $(k \geq 2)$

$$r : U(b\Omega) \longrightarrow \mathbb{R}.$$

Consider an arbitrary boundary point z_0. After an affine change of coordinates the Taylor development of r around z_0 becomes

$$r(z) = z^1 + \overline{z}^1 + \sum a_{ij} z^i z^j + \sum \overline{a}_{ij} \overline{z}^i \overline{z}^j + \sum c_{i\overline{j}} z^i \overline{z}^j + o(\|z\|^2) \qquad (2.1)$$

(so we have assumed $z_0 = 0$). The quadratic coordinate change

$$\begin{aligned} w^1 &= z^1 + \sum a_{ij} z^i z^j \\ w^k &= z^k, \quad k = 2 \ldots n \end{aligned} \qquad (2.2)$$

transforms (2.1) into

$$r(w) = w^1 + \overline{w}^1 + \sum c_{i\overline{j}} w^i \overline{w}^j + o(\|w\|^2). \qquad (2.3)$$

The Levi matrices of (2.1) and (2.3) coincide because of the transformation rule (1.4). A final unitary coordinate change transforms (2.3) into

$$r(u) = \sum_{i=1}^{n} a_i u^i + \sum_{i=1}^{n} \overline{a}_i \overline{u}^i + \sum_{i=1}^{n} \lambda_i |u^i|^2 + o(\|u\|^2), \qquad (2.4)$$

where the λ_i are positive. But this means that the Hesse matrix of r (in the u-coordinates) is positive definite. So the same was true for (2.3) and we have

Proposition 2.5 *If z_0 is a boundary point of the strictly pseudoconvex domain Ω, there is a neighbourhood U_0 of z_0 and a biholomorphic transformation $T : U_0 \xrightarrow{\sim} V_0 \subset \mathbb{C}^n$, such that $T(\Omega \cap U_0)$ is strictly convex and $T(b\Omega \cap U_0)$ is a closed strictly convex hypersurface in V_0. – T can be chosen as a quadratic transformation.*

Conversely a *strictly convex function r is* easily seen to be *strictly plurisubharmonic*: we may again assume that the Levi matrix L of r is in diagonal form; then the diagonal elements are

$$\lambda_j = \frac{1}{4} \left(\frac{\partial^2 r}{(\partial x^j)^2} + \frac{\partial^2 r}{(\partial y^j)^2} \right)$$

Since the Hesse matrix of r is positive definite one must have

$$\frac{\partial^2 r}{(\partial x^j)^2}, \ \frac{\partial^2 r}{(\partial y^j)^2} > 0,$$

and so the $\lambda_j > 0$.

As upshot of the above discussion strict pseudoconvexity is characterised as *strict Euclidean convexity localised and made biholomorphically invariant*. Now a strictly convex Euclidean hypersurface is supported by its tangent planes; in fact, in the situation (2.3) the plane

$$M = \{w : w^1 + \overline{w}^1 = 0\}$$

lies locally on one side of the surface

$$S = \{w : r(w) = 0\}$$

because we have on M

$$r(w) \geq 0$$
$$r(w) = 0 \quad \text{if and only if } w = 0.$$

Transforming back to (2.1) we can state

Proposition 2.6 *For each $\zeta \in b\Omega$ there is a holomorphic quadratic polynomial in z,*

$$F(\zeta, z) = \sum_j a_j(\zeta)(z^j - \zeta^j) + \sum_{j,k} a_{jk}(\zeta)(z^j - \zeta^j)(z^k - \zeta^k),$$

and a neighbourhood U_0 of ζ, such that

$$\{z \in U_0 : r(z) \leq 0, \operatorname{Re} F(\zeta, z) = 0\} = \{\zeta\}.$$

In other words,

$$M_\zeta = \{z : \operatorname{Re} F(\zeta, z) = 0\}$$

is a *local pluriharmonic supporting hypersurface for $S = b\Omega$.*

2. We need a quantitative version of 2.6. The Taylor development of the given defining function r around an arbitrary point ζ in U is

$$r(z) = r(\zeta) - \sum r_j(\zeta)(\zeta^j - z^j) + \frac{1}{2} \sum r_{jk}(\zeta)(\zeta^j - z^j)(\zeta^k - z^k)$$
$$- \sum r_{\bar{j}}(\zeta)(\overline{\zeta}^j - \overline{z}^j) + \frac{1}{2} \sum r_{\bar{j}\bar{k}}(\zeta)(\overline{\zeta}^j - \overline{z}^j)(\overline{\zeta}^k - \overline{z}^k) \qquad (2.7)$$
$$+ \sum r_{j\bar{k}}(\zeta)(\zeta^j - z^j)(\overline{\zeta}^k - \overline{z}^k) + o(\|\zeta - z\|^2).$$

Definition 2.8 *The polynomial*

$$F(\zeta, z) = \sum_{j=1}^n r_j(\zeta)(\zeta^j - z^j) - \frac{1}{2} \sum_{j,k=1}^n r_{jk}(\zeta)(\zeta^j - z^j)(\zeta^k - z^k)$$

is the Levi polynomial of the function r at ζ.

With this definition (2.7) becomes

$$r(z) = r(\zeta) - 2 \operatorname{Re} F(\zeta, z) + L_r(\zeta; \zeta - z) + o(\|\zeta - z\|^2). \tag{2.9}$$

As L_r is positive definite, there is a positive c such that

$$L_r(\zeta; \zeta - z) \geq 2c\|\zeta - z\|^2;$$

c can be chosen independent of ζ (restricted to a neighbourhood $V \subset\subset U$ of $b\Omega$). There is, moreover, another positive constant ε which can again be found to be independent of ζ such that

$$o(\|\zeta - z\|^2) \leq c\|\zeta - z\|^2$$

for $\|\zeta - z\| \leq \varepsilon$. From (2.9) we then deduce

$$2 \operatorname{Re} F(\zeta, z) \geq r(\zeta) - r(z) + c\|\zeta - z\|^2. \tag{2.10}$$

To sum up we have

Theorem 2.11 *i. Let the strictly pseudoconvex domain Ω be given by a strictly plurisubharmonic defining function r on a neighbourhood U of $b\Omega$. Then there is a neighbourhood V of $b\Omega$ in U and there are positive constants c and ε such that the Levi polynomial of r satisfies the estimate*

$$2 \operatorname{Re} F(\zeta, z) \geq r(\zeta) - r(z) + c\|\zeta - z\|^2$$

on $W_\varepsilon = \{(\zeta, z) : \zeta \in V, \|\zeta - z\| < \varepsilon\}$.

ii. If $\zeta \in b\Omega$, then for

$$M_\zeta = \{z : \operatorname{Re} F(\zeta, z) = 0\}$$

one has

$$M_\zeta \cap \overline{\Omega} \cap \{z : \|\zeta - z\| < \varepsilon\} = \{\zeta\}.$$

Statement *ii.* follows from *i.* (and contains less information.)

We shall need later the following additional properties of the Levi polynomial:

Proposition 2.12 *Suppose r is of class C^k, $k \geq 3$.*

i. $F(\zeta, z) = \sum F_j(\zeta, z)(\zeta^j - z^j)$, where the functions

$$F_j(\zeta, z) = r_j(\zeta) - \frac{1}{2} \sum r_{jk}(\zeta)(\zeta^k - z^k)$$

are in C^{k-2} and holomorphic in z.

ii. For all z in V,

$$d_\zeta r(\zeta) \wedge d_\zeta \operatorname{Im} F(\zeta, z) = 2i\, \partial r(\zeta) \wedge \overline{\partial} r(\zeta) + O(\|\zeta - z\|).$$

iii. $F(\zeta, z) - r(\zeta) = \overline{F(z, \zeta)} - r(z) + O(\|\zeta - z\|^3).$

Proof   *i.* follows from the construction. For *ii.* one notes

$$i\, d_\zeta(\text{Im } F(\zeta, z)) = \partial r(\zeta) - \overline{\partial} r(\zeta) + O(\|\zeta - z\|).$$

To prove *iii.* we develop $\overline{F}(z, \zeta) - r(z)$ around ζ and compare with the left-hand side: all terms of order ≤ 2 cancel.   □

Part *ii.* of the proposition tells us that $r(\zeta)$ and Im $F(\zeta, z)$ can serve as part of a local C^{k-2} coordinate system around z. – The symmetry property *iii.* will play a decisive role later.

3.   The preceding pages have shown that strictly pseudoconvex boundaries have a very simple complex geometry: after a biholomorphic transformation they turn out to be (locally) convex. Moreover, we have the precise estimate (2.10). Given the importance of this information in all that follows, it is natural to ask whether these properties are shared by general pseudoconvex boundaries. There is an immediate negative answer given by Kohn and Nirenberg:

Proposition 2.13   *There is a pseudoconvex domain in $\mathbb{C}^2$ with the following properties: $b\Omega$ is smooth and real analytic; $0 \in b\Omega$; there is no local holomorphic function h with $h(0) = 0$ and $\Omega \cap \{z : h(z) = 0\} = \emptyset$. In particular, Ω is not locally convexifiable by a biholomorphic map.*

For a proof see [ForS 87]. This means that the method of holomorphic support functions which we are going to use and which is based on the properties of the Levi polynomial proved above, cannot work in the general pseudoconvex case. But even if a holomorphic support function h_ζ exists for a boundary point ζ, it may happen that the contact set

$$A_\zeta = b\Omega \cap \{z : h_\zeta(z) = 0\}$$

contains more than just the point ζ. A result of Diederich and Fornæss [DiF 78] shows that in some cases A_ζ cannot contain complex analytic germs:

Theorem 2.14   *Let Ω be a bounded pseudoconvex domain in $\mathbb{C}^n$ with real analytic boundary $b\Omega$. Then $b\Omega$ does not contain any germ of a complex analytic set (of dimension > 0).*

In this case, analytic germs must have finite order of contact with $b\Omega$, because of the real analyticity of the boundary. It is this more stringent condition which seems to be essential:

Definition 2.15   *A bounded pseudoconvex domain Ω with smooth boundary is called a finite type domain if the order of contact between complex analytic curves and the boundary of Ω is bounded.*

So both real analytic pseudoconvex domains $\Omega \subset\subset \mathbb{C}^n$ and strictly pseudoconvex domains are of finite type. The precise notion of type underlying the above definition is still a topic of research; essential contributions are due to Catlin and D'Angelo, see [DAn 82], [Cat 84].

For a finite type domain holomorphic support functions, if they exist, may still admit non-trivial contact sets A which are, it is true, not too large. Nevertheless their existence would be an obstruction to good estimates for integral kernels constructed from these support functions. But one might hope to construct better support functions... This problem gives rise to many interesting open questions. We shall now discuss a case where a solution has recently been found.

4.   Trivially, convex domains admit holomorphic support functions given by the equation of the holomorphic tangent spaces

$$h_\zeta(z) = \sum_{j=1}^{n} \frac{\partial r}{\partial \zeta_j}(\zeta)(\zeta_j - z_j) = 0 \,.$$

This function $h_\zeta(z)$ in general has all the bad features discussed above: A_ζ is much too large. But if the domain is in addition of finite type, we have an analogue of the Levi polynomial which has been constructed recently by Diederich and Fornæss:

Theorem 2.16 *There are positive constants p, c, R and a smooth function $F(\zeta, z)$ on $b\Omega \times \mathbb{C}^n$, holomorphic in z, with $F(\zeta, \zeta) = 0$, such that*

$$\text{Re } F(\zeta, z) \geq c\|\zeta - z\|^p$$

for $\|\zeta - z\| \leq R$.

(This is a simplified version of the more precise result in [DiF 99].) F is constructed using the Taylor development of a convex defining function r, as in the construction of the Levi polynomial. But, as opposed to the strictly pseudoconvex case, this construction requires a deep understanding of the complex geometry of finite type boundaries which has been gradually built up in the work of the above mentioned authors over the last 25 years. In fact, more precise information on F as constructed above, is available.

We will not explain, in this book, the geometry of weakly pseudoconvex boundaries, but once more refer to [DAn 82], [Cat 84]. We shall, however, come back to an application of Theorem 2.16 in the framework of integral formulae in §8 and in Chapter VIII, §§1 and 2.

As to convex domains of infinite type — they seem to be completely out of reach, at present.

§3   The Basic Homotopy Formula for the Ball

0.   Let Ω be a strictly pseudoconvex domain in $\mathbb{C}^n$ given by a strictly plurisubharmonic defining function

$$r : U(b\Omega) \longrightarrow \mathbb{R}$$

with Levi polynomial

$$F(\zeta, z) = \sum_{j=1}^{n} F_j(\zeta, z)(\zeta^j - z^j)$$

$$F_j(\zeta, z) = r_j(\zeta) - \frac{1}{2} \sum_{k=1}^{n} r_{jk}(\zeta)(\zeta^k - z^k)$$

(Recall the notation

$$r_j(\zeta) = \frac{\partial r(\zeta)}{\partial \zeta^j},$$

etc.) F gives rise to a Leray form

$$k(\zeta, z) = \frac{1}{F(\zeta, z)} \sum_{j=1}^{n} F_j(\zeta, z) \, d\zeta^j$$

and corresponding CF kernels

$$K_q(\zeta, z) = \Omega_q(k), q = 0, \ldots, n.$$

Since k is holomorphic in z we have $K_q(\zeta, z) \equiv 0$ for $q \geq 1$, whereas $K_0(\zeta, z)$, which is $\not\equiv 0$, is still holomorphic in z. Our aim is to replace, in the Bochner-Martinelli integral representation for $(0, q)$-forms f:

$$f(z) = \int_{b\Omega} f(\zeta) \wedge B_{nq}(\zeta, z) - \int_{\Omega} \bar{\partial}f(\zeta) \wedge B_{nq}(\zeta, z) - \bar{\partial}_z \int_{\Omega} f(\zeta) \wedge B_{n,q-1}(\zeta, z),$$

the boundary integral by the corresponding boundary integral over K_q and integrals involving the transition kernels $A_q(k, b)$ between K_q and B_q. This has the advantage that we are left with an integral representation which contains, beside $\bar{\partial}$-exact respectively holomorphic terms, only integrals involving $\bar{\partial}f$; in particular, one obtains explicit solution operators to the CR equations.

Unfortunately, things are not so easy. In fact, the kernels $K_q(\zeta, z)$ are not defined on the whole set $b\Omega \times \Omega$ (which would be necessary for our procedure to work), but only close to the boundary diagonal. Therefore they have to be patched together with the BMK kernel: this technical difficulty causes the kernels $K_q(\zeta, z)$ to have the crucial properties of being holomorphic or zero only near the boundary diagonal. They do not immediately yield solution operators to the $\bar{\partial}$-equation; we shall see that the corresponding integrals are still "approximate" solution operators ("parametrices") for the CR system – which has to be (and will be) sufficient for our purposes.

For technical reasons it is also useful to replace all boundary integrals by volume integrals; this requires an additional modification of the kernels which, however, leaves their essential properties unchanged.

Before embarking on the general construction valid on arbitrary strictly pseudoconvex domains, we want to study one case where the above mentioned difficulties do not occur: *the case of the ball.*

1.   The *unit ball*

$$\mathbb{D} = \{\zeta \in \mathbb{C}^n : \|\zeta\|^2 < 1\}$$

is given by the strictly plurisubharmonic boundary function

$$r(\zeta) = \|\zeta\|^2 - 1.$$

Its Levi polynomial is

$$F(\zeta, z) = \sum_{j=1}^{n} \bar\zeta^j (\zeta^j - z^j) = \langle \zeta, \zeta - z \rangle, \tag{3.1}$$

where $\langle, \rangle$ denotes the hermitian scalar product on $\mathbb{C}^n$. Let us introduce the function

$$\Phi(\zeta, z) = F(\zeta, z) - r(\zeta) = 1 - \langle \zeta, z \rangle. \tag{3.2}$$

An immediate computation gives

Lemma 3.3   i. $\Phi(\zeta, z) = \overline{\Phi(z, \zeta)}$

ii. $2 \operatorname{Re} \Phi(\zeta, z) = -r(\zeta) - r(z) + \|\zeta - z\|^2$

This implies that the form

$$k_0(\zeta, z) = \frac{1}{\Phi(\zeta, z)} \partial r(\zeta) \tag{3.4}$$

is smooth on $\overline{\mathbb{D}} \times \mathbb{D}$. Let us also introduce the form

$$b_0(\zeta, z) = \frac{1}{P(\zeta, z)} \sum (\bar\zeta^j - \bar z^j) \, d\zeta^j \tag{3.5}$$

with

$$P(\zeta, z) = \|\zeta - z\|^2 + r(\zeta) r(z) \tag{3.6}$$

and the kernels

$$K_q(\zeta, z) = \Omega_q(k_0) \tag{3.7}$$
$$A_q(\zeta, z) = A_q(k_0, b_0) \tag{3.8}$$

(given in definition II.1.36 and II.1.37).

They have the following properties:

Lemma 3.9   i. $K_q(\zeta, z)$ and $A_q(\zeta, z)$ are regular on $\overline{\mathbb{D}} \times \mathbb{D}$.

ii. $K_0(\zeta, z)$ is holomorphic in z, $K_q(\zeta, z) \equiv 0$ for $q \geq 1$.

Now, for $r(\zeta) = 0$, we have the identities

$$k_0(\zeta, z) = k(\zeta, z) = \frac{1}{F(\zeta, z)} \partial r(\zeta)$$

$$b_0(\zeta, z) = b(\zeta, z) = \frac{1}{\|\zeta - z\|^2} \sum (\bar{\zeta}^j - \bar{z}^j) \, d\zeta^j;$$

consequently the forms $\Omega_q(k_0)$ and $\Omega_q(k)$, resp. $A_q(k_0, b_0)$ and $A_q(k, b)$, have the same pull-back to $b\mathbb{D} \times \mathbb{D}$. But k and b are *Leray forms*, the latter generating the Bochner-Martinelli-Koppelman kernel. Koppelman's homotopy formula can therefore be applied and yields

Lemma 3.10 *On $b\mathbb{D} \times \mathbb{D}$ – i. e.for the pull-back of the forms in question – one has*

$$B_{nq} = K_q + (-1)^{q+1} \bar{\partial}_\zeta A_q + \bar{\partial}_z A_{q-1}.$$

2.   We now use Lemma 3.10 in order to pursue the programme sketched above. If f is a C^1-$(0, q)$-form on $\overline{\mathbb{D}}$, the BMK formula reads

$$f(z) = \int\limits_{b\mathbb{D}} f(\zeta) \wedge B_{nq}(\zeta, z) - \int\limits_{\mathbb{D}} \bar{\partial} f(\zeta) \wedge B_{nq}(\zeta, z) - \bar{\partial}_z \int\limits_{\mathbb{D}} f(\zeta) \wedge B_{n,q-1}(\zeta, z). \quad (3.11)$$

We replace B_{nq} in the boundary integral by the right hand side of 3.10 and apply Stokes' theorem.

a) $q = 0$

$$\int\limits_{b\mathbb{D}} f(\zeta) B_{n0}(\zeta, z) = \int\limits_{b\mathbb{D}} f(\zeta) K_0(\zeta, z) - \int\limits_{b\mathbb{D}} f(\zeta) \bar{\partial}_\zeta A_0(\zeta, z)$$

$$= \int\limits_{\mathbb{D}} \bar{\partial} f(\zeta) \wedge K_0(\zeta, z) + \int\limits_{\mathbb{D}} f(\zeta) \bar{\partial}_\zeta K_0(\zeta, z) - \int\limits_{\mathbb{D}} \bar{\partial} f(\zeta) \wedge \bar{\partial}_\zeta A_0(\zeta, z)$$

$$(3.12)$$

(note that the integrands are of type $(n, \cdot)$ in ζ, so $d_\zeta = \bar{\partial}_\zeta$. The same remark applies also in the sequel).

b) $q \geq 1$

$$\int\limits_{b\mathbb{D}} f(\zeta) \wedge B_{nq}(\zeta, z) = (-1)^{q+1} \int\limits_{b\mathbb{D}} f(\zeta) \wedge \bar{\partial}_\zeta A_q(\zeta, z) + \bar{\partial}_z \int\limits_{b\mathbb{D}} f(\zeta) \wedge A_{q-1}(\zeta, z)$$

$$(3.13)$$

$$= (-1)^{q+1} \int\limits_{\mathbb{D}} \bar{\partial} f(\zeta) \wedge \bar{\partial}_\zeta A_q(\zeta, z) + \int\limits_{\mathbb{D}} \bar{\partial} f(\zeta) \wedge \bar{\partial}_z A_{q-1}(\zeta, z) +$$

$$(-1)^q \bar{\partial}_z \int\limits_{\mathbb{D}} f(\zeta) \wedge \bar{\partial}_\zeta A_{q-1}(\zeta, z).$$

Combining (3.12) resp. (3.13) with (3.11) we obtain our main result

Theorem 3.14 (Homotopy formula for the ball)   *i. If f is a C^1 function on $\overline{\mathbb{D}}$, then for $z \in \mathbb{D}$,*

$$f(z) = \int_{\mathbb{D}} f(\zeta)\overline{\partial}_\zeta K_0(\zeta, z) + \int_{\mathbb{D}} \overline{\partial} f(\zeta) \wedge \left(K_0(\zeta, z) - \overline{\partial}_\zeta A_0(\zeta, z) - B_{n0}(\zeta, z)\right),$$

where the kernel $\overline{\partial}_\zeta K_0(\zeta, z)$ is holomorphic in z.

ii. Let $q \geq 1$ and f a C^1-$(0, q)$-form on $\overline{\mathbb{D}}$. Then for all $z \in \mathbb{D}$,

$$f(z) = \int_{\mathbb{D}} \overline{\partial} f(\zeta) \wedge \left((-1)^{q+1}\overline{\partial}_\zeta A_q(\zeta, z) + \overline{\partial}_z A_{q-1}(\zeta, z) - B_{nq}(\zeta, z)\right)$$

$$+ \overline{\partial}_z \int_{\mathbb{D}} f(\zeta) \wedge \left((-1)^q \overline{\partial}_\zeta A_{q-1}(\zeta, z) - B_{n,q-1}(\zeta, z)\right)$$

We abbreviate the kernels in the above integrals by

$$
\begin{aligned}
P(\zeta, z) &= \overline{\partial}_\zeta K_0(\zeta, z) \\
T_0(\zeta, z) &= K_0(\zeta, z) - \overline{\partial}_\zeta A_0(\zeta, z) - B_0(\zeta, z) \\
T_q(\zeta, z) &= (-1)^{q+1}\overline{\partial}_\zeta A_q(\zeta, z) + \overline{\partial}_z A_{q-1}(\zeta, z) - B_q(\zeta, z) \\
S_{q-1}(\zeta, z) &= (-1)^q \overline{\partial}_\zeta A_{q-1}(\zeta, z) - B_{q-1}(\zeta, z)
\end{aligned}
\tag{3.15}
$$

and denote the corresponding operators by the same letters in boldface. Then Theorem 3.14 can be written more concisely as

Theorem 3.14′ (Homotopy formula)   *In the situation of 3.14, for a $(0, q)$-form f,*

$$f = \boldsymbol{P}f + \boldsymbol{T}_0\overline{\partial} f, \quad q = 0$$
$$f = \boldsymbol{T}_q\overline{\partial} f + \overline{\partial}\boldsymbol{S}_{q-1}f, \quad q \geq 1$$

with $\boldsymbol{P}f(z) = \int_{\mathbb{D}} f(\zeta) \wedge P(\zeta, z)$, etc. The function $\boldsymbol{P}f$ is holomorphic on $\mathbb{D}$.

3.   Before we can give some applications of the above theorem it is convenient to note the simplest regularity properties of the arising operators.

Proposition 3.16 *The operators $\boldsymbol{P}$, $\boldsymbol{T}_q$, $\boldsymbol{S}_{q-1}$ are linear continuous maps between the following spaces:*

$$
\begin{aligned}
&i. \boldsymbol{P} : L^p(\mathbb{D}) \longrightarrow L^p(\mathbb{D}), && 1 < p < \infty \\
&ii. \boldsymbol{T}_q : L^p_{0,q+1}(\mathbb{D}) \longrightarrow L^r_{0,q}(\mathbb{D}) \\
&iii. \boldsymbol{S}_{q-1} : L^p_{0,q}(\mathbb{D}) \longrightarrow L^r_{0,q-1}(\mathbb{D})
\end{aligned}
\left.\begin{aligned}\\ \\ \end{aligned}\right\} 1 \leq p \leq \infty, \ \tfrac{1}{r} > \tfrac{1}{p} - \tfrac{1}{2n+2}
$$

If, moreover, f is in C^k, so is $\boldsymbol{P}f$, $\boldsymbol{S}_q f$, $\boldsymbol{T}_q f$.

The last statement of the proposition follows from the construction of the operators and the corresponding property of the BMK kernel. The other statements will be proved in greater generality and precision in §5.

Now suppose $f \in L^1(\mathbb{D}) \cap C^1$ and $\bar{\partial} f \in L^1$. We can easily (cf. V.2.6) find a sequence $f_j \in C^1(\overline{\mathbb{D}})$ with

$$f_j \xrightarrow[L^1]{} f, \quad \bar{\partial} f_j \xrightarrow[L^1]{} \bar{\partial} f$$

Applying Theorem 3.14' to the f_j and using Proposition 3.16 we obtain a more general version of our theorem:

Theorem 3.14″ (Homotopy formula)   *i. If $f \in L^1(\mathbb{D}) \cap C^1$ and $\bar{\partial} f \in L^1_{0,1}$, for $1 \le p \le \infty$, then*

$$f = Pf + T_0 \bar{\partial} f$$

ii. If $f \in L^1_{0,q}(\mathbb{D}) \cap C^1(\mathbb{D})$ and $\bar{\partial} f \in L^1_{0,q+1}(\mathbb{D})$, then

$$f = T_q \bar{\partial} f + \bar{\partial} S_{q-1} f.$$

If, moreover, f and $\bar{\partial} f$ are in L^p, then all the above forms are in L^p.

(One can even drop the local regularity assumption $f \in C^1$ by working with the distribution derivatives of f).

Proof   We first prove *ii.*. By 3.14' one has

$$f_j = T_q \bar{\partial} f_j + \bar{\partial} S_{q-1} f_j \tag{3.17}$$

Since $f_j \to f$ and $\bar{\partial} f_j \to \bar{\partial} f$ in L^1, the first two terms in (3.17) converge to f and $T_q \bar{\partial} f$ in L^1, because of 3.16. The last term converges to $\bar{\partial} S_{q-1} f$ in the sense of distributions; (3.17) now shows the L^1-convergence of the last term as well, and the required identity.

Statement *i.* requires a more subtle argument. We have for the f_j by 3.14

$$f_j = Pf_j + T_0 \bar{\partial} f_j.$$

Taking limits in L^1 we get

$$f = \lim_{j \to \infty} Pf_j + T_0 \bar{\partial} f.$$

But for $z \in \Omega$ fixed,

$$\lim_{j \to \infty} Pf_j(z) = \lim_{j \to \infty} \int_\Omega f_j(\zeta) \wedge \bar{\partial}_\zeta K_0(\zeta, z)$$

$$= \int_\Omega f(\zeta) \wedge \bar{\partial}_\zeta K_0(\zeta, z)$$

$$= Pf(z).$$

So $\lim_{j \to \infty} Pf_j = Pf$ in L^1, as required. – The case $p > 1$ now follows easily.   □

4.   The main homotopy formula contains two series of solution operators to the CR equations on the ball.

Theorem 3.18 *Let f be a $\bar{\partial}$-closed $(0,q)$-form on $\mathbb{D}$, $f \in C^k(\mathbb{D}) \cap L^p_{0,q}(\mathbb{D})$, $k \geq 1$, $q \geq 1$, $1 \leq p \leq \infty$. Then the CR equation*

$$\bar{\partial}u = f, \quad \bar{\partial}f = 0$$

has a solution $u \in C^k(\mathbb{D}) \cap L^r_{0,q}(\mathbb{D})$, $\frac{1}{r} > \frac{1}{p} - \frac{1}{2n+2}$.

In fact, $u = S_{q-1}f$ is such a solution. If we apply Theorem 3.14 i and ii to this solution u, we immediately see that $T_{q-1}f$ *is also a solution with the required properties*. The two solution operators T_{q-1} and S_{q-1} differ by a $\bar{\partial}$-exact form respectively a holomorphic function, namely

$$\bar{\partial}_z \int\limits_{\mathbb{D}} f(\zeta) \wedge A_{q-2}(\zeta, z) \text{ resp. } \int\limits_{\mathbb{D}} f(\zeta) \wedge K_0(\zeta, z).$$

We shall see later (Chapter VI, §6) that the operators T_q are "more natural". For $q = 1$ we can make this statement precise at the end of this paragraph. But let us first note a qualitative corollary of the above.

Theorem 3.19 *Any C^∞-smooth $\bar{\partial}$-closed $(0,q)$-form f on a complex manifold X is locally $\bar{\partial}$-exact: given any point $x \in X$ there is a neighbourhood U of x and a smooth $(0, q-1)$-form u on U with $\bar{\partial}u = f$.*

5.   We finish this paragraph with a study of the operator

$$Pf(z) = \int\limits_{\mathbb{D}} f(\zeta)\bar{\partial}_\zeta K_0(\zeta, z) = \int\limits_{\mathbb{D}} f(\zeta)P(\zeta, z). \tag{3.20}$$

The kernel $P(\zeta, z)$ can be explicitly calculated:

$$P(\zeta, z) = \left(\frac{1}{2\pi i}\right)^n \bar{\partial}_\zeta \left[\frac{1}{\Phi^n} \partial r \wedge (\bar{\partial}\partial r)^{n-1}\right] \tag{3.21}$$

$$= \left(\frac{1}{2\pi i}\right)^n \left[\frac{(\bar{\partial}\partial r)^n}{\Phi^n} - n\frac{\bar{\partial}\Phi \wedge \partial r \wedge (\bar{\partial}\partial r)^{n-1}}{\Phi^{n+1}}\right]$$

Now we have

$$\bar{\partial}\Phi = -\sum_{j=1}^n z^j \, d\bar{\zeta}^j$$

$$\partial r = \sum \bar{\zeta}^j \, d\zeta^j \tag{3.22}$$

$$\bar{\partial}\partial r = \sum d\bar{\zeta}^j \wedge d\zeta^j$$

Insertion of (3.22) into (3.21) leads to

$$P(\zeta,z) = \frac{n!}{(2\pi i)^n}\frac{1}{\Phi^{n+1}}\left[\Phi + \langle\zeta,z\rangle\right](d\bar\zeta \wedge d\zeta)^N \tag{3.23}$$

$$= \frac{n!}{(2\pi i)^n}\frac{1}{\Phi^{n+1}}(d\bar\zeta \wedge d\zeta)^N$$

$$= \frac{n!}{\pi^n}\frac{1}{\Phi^{n+1}}\,dV(\zeta)$$

The operator P with kernel $P(\zeta,z)$ maps L^2 continuously into $L^2\cap\mathcal{O}$, the *space of square integrable holomorphic functions*. Its adjoint operator (with respect to the L^2 scalar product) is given by

$$P^*f(\zeta) = \frac{n!}{\pi^n}\int_{\mathbb{D}} f(\zeta)\frac{1}{\overline{\Phi(z,\zeta)}^{n+1}}\,dV(\zeta) \tag{3.24}$$

(use Fubini's theorem, see also Chapter V, where the above subject is discussed in more detail). This implies

$$P = P^*, \tag{3.25}$$

because of Lemma 3.3, that is, P *is self-adjoint*. Finally, it follows from 3.14.*i* that

$$PPf = Pf, \tag{3.26}$$

because Pf is holomorphic. Summing up, we have proved

Theorem 3.27 *The operator*

$$Pf(z) = \int_{\mathbb{D}} f(\zeta)P(\zeta,z)$$

is the orthogonal projection from L^2 onto $L^2\cap\mathcal{O}$. Its kernel is (3.23)

$$P(\zeta,z) = \frac{n!}{\pi^n}\frac{1}{(1-\langle\zeta,z\rangle)^{n+1}}\,dV(\zeta).$$

It is more consistent – following Chapter I – to define the function

$$\mathcal{P}(\zeta,z) = \frac{n!}{\pi^n}\frac{1}{(1-\langle z,\zeta\rangle)^{n+1}} \tag{3.28}$$

as the kernel of P. Then P is given as a scalar product

$$Pf(z) = (f(\cdot),\mathcal{P}(\cdot,z))_{\mathbb{D}}$$

as in Chapter I.

Definition 3.29

$$\mathcal{P}(\zeta,z) = \frac{n!}{\pi^n}\frac{1}{(1-\langle z,\zeta\rangle)^{n+1}}$$

is called the Bergman kernel function of $\mathbb{D}$, P the Bergman projection.

We now turn again to the solution operator $\boldsymbol{T}_0$. If f is a $\overline{\partial}$-closed $(0,1)$-form in $L^2 \cap C^1$, then $u = \boldsymbol{T}_0 f$ is a square integrable function satisfying $\overline{\partial} u = f$ and

$$u = \boldsymbol{P}u + \boldsymbol{T}_0 \overline{\partial} u = \boldsymbol{P}u + u;$$

so $\boldsymbol{P}u = 0$, and we recognise $\boldsymbol{T}_0 f$ as the solution of minimal L^2-norm (that is orthogonal to the holomorphic functions) of the CR equation.

§4   The Basic Integral Representation

1.   We work with an arbitrary hermitian metric

$$ds^2 = \sum_{j,k=1}^{n} g_{j\bar{k}}(\zeta)\, d\zeta^j\, d\overline{\zeta}^k$$

on $\mathbb{C}^n$, associate with it the square norm function

$$R^2(\zeta, z) = \sum_{j,k=1}^{n} g_{j\bar{k}}(\zeta)(\zeta^j - z^j)(\overline{\zeta}^k - \overline{z}^k),$$

the Leray form

$$h(\zeta, z) = \frac{1}{R^2(\zeta, z)} \sum_{j=1}^{n} \left(\sum_{k=1}^{n} g_{j\bar{k}}(\zeta)(\overline{\zeta}^k - \overline{z}^k) \right) d\zeta^j$$

on $\mathbb{C}^n \times \mathbb{C}^n - \Delta$ (where Δ is the diagonal) and the Cauchy-Fantappiè kernels

$$H_q(\zeta, z) = \Omega_q(h); \tag{4.1}$$

see II(2.1) – (2.5). For an arbitrary domain $\Omega \subset\subset \mathbb{C}^n$ with piecewise smooth boundary and any form $f \in C^1_{0,q}(\overline{\Omega})$ we have, by Theorem II.2.9, the identity

$$f(z) = \int_{b\Omega} f(\zeta) \wedge H_q(\zeta, z) - \int_{\Omega} \overline{\partial} f(\zeta) \wedge H_q(\zeta, z) - \overline{\partial} \int_{\Omega} f(\zeta) \wedge H_{q-1}(\zeta, z), \quad z \in \Omega. \tag{4.2}$$

Isotropic kernels in the sense of I §5 will be defined in terms of the norm function R^2. Then the kernels H_q are isotropic of type 1 (or order $1 - 2n$).

2.   In all that follows Ω will be a strictly pseudoconvex domain in $\mathbb{C}^n$ given by a strictly plurisubharmonic defining function

$$r : U(b\Omega) \longrightarrow \mathbb{R};$$

$$F(\zeta, z) = \sum_{j=1}^{n} r_j(\zeta)(\zeta^j - z^j) - \frac{1}{2} \sum_{j,k=1}^{n} r_{jk}(\zeta)(\zeta^j - z^j)(\zeta^k - z^k)$$

$$= \sum_{j=1}^{n} F_j(\zeta, z)(\zeta^j - z^j)$$

is its Levi polynomial. The results of §2 provide us with positive numbers δ, ε and c, such that for

$$|r(\zeta)| < 2\delta, \quad R^2(\zeta, z) < \varepsilon^2 \tag{4.3}$$

the estimate

$$2 \operatorname{Re} F(\zeta, z) \geq r(\zeta) - r(z) + cR^2(\zeta, z)$$

holds. This means that

$$2 \operatorname{Re} (F(\zeta, z) - r(\zeta)) \geq -r(\zeta) - r(z) + cR^2(\zeta, z) \tag{4.4}$$

on the set $W_{\delta,\varepsilon}$ defined by the inequalities (4.3).

Choose a smooth patching function $\varphi(\zeta, z)$ on $\mathbb{C}^n \times \mathbb{C}^n$ such that

$$\varphi(\zeta, z) = \begin{cases} 1 & \text{for } R(\zeta, z) \leq \frac{\varepsilon}{2} \\ 0 & \text{for } R(\zeta, z) \geq \frac{3}{4}\varepsilon \end{cases}$$

and introduce

$$\Phi(\zeta, z) = \varphi(\zeta, z) \left(F(\zeta, z) - r(\zeta) \right) + (1 - \varphi(\zeta, z)) R^2(\zeta, z). \tag{4.5}$$

This function is defined on the set

$$W_\delta = \{(\zeta, z) : |r(\zeta)| < 2\delta\} \subset \mathbb{C}^n \times \mathbb{C}^n$$

and has the following properties:

   i.
$$\overline{\partial}_z \Phi(\zeta, z) = 0 \quad \text{for } R(\zeta, z) < \frac{\varepsilon}{2} \tag{4.6}$$

   ii.
$$\Phi(\zeta, \zeta) = 0 \quad \text{for } \zeta \in b\Omega \tag{4.7}$$

   iii. *There is a positive constant γ such that*

$$|\Phi(\zeta, z)| \geq \gamma(-r(\zeta) - r(z) + R^2(\zeta, z) + |\operatorname{Im} \Phi(\zeta, z)|) \tag{4.8}$$

   iv. *Given any $z \in b\Omega$, the functions $r(\zeta)$ and $\operatorname{Im} \Phi(\zeta, z)$ are part of a smooth local coordinate system around z.*

   v.
$$\Phi(\zeta, z) - \overline{\Phi(z, \zeta)} = \mathcal{E}_3 \tag{4.9}$$

All these statements follow from (4.4) and Propositions 2.11 and 2.12. – It is convenient to further introduce the function

$$\widetilde{F}(\zeta, z) = \varphi(\zeta, z)F(\zeta, z) + (1 - \varphi(\zeta, z))R^2(\zeta, z) \qquad (4.10)$$

which coincides with Φ for $\zeta \in b\Omega$. Then $\widetilde{F}$ decomposes as

$$\widetilde{F}(\zeta, z) = \sum_{j=1}^{n} P_j(\zeta, z)(\zeta^j - z^j)$$

$$P_j(\zeta, z) = \varphi(\zeta, z)F_j(\zeta, z) + (1 - \varphi(\zeta, z))R_j(\zeta, z)$$

$$F_j(\zeta, z) = r_j(\zeta) - \frac{1}{2}\sum_{k=1}^{n} r_{jk}(\zeta)(\zeta^k - z^k)$$

$$R_j(\zeta, z) = \sum_{k=1}^{n} g_{j\bar{k}}(\zeta)(\overline{\zeta}^k - \overline{z}^k)$$

This leads us to defining the following forms:

$$k_0(\zeta, z) = \frac{1}{\Phi(\zeta, z)} \sum_{j=1}^{n} P_j(\zeta, z)\, d\zeta^j, \qquad (4.11)$$

$$\tilde{k}(\zeta, z) = \frac{1}{\widetilde{F}(\zeta, z)} \sum_{j=1}^{n} P_j(\zeta, z)\, d\zeta^j, \qquad (4.12)$$

$$h(\zeta, z) = \frac{1}{R^2(\zeta, z)} \sum_{j=1}^{n} R_j(\zeta, z)\, d\zeta^j, \qquad (4.13)$$

$$h_0(\zeta, z) = \frac{1}{\mathrm{P}(\zeta, z)} \sum_{j=1}^{n} R_j(\zeta, z)\, d\zeta^j, \qquad (4.14)$$

where we set

$$\mathrm{P}(\zeta, z) = R^2(\zeta, z) + 2r(\zeta)r(z).$$

The form $k_0(\zeta, z)$ is defined on W_δ. We can choose a final patching function $\chi(\zeta)$ depending on ζ alone, with

$$\chi(\zeta) = \begin{cases} 1 & \text{for } |r(\zeta)| \le \delta \\ 0 & \text{for } |r(\zeta)| \ge \frac{3}{2}\delta \end{cases}$$

and extend k_0 to k defined on $\overline{\Omega} \times \overline{\Omega} - \Lambda$:

$$k(\zeta, z) = \chi(\zeta)k_0(\zeta, z). \qquad (4.15)$$

Λ denotes the boundary diagonal.

Definition 4.16 *The CF kernels*

$$K_q(\zeta, z) = \Omega_q(k), \quad q = 0, \ldots, n - 1$$

are called boundary kernels associated with the strictly pseudoconvex domain Ω (and its defining function r).

In explicit terms,

$$K_q(\zeta, z) = (-1)^{q(q-1)/2} \binom{n-1}{q} \left(\frac{1}{2\pi i}\right)^n k \wedge (\bar{\partial}_\zeta k)^{n-q-1} \wedge (\bar{\partial}_z k)^q \qquad (4.17)$$

In addition to the K_q we consider the *transition kernels* between K_q and $\Omega_q(h_0)$ (cf. II.1.37):

$$A_q(\zeta, z) = A_q(k, h_0). \qquad (4.18)$$

Now it is crucial to note that the forms k and h_0 coincide on $b\Omega \times \bar{\Omega}$ with $\tilde{k}$ respectively h, and these latter forms are Leray forms! Moreover, $\Omega_q(h_0) = H_q(\zeta, z)$ for $r(\zeta) = 0$. Consequently, Koppelman's homotopy formula yields the relation

$$H_q(\zeta, z) = K_q(\zeta, z) + (-1)^{q+1} \bar{\partial}_\zeta A_q(\zeta, z) + \bar{\partial}_z A_{q-1}(\zeta, z), \qquad (4.19)$$

valid for the pull-back of these forms to $b\Omega \times \Omega$. To sum up:

Proposition 4.20 *The boundary kernels $K_q(\zeta, z)$ and the transition kernels $A_q(\zeta, z)$ are smooth on $\bar{\Omega} \times \bar{\Omega} - \Lambda$ and satisfy, on $b\Omega \times \bar{\Omega} - \Lambda$, the homotopy formula (4.19).*

3.   We can now execute the programme described in §3. The computations are completely parallel to (3.11) – (3.14).

Let $f \in C_{0,q}^1(\bar{\Omega})$. Then the boundary integral in (4.2) can be replaced using (4.19)

$$\int_{b\Omega} f(\zeta) \wedge H_q(\zeta, z) = \int_{b\Omega} f(\zeta) \wedge K_q(\zeta, z)$$
$$+ (-1)^{q+1} \int_{b\Omega} f(\zeta) \wedge \bar{\partial}_\zeta A_q(\zeta, z) + \bar{\partial}_z \int_{b\Omega} f(\zeta) \wedge A_{q-1}(\zeta, z)$$

For $z \in \Omega$ all the above integrands are regular in $\zeta \in \bar{\Omega}$; we can therefore transform them into volume integrals and get

$$\int_{b\Omega} f(\zeta) \wedge H_q(\zeta, z) = \int_\Omega \bar{\partial} f(\zeta) \wedge K_q(\zeta, z) + (-1)^q \int_\Omega f(\zeta) \wedge \bar{\partial}_\zeta K_q(\zeta, z)$$
$$+ (-1)^{q+1} \int_\Omega \bar{\partial} f(\zeta) \wedge \bar{\partial}_\zeta A_q(\zeta, z) + \bar{\partial}_z \int_\Omega \bar{\partial} f(\zeta) \wedge A_{q-1}(\zeta, z)$$
$$+ (-1)^q \bar{\partial}_z \int_\Omega f(\zeta) \wedge \bar{\partial}_\zeta A_{q-1}(\zeta, z) \quad (4.21)$$

Inserting (4.21) in (4.2) gives our main result:

Theorem 4.22 (Homotopy formula) *Let the kernels H_q, K_q and A_q be defined as above in (4.1), (4.16) and (4.17). If $f \in C_{0,q}^1(\overline{\Omega})$ and $z \in \Omega$, the following integral representation holds:*

$$f(z) = (-1)^q \int\limits_{\Omega} f(\zeta) \wedge \overline{\partial}_\zeta K_q(\zeta, z) + \int\limits_{\Omega} \overline{\partial} f(\zeta) \wedge \left[K_q(\zeta, z) + (-1)^{q+1} \overline{\partial}_\zeta A_q(\zeta, z) \right.$$

$$\left. + \overline{\partial}_z A_{q-1}(\zeta, z) - H_q(\zeta, z) \right] + \overline{\partial}_z \int\limits_{\Omega} f(\zeta) \wedge \left[(-1)^q \overline{\partial}_\zeta A_{q-1}(\zeta, z) - H_{q-1}(\zeta, z) \right]$$

Let us again abbreviate the above kernels

$$P_q(\zeta, z) = (-1)^q \overline{\partial}_\zeta K_q(\zeta, z)$$
$$T_q(\zeta, z) = K_q(\zeta, z) + (-1)^{q-1} \overline{\partial}_\zeta A_q(\zeta, z) + \overline{\partial}_z A_{q-1}(\zeta, z) - H_q(\zeta, z) \qquad (4.23)$$
$$S_{q-1}(\zeta, z) = (-1)^q \overline{\partial}_\zeta A_{q-1}(\zeta, z) - H_{q-1}(\zeta, z)$$

and denote the corresponding operators by the same letter in boldface. Then Theorem 4.22 reads as

Theorem 4.22′ (Homotopy formula) *For $f \in C_{0,q}^1(\overline{\Omega})$, the identity*

$$f = \boldsymbol{P}_q f + \boldsymbol{T}_q \overline{\partial} f + \overline{\partial} \boldsymbol{S}_{q-1} f$$

holds on Ω. The operators are defined by integration against the kernels (4.23).

We shall prove in the next paragraph

Proposition 4.24 *The operators $\boldsymbol{P}_0$, $\boldsymbol{T}_q$, $\boldsymbol{S}_{q-1}$ are linear continuous maps between the following spaces:*

> *i.* $\boldsymbol{P}_0 : L^p(\Omega) \longrightarrow L^p(\Omega)$, $1 < p < \infty$
> *ii.* $\boldsymbol{T}_q : L_{0,q+1}^p(\Omega) \longrightarrow L_{0,q}^r(\Omega)$ $\left. \right\} 1 \leq p \leq \infty, \; \frac{1}{r} > \frac{1}{p} - \frac{1}{2n+2}$
> *iii.* $\boldsymbol{S}_{q-1} : L_{0,q}^p(\Omega) \longrightarrow L_{0,q-1}^r(\Omega)$

If, moreover, f is in C^k, so is $\boldsymbol{P}_0 f$, $\boldsymbol{S}_{q-1} f$, $\boldsymbol{T}_q f$. In fact, $\boldsymbol{P}_0 f$ is smooth on Ω for all $f \in L^1$

Let us now look more closely at the operators $\boldsymbol{P}_q$ for $q \geq 1$. Since $k(\zeta, z)$ is holomorphic in z near the boundary diagonal Λ, (4.16) shows that $K_q(\zeta, z)$ vanishes near Λ for $q \geq 1$ (it is holomorphic near Λ in case $q = 0$). This means that the kernels $K_q(\zeta, z)$ are smooth on the whole set $\overline{\Omega} \times \overline{\Omega}$, and we have the crucial

Theorem 4.25 *If $q \geq 1$, then the operators $\boldsymbol{P}_q$ are continuous maps from L^1 to $C^\infty(\overline{\Omega})$:*

$$\boldsymbol{P}_q : L_{0,q}^1(\Omega) \longrightarrow C_{0,q}^\infty(\overline{\Omega}),$$

where $C^\infty(\overline{\Omega})$ carries the usual Fréchet topology of uniform convergence with all derivatives. For $q = 0$, the form $\overline{\partial} \boldsymbol{P}_0 f$ is smooth on $\overline{\Omega}$.

The last statement follows by the same argument as the assertion for $q \geq 1$. We have tacitly assumed $r \in C^\infty$. If r is in C^k, $k \geq 4$, the theorem still holds (by an analysis of the above argument).

Because of Proposition 4.24, we can apply the representation formula 4.22' to a larger class of forms:

Theorem 4.22″ (Homotopy formula) *If $f \in L^1_{0,q}$ and $\overline{\partial} f \in L^1_{0,q+1}$, then*

$$f = P_q f + T_q \overline{\partial} f + \overline{\partial} S_{q-1} f$$

If, moreover, f and $\overline{\partial} f$ are in L^p, for $1 \leq p \leq \infty$, then each term on the right hand side is in L^p.

Proof   We choose forms $f_j \in C^1_{0,q}(\overline{\Omega})$ such that $f_j \xrightarrow[L^1]{} f$ and $\overline{\partial} f_j \xrightarrow[L^1]{} \overline{\partial} f$ (A detailed construction of such a sequence is exhibited in V.2.6). By 4.22,

$$f_j = P_q f_j + T_q \overline{\partial} f_j + \overline{\partial} S_{q-1} f_j$$

(4.25) implies the convergence in L^p of $P_q f_j \longrightarrow P_q f$, $T_q \overline{\partial} f_j \longrightarrow T_q \overline{\partial} f$, $S_{q-1} f_j \longrightarrow S_{q-1} f$. Therefore $\overline{\partial} S_{q-1} f_j \xrightarrow[\mathcal{D}']{} \overline{\partial} S_{q-1} f$ in the sense of distributions. Moreover, we have in the limit the desired equality. Since three of the four terms belong to L^1, so does the fourth, and $\overline{\partial} S_{q-1} f_j \longrightarrow \overline{\partial} S_{q-1} f$ in L^1. – The case $q = 0$ is treated as in 3.14″. $\square$

The far reaching consequences of these results will be investigated in the next paragraphs, once we have at our disposition the necessary regularity properties of the above operators.

§5   Admissible Kernels and L^p-Estimates

1.   The operators P_q, T_q, S_q having come up in the preceding paragraphs belong to a new class of operators whose kernels and regularity properties will now be described. Let us first recall their main building blocks. We keep the notations of the last paragraph, in particular, $r(\zeta)$ is a *strictly plurisubharmonic defining function* for a strictly pseudoconvex domain Ω, and $F(\zeta, z)$ its *Levi polynomial*.

Definition 5.1   *The function*

$$\Phi(\zeta, z) = \varphi(\zeta, z)(F(\zeta, z) - r(\zeta)) + (1 - \varphi(\zeta, z))R^2(\zeta, z)$$

is called the extended Levi polynomial of $r(\zeta)$.

Its main properties have been described in §4 – we recall them here as

Proposition 5.2 $\Phi(\zeta, z)$ *is smooth on* $W_\delta = \{(\zeta, z) \in \mathbb{C}^n \times \mathbb{C}^n : |r(\zeta)| < \delta\}$, *it is holomorphic in* z *if* $R(\zeta, z) < \frac{\varepsilon}{2}$, *it vanishes on the boundary diagonal* $\Lambda = \{(\zeta, \zeta) : r(\zeta) = 0\}$. *The functions* $r(\zeta)$ *and* Im $\Phi(\zeta, z)$ *are part of a smooth local coordinate system around any* $z \in b\Omega$. Φ *is almost symmetric, meaning that the relation*

$$\Phi(\zeta, z) - \overline{\Phi(z, \zeta)} = \mathcal{E}_3(\zeta, z) \tag{5.3}$$

holds, and it satisfies the estimate

$$|\Phi(\zeta, z)| \geq \gamma(-r(\zeta) - r(z) + R^2(\zeta, z) + |\text{Im } \Phi(\zeta, z)|) \tag{5.4}$$

(Here δ, ε, γ *are sufficiently small positive constants).*

The estimate (5.4) is crucial. If z is a boundary point, we can choose coordinates $x_1, \ldots,$ x_{2n} such that $x_j(z) = 0$ and

$$\begin{aligned} x_1(\zeta) &= -r(\zeta) \\ x_2(\zeta) &= \text{Im } \Phi(\zeta, z) \end{aligned} \tag{5.5}$$

These coordinates have the following geometrical significance: the plane $x_1 = 0$ is the (real) tangent plane to the boundary (at $z = 0$); within that plane, the plane $x_2 = 0$ defines the holomorphic tangent plane to the boundary (because Im $\Phi(\zeta, z) = $ Im $\sum_j r_j(\zeta)(\zeta^j - z^j) + O(\|\zeta - z\|^2)$. Now (5.4) reads as

$$|\Phi(\zeta, z)| \geq \gamma(x_1 + |x_2| + \|x\|^2); \tag{5.4'}$$

so $|\Phi|$ *is estimated from below by the squared norm in all directions, but by a first power of the coordinates in the* x_1- *and* x_2-*directions.* We refer to these estimates as *anisotropic estimates.*

The function

$$P(\zeta, z) = R^2(\zeta, z) + 2r(\zeta)r(z) \tag{5.6}$$

will be called the *extended norm function.* It is positive on $\overline{\Omega} \times \overline{\Omega} - \Lambda$ and satisfies near Λ, the estimate

$$P(\zeta, z) \geq R^2(\zeta, z) \tag{5.7}$$

We shall need the following notation

Definition 5.8 *For a double differential form* $\mathcal{K}(\zeta, z)$ *the adjoint form is given as*

$$\mathcal{K}^*(\zeta, z) = \overline{\mathcal{K}(z, \zeta)}$$

Consequently, if $\mathcal{K}$ is of double type $(p, q; r, s)$, then $\mathcal{K}^*$ is of double type $(s, r; q, p)$.

2.   We can now introduce the class of kernels and operators hinted at the beginning of this paragraph.

Definition 5.9 *A double differential form $\mathcal{A}(\zeta, z)$ on $\overline{\Omega} \times \overline{\Omega}$ is an admissible kernel, if it has the following properties:*

    i. $\mathcal{A}$ is smooth on $\overline{\Omega} \times \overline{\Omega} - \Lambda$

    ii. For any point $(\zeta_0, \zeta_0) \in \Lambda$ there is a neighbourhood $U \times U$ and a representation

$$\mathcal{A} = \mathcal{E}_j P^{-t_0} \Phi^{t_1} \overline{\Phi}^{t_2} \Phi^{*t_3} \overline{\Phi}^{*t_4} r^k r^{*l} \tag{5.10}$$

    with $j, t_0, \ldots, l$ integers and $j, t_0, k, l \geq 0$, $-t = t_1 + \cdots t_4 \leq 0$.

Of course $r^*(\zeta) = r(z)$.

Remark.   We will explain in Chapter IV, 5.49 that the condition $t \geq 0$ is not really a restriction.

Definition 5.11 *The type of a representation (5.10) is the integer*

$$\lambda = 2n + j + \min(2, t - k - l) - 2(t_0 + t - k - l)$$

where again $-t = t_1 + t_2 + t_3 + t_4 \leq 0$. – $\mathcal{A}$ is called of type $\geq \lambda$ if $\mathcal{A}$ has at each point of Λ a representation of type $\geq \lambda$.

The complicated definition takes into account the fundamental anisotropic estimate (5.4) for Φ – this explains the term $\min(2, t - k - l)$ – and the comparison between Φ, P, r, R^2 and their adjoints as expressed in 5.16 – 5.18. Kernels of positive type will be seen to have regularising properties – this is the content of Proposition 5.19 –, whereas kernels of type 0 define at best L^p-continuous operators.

Definition 5.12   *i. If $\mathcal{A}$ is an admissible kernel of type $\geq \lambda$ then the operator*

$$Af(z) = (f(\cdot), \mathcal{A}(\cdot, z)) = \int_\Omega f(\zeta) \wedge *\overline{\mathcal{A}(\zeta, z)}$$

    is called an admissible operator of type $\geq \lambda$.

    ii. The sum of admissible operators A_λ and A_μ of type $\geq \lambda$ resp. μ is called admissible of type $\geq \min(\lambda, \mu)$.

The symmetry of conditions 5.10 and 5.11 implies: if the operator A with kernel $\mathcal{A}$ is admissible of type $\geq \lambda$, the same holds for the operator A^* with kernel $\mathcal{A}^*$ (the formal adjoint of A) and the operator

$$Af(z) = \int_\Omega f(\zeta) \wedge \mathcal{A}(\zeta, z)$$

(defined for the appropriate types of differential forms). We shall return to these notions in Chapter IV. 5.49 in greater generality, namely on complex manifolds. Moreover, Chapter VII contains a fairly complete regularity theory of admissible operators.

The most important examples of admissible operators and kernels have already been introduced in the last two paragraphs:

Proposition 5.13   i. *The operator P_0 is admissible of type ≥ 0.*

    ii. *The operators T_q and S_q are sums of admissible operators of type ≥ 1 and isotropic type 1 operators.*

Proof   We have to analyse the kernels $\bar{\partial}_\zeta K_0$, $\bar{\partial}_\zeta A_q$ and $\bar{\partial}_z A_q$ near the boundary diagonal Λ.

a) $K_0 = \text{const} \cdot k \wedge (\bar{\partial}_\zeta k)^{n-1} = \frac{\mathcal{E}_0}{\Phi^n}$,

so

$$\bar{\partial}_\zeta K_0 = \frac{\mathcal{E}_0}{\Phi^n} + \frac{\mathcal{E}_0}{\Phi^{n+1}}.$$

The summands on the right-hand side are of type ≥ 2 resp. ≥ 0.

b) A_q is a sum of kernels of the form

$$\text{const} \cdot k \wedge h_0 \wedge (\bar{\partial}_\zeta k)^a \wedge (\bar{\partial}_\zeta h_0)^b \wedge (\bar{\partial}_z h_0)^c$$

with $a + b + c = n - 2$. This leads to a representation

$$A_q = \sum_t \frac{\mathcal{E}_1}{P^{n-t}\Phi^t}, \quad 1 \leq t \leq n - 1.$$

Now

$$\bar{\partial}_\zeta P = \bar{\partial}_\zeta (R^2 + 2rr^*) = \mathcal{E}_1 + \mathcal{E}_0 r^*$$
$$\bar{\partial}_z P \qquad\qquad\;\; = \mathcal{E}_1 + \mathcal{E}_0 r$$
$$\bar{\partial}_\zeta \Phi = \mathcal{E}_0$$
$$\bar{\partial}_z \Phi = 0$$

(near the boundary diagonal). This gives

$$\bar{\partial}_\zeta A_q = \sum_t \left(\frac{\mathcal{E}_0}{P^{n-t}\Phi^t} + \frac{\mathcal{E}_2}{P^{n-t+1}\Phi^t} + \frac{\mathcal{E}_1 r^*}{P^{n-t+1}\Phi^t} + \frac{\mathcal{E}_1}{P^{n-t}\Phi^{t+1}} \right)$$
$$\bar{\partial}_z A_q = \sum_t \left(\frac{\mathcal{E}_0}{P^{n-t}\Phi^t} + \frac{\mathcal{E}_2}{P^{n-t+1}\Phi^t} + \frac{\mathcal{E}_1 r}{P^{n-t+1}\Phi^t} \right),$$

and all terms are easily seen to be of type ≥ 1. $\square$

3. We are now going to estimate admissible kernels near the diagonal. To simplify the notation, we use the following conventions for positive functions f and g:

$$f \lesssim g \tag{5.14}$$

means: *there is a positive constant γ with*

$$f \leq \gamma g;$$

$$f \approx g \tag{5.15}$$

means: $f \lesssim g$ and $g \lesssim f$. The following information is now obvious or follows from (5.4):

Lemma 5.16   *i.* $|\Phi| \gtrsim |r|$, $|\Phi| \gtrsim |r^*|$, $|\Phi| \gtrsim R^2$

ii. $P \gtrsim R^2$

and i. also holds for $|\Phi^*|$.

Slightly more delicate is

Lemma 5.17

$$\left|\frac{\Phi}{\Phi^*}\right| \approx 1$$

Proof

$$\frac{\Phi}{\Phi^*} = \frac{\Phi^* + \mathcal{E}_3}{\Phi^*} = 1 + \frac{\mathcal{E}_3}{\Phi^*}$$

in view of (5.3). But $|\Phi^*| \gtrsim R^2$, therefore

$$\left|\frac{\mathcal{E}_3}{\Phi^*}\right| \lesssim 1,$$

and we have $|\frac{\Phi}{\Phi^*}| \lesssim 1$. The estimate $|\frac{\Phi^*}{\Phi}| \lesssim 1$ is proved in the same way. $\square$

We also need

Lemma 5.18 $P(\zeta, z) \gtrsim r(\zeta)^2 + r(z)^2$

Proof   Let $\zeta_0 \in b\Omega$ and choose a local coordinate system near ζ_0 such that $r(\zeta)$ is one of the coordinates. Then for ζ and z in the realm of that coordinate system,

$$P(\zeta, z) \gtrsim (r(\zeta) - r(z))^2 + 2r(\zeta)r(z) = r(\zeta)^2 + r(z)^2.$$

A compactness argument completes the proof. $\square$

The above lemmas will be put together to yield the following fundamental estimates for admissible kernels:

Proposition 5.19 *Let $\mathcal{A}$ be an admissible kernel of type λ with $0 < \lambda < 2n$. Then the integrals*

$$\int_\Omega |\mathcal{A}(\zeta, z)|^\gamma dV(\zeta) \text{ and } \int_\Omega |\mathcal{A}(\zeta, z)|^\gamma dV(z)$$

are uniformly bounded on Ω, if

$$1 \leq \gamma < \frac{2n+2}{2n+2-\lambda}.$$

We have used the following convention: $|\mathcal{A}|$ means the absolute value of an arbitrary coefficient of the double form $\mathcal{A}$; so $|\mathcal{A}|$ is a positive function on $\overline{\Omega} \times \overline{\Omega} - \Lambda$ which can be integrated with respect to ζ or z. A more general situation will be studied in Chapter VII, §2. Lemma VII.2.1, in fact, immediately gives our proposition. But we include here a direct proof following somewhat different lines in order to make the reader's task easier.

Proof a) Choose an open covering of $b\Omega$ by balls U_i such that $\mathcal{A}$ has a representation (5.10) of type $\lambda > 0$ on $U_i \times U_i$; let $V_i \subset\subset U_i$ be slightly smaller balls which still cover $b\Omega$. If $z \in \Omega - \bigcup V_i$, then

$$\int_\Omega |\mathcal{A}(\zeta, z)|^\gamma dV(\zeta)$$

is bounded independently of z (and γ can be arbitrary). If $z \in V_i$, then

$$\int_\Omega |\mathcal{A}(\zeta, z)|^\gamma dV(\zeta) = \int_{\Omega - U_i} |\mathcal{A}(\zeta, z)|^\gamma dV(\zeta) + \int_{U_i} |\mathcal{A}(\zeta, z)|^\gamma dV(\zeta).$$

The first integral on the right hand side is uniformly bounded for $z \in V_i$ (and arbitrary γ); so we have to estimate the last integral. By repeatedly using 5.16 and 5.17 we see that

$$|\mathcal{A}(\zeta, z)| \lesssim \left| \frac{1}{\mathbf{P}^a \overline{\Phi} t} r^k r^{*l} \right|. \tag{5.20}$$

We distinguish the cases $t \geq k + l$ and $t \leq k + l$.

b) Let $t \geq k + l$. Then, in view of 5.16,

$$|\mathcal{A}(\zeta, z)| \lesssim \frac{1}{\mathbf{P}^a |\Phi|^{t-k-l}}. \tag{5.21}$$

The types of the right-hand sides of (5.20) and (5.21) are still λ. Introducing λ we arrive at three different cases

$$|\mathcal{A}(\zeta, z)| \lesssim \frac{1}{\mathbf{P}^{n-\lambda/2}} \tag{5.22}$$

$$|\mathcal{A}(\zeta, z)| \lesssim \frac{1}{\mathbf{P}^{n-1/2-\lambda/2}|\Phi|} \tag{5.23}$$

$$|\mathcal{A}(\zeta, z)| \lesssim \frac{1}{\mathbf{P}^{n+1-m-\lambda/2}|\Phi|^m}, \quad m \geq 2 \tag{5.24}$$

c) We only treat (5.24); the cases (5.22) and (5.23) are treated similarly and are easier. It is sufficient to take $m = 2$ because the other cases can be reduced to this one. So we have to estimate

$$\int_{U_i} \frac{dV(\zeta)}{(P^{n-1-\lambda/2}|\Phi|^2)^\gamma}. \qquad (5.25)$$

The following lemma, which follows from the construction of Φ, will be frequently used.

d) **Lemma 5.26** *There is a coordinate system* $x_1, \dots, x_{2n}$ *with*

$$x_1(\zeta) = -r(\zeta)$$
$$x_2(\zeta) = \operatorname{Im} \Phi(\zeta, z)$$
$$x(z) = (-r(z), 0, \dots, 0).$$

The system is defined for $z \in U_i$ *on all of* U_i *provided* U_i *is sufficiently small.*

(Note that (5.5) is a special case of 5.26.)

Now the estimates 5.4, 5.16 and 5.18 can be expressed in the coordinates 5.26 as follows:

$$P(\zeta, z) \gtrsim (\|\tilde{x}\|^2 + x_1^2 + x_2^2 + |r(z)|^2)$$
$$\gtrsim (\|\tilde{x}\|^2 + x_1^2 + x_2^2)$$
$$|\Phi(\zeta, z)| \gtrsim (\|\tilde{x}\|^2 + x_1 + |x_2| + |r(z)|)$$
$$\gtrsim (\|\tilde{x}\|^2 + x_1 + |x_2|)$$

with $\tilde{x} = (x_3, x_4, \dots, x_{2n})$. (5.25) can therefore be estimated by the integral

$$\int_{\tilde{U}_i} \frac{dV(x)}{(\|\tilde{x}\|^2 + x_1^2 + x_2^2)^{\gamma(n-1-\lambda/2)}(\|\tilde{x}\|^2 + x_1 + |x_2|)^{2\gamma}}, \qquad (5.27)$$

with $\tilde{U}_i = \{(x_1, x_2, \tilde{x}) : \|\tilde{x}\| \leq 1, 0 < x_1 \leq 1, |x_2| \leq 1\}$.

e) Let us state, also for frequent later use, another elementary inequality:

Lemma 5.28 *For* $a, b > 0$, $k \in \mathbb{N}$, $0 < \varepsilon < k$, *one has*

$$(a + b)^k \gtrsim a^{k-\varepsilon} b^\varepsilon$$

f) We apply (5.28) to the denominator in (5.27):

$$(\|\tilde{x}\|^2 + x_1 + |x_2|)^{2\gamma} \gtrsim \|\tilde{x}\|^{2\varepsilon}(x_1 + |x_2|)^{2\gamma-\varepsilon}.$$

Then the denominator is estimated from below by

$$\|\tilde{x}\|^{2\gamma(n-1-\lambda/2)+2\varepsilon}(x_1 + |x_2|)^{2\gamma-\varepsilon}.$$

If it is possible to choose $\varepsilon > 0$ such that

$$2\gamma(n - 1 - \lambda/2) + 2\varepsilon < 2n - 2$$
$$2\gamma - \varepsilon < 2,$$

then the integral exists. But this amounts to choosing

$$\gamma < \frac{2n + 2}{2n + 2 - \lambda},$$

which is the estimate of γ given in the claim.

g)   The case $t \leq k + l$ is easily reduced, by repeatedly applying 5.16 and 5.18, to the integral of $\frac{1}{P^{n-\lambda/2}}$ which has already been treated.

h)   As all the estimates are symmetric in ζ and z, the second integral in (5.19) can be treated in the same way as the first; this concludes the proof. $\square$

The case of kernels of type 0 is much more delicate: in fact, an estimate (5.19) does not hold in this case. We need an additional definition.

Definition 5.29   *i. An admissible kernel $\mathcal{A}$ of type ≥ 0 is special admissible, if, near each point $(\zeta_0, \zeta_0) \in \Lambda \subset b\Omega \times b\Omega$, it can be estimated as*

$$|\mathcal{A}(\zeta, z)| \lesssim \frac{1}{P^{t_0}|\Phi|^t} \text{ with } t \neq 2 \text{ and } 2n + \min(2, t) - 2(t + t_0) \geq 0.$$

ii. An operator with special admissible kernel is called special admissible.

Since $\mathcal{A}$ is assumed to be of type ≥ 0, we have the following different cases:

$$|\mathcal{A}(\zeta, z)| \lesssim \frac{1}{P^{n-t/2}|\Phi|^t}, \quad 0 \leq t < 2 \tag{5.30}$$

$$|\mathcal{A}(\zeta, z)| \lesssim \frac{1}{P^{n+1-t}|\Phi|^t}, \quad t > 2 \tag{5.31}$$

Proposition 5.32 *Let $\mathcal{A}$ be a special admissible kernel and ε be a positive number. Then there is a constant C (depending on ε) such that*

$$\int_\Omega |\mathcal{A}(\zeta, z)||r(\zeta)|^{-\varepsilon} dV(\zeta) \leq C|r(z)|^{-\varepsilon}$$

$$\int_\Omega |\mathcal{A}(\zeta, z)||r(z)|^{-\varepsilon} dV(z) \leq C|r(\zeta)|^{-\varepsilon}$$

The result can again be deduced from Chapter VII, Lemma 2.1; but we still want to proceed more directly.

Proof   We only have to look at points z close to the boundary and $\zeta \in U$ close to z, and can therefore work with the coordinates 5.26. This leads to the integrals

$$\int\limits_{\widetilde{U}} \frac{dV(\zeta)}{(\|\widetilde{x}\|^2 + x_1^2 + x_2^2 + r(z)^2)^{n-t/2}(\|\widetilde{x}\|^2 + x_1 + |x_2| + |r(z)|)^t x_1^\varepsilon}, \ t < 2, \quad (5.33)$$

resp.

$$\int\limits_{\widetilde{U}} \frac{dV(\zeta)}{(\|\widetilde{x}\|^2 + x_1^2 + x_2^2 + r(z)^2)^{n+1-t}(\|\widetilde{x}\|^2 + x_1 + |x_2| + |r(z)|)^t x_1^\varepsilon}, \ t > 2. \quad (5.34)$$

$\widetilde{U}$ is a neighbourhood of 0 in the space $\mathbb{R}^+ \times \mathbb{R} \times \mathbb{R}^{2n-2} = \{(x_1, x_2, \widetilde{x})\}$. We abbreviate $r = |r(z)|$ and want to transform the coordinates in dependence of r. This will not leave $\widetilde{U}$ invariant unless $\widetilde{U}$ is the whole space; so we take as $\widetilde{U}$:

$$\widetilde{U} = \mathbb{R}^+ \times \mathbb{R} \times \mathbb{R}^{2n-2}.$$

From now on we have to treat (5.33) and (5.34) separately.

Case (5.33). Call the integral J. Then

$$J \leq \int\limits_{\mathbb{C}^{n-1}} \int\limits_{\mathbb{R}} \int\limits_{\mathbb{R}^+} \frac{dV(x_1, x_2, \widetilde{x})}{(\|\widetilde{x}\|^2 + x_1^2 + x_2^2 + r^2)^{n-t/2}(x_1 + |x_2| + r)^t x_1^\varepsilon}$$

$$= \frac{1}{r^\varepsilon} \int\limits_{\mathbb{C}^{n-1}} \int\limits_{\mathbb{R}} \int\limits_{\mathbb{R}^+} \frac{dV(x_1, x_2, \widetilde{x})}{(\|\widetilde{x}\|^2 + x_1^2 + x_2^2 + 1)^{n-t/2}(x_1 + |x_2| + 1)^t x_1^\varepsilon},$$

where we have used the transformation

$$(x_1, x_2, \widetilde{x}) \longmapsto (rx_1, rx_2, r\widetilde{x}).$$

We integrate over the 8 domains defined by the inequalities $\|\widetilde{x}\| \gtrless 1$, $x_1 \gtrless 1$, $|x_2| \gtrless 1$. Whenever one of the 3 variables remains in a bounded domain, that is in 7 of the 8 cases, the corresponding integral is easily seen to be finite. It remains the case

$$J_\infty = \frac{1}{r^\varepsilon} \int\limits_{\|\widetilde{x}\| \geq 1} \int\limits_{|x_2| \geq 1} \int\limits_{x_1 \geq 1} \cdots$$

We use 5.28:

$$(\|\widetilde{x}\|^2 + x_1^2 + x_2^2 + 1)^{n-t/2}(x_1 + |x_2| + 1)^t x_1^\varepsilon$$

$$\geq (\|\widetilde{x}\|^2 + x_1^2 + x_2^2)^{n-t/2}(x_1 + |x_2|)^t x_1^\varepsilon$$

$$= (\|\widetilde{x}\|^2 + x_1^2 + x_2^2)^{n-1+\alpha/2}(x_1 + |x_2|)^{2-\alpha} x_1^\varepsilon \quad (\text{with } t = 2 - \alpha)$$

$$\geq \|\widetilde{x}\|^{2(n-1+\alpha/2-\delta)}(x_1 + |x_2|)^{2-\alpha+2\delta} x_1^\varepsilon$$

$$\geq \|\widetilde{x}\|^{2n-2+\alpha-2\delta}|x_2|^{2-\alpha+2\delta-\eta} x_1^{\eta+\varepsilon}$$

for $\delta, \eta > 0$. We choose

$$\eta = 1 - \frac{\varepsilon}{2}; \; \frac{\alpha}{2} - \frac{\varepsilon}{4} < \delta < \frac{\alpha}{2}.$$

This implies

$$\eta + \varepsilon > 1$$
$$2 - \alpha + 2\delta - \eta > 1$$
$$2n - 2 + \alpha - 2\delta > 2n - 2,$$

and yields

$$J_\infty \lesssim \frac{1}{r^\varepsilon} \int\limits_{\substack{\|\widetilde{x}\| \geq 1 \\ |x_2| \geq 1 \\ x_1 \geq 1}} \frac{dV(\widetilde{x}) \, dx_2 \, dx_1}{\|\widetilde{x}\|^{2n-2+\alpha-2\delta} |x_2|^{2-\alpha+2\delta-\eta} x_1^{\eta+\varepsilon}} = C r^{-\varepsilon}.$$

Let us now treat case (5.34). Let J be that integral. Then

$$J \lesssim \int\limits_{\mathbb{C}^{n-1}} \int\limits_{\mathbb{R}} \int\limits_{\mathbb{R}^+} \frac{dV(x_1, x_2, \widetilde{x})}{(\|\widetilde{x}\|^2 + x_1^2 + x_2^2 + r^2)^{n+1-t} (\|\widetilde{x}\|^2 + x_1 + |x_2| + r)^t x_1^\varepsilon}$$

$$\lesssim \int\limits_{\mathbb{C}^{n-1}} \int\limits_{\mathbb{R}} \int\limits_{\mathbb{R}^+} \frac{dV(x_1, x_2, \widetilde{x})}{\|\widetilde{x}\|^{2n+2-2t} (\|\widetilde{x}\|^2 + x_1 + |x_2| + r)^t x_1^\varepsilon}$$

in the coordinates (5.26), with $r = |r(z)|$. We transform

$$(x_1, x_2, \widetilde{x}) \longmapsto (r x_1, r x_2, \sqrt{r} \widetilde{x})$$

and obtain

$$J \lesssim \frac{1}{r^\varepsilon} \int\limits_{\mathbb{C}^{n-1}} \int\limits_{\mathbb{R}} \int\limits_{\mathbb{R}^+} \frac{dV(x_1, x_2, \widetilde{x})}{\|\widetilde{x}\|^{2n+2-2t} (\|\widetilde{x}\|^2 + x_1 + |x_2| + 1)^t x_1^\varepsilon}$$

As before, we have only to study integration over a neighbourhood of ∞, the other integrals being easier to treat (because $t > 2$):

$$J_\infty = \frac{1}{r^\varepsilon} \int\limits_{\substack{\|\widetilde{x}\| \geq 1 \\ |x_2| \geq 1 \\ x_1 \geq 1}} \frac{dV(x_1, x_2, \widetilde{x})}{\|\widetilde{x}\|^{2n+2-2t} (\|\widetilde{x}\|^2 + x_1 + |x_2|)^t x_1^\varepsilon}$$

(We could of course drop the 1.) Using (5.28) we get

$$\|\widetilde{x}\|^{2n+2-2t} (\|\widetilde{x}\|^2 + x_1 + |x_2|)^t x_1^\varepsilon$$
$$\geq \|\widetilde{x}\|^{2n+2-2t} \|\widetilde{x}\|^{2t-2\delta} (x_1 + |x_2|)^\delta x_1^\varepsilon$$
$$\geq \|\widetilde{x}\|^{2n+2-2\delta} |x_2|^{\delta-\eta} x_1^{\eta+\varepsilon}.$$

We can choose

$$\eta = 1 - \frac{\varepsilon}{2}, \qquad \delta = 2 - \frac{\varepsilon}{4}$$

and get

$$J_\infty \le C \cdot r^{-\varepsilon},$$

as required. – If $t = 2$ none of the above procedures works.

Interchanging the roles of ζ and z yields the second claim in (5.32). $\qquad\square$

To apply the above estimates we need the following information:

Proposition 5.35 (Young inequality) *Let (X, μ) and (Y, ν) be measure spaces, $\mathcal{K}(x, y)$ a $\mu \times \nu$-measurable function on $X \times Y$ satisfying*

$$\int_X |\mathcal{K}(x, y)|^s \, d\mu(x) \le M^s \quad \text{for almost all } y \in Y,$$

$$\int_Y |\mathcal{K}(x, y)|^s \, d\nu(y) \le M^s \quad \text{for almost all } x \in X,$$

$M < \infty$, $s \ge 1$. *Then the operator K defined by*

$$Kf(y) = \int_X f(x)\mathcal{K}(x, y) \, d\mu(x)$$

is bounded from $L^p(X)$ to $L^r(Y)$ with norm $\le M$ for all $1 \le p \le \infty$ and r such that

$$\frac{1}{r} = \frac{1}{p} + \frac{1}{s} - 1.$$

For the proof see [Ran 86].

Proposition 5.36 *Let $\mathcal{A}$ be an admissible kernel such that for any ε with $0 < \varepsilon < 1$ there is a constant C with*

$$\int_\Omega |\mathcal{A}(\zeta, z)| |r(\zeta)|^{-\varepsilon} \, dV(\zeta) \le C |r(z)|^{-\varepsilon}$$

$$\int_\Omega |\mathcal{A}(\zeta, z)| |r(z)|^{-\varepsilon} \, dV(z) \le C |r(\zeta)|^{-\varepsilon}.$$

Then the operator A with

$$Af(z) = \int_\Omega f(\zeta)\mathcal{A}(\zeta, z) \, dV(\zeta)$$

is continuous from $L^p \to L^p$ for each p, $1 < p < \infty$.

Proof   Let s be the dual exponent: $\frac{1}{p} + \frac{1}{s} = 1$. We write

$$|Af(z)| \leq \int_{\Omega} |\mathcal{A}(\zeta,z)|^{\frac{1}{p}} |f(\zeta)| |r(\zeta)|^{\frac{\varepsilon}{s}} |\mathcal{A}(\zeta,z)|^{\frac{1}{s}} |r(\zeta)|^{-\frac{\varepsilon}{s}} dV(\zeta)$$

and apply Hölder's inequality:

$$|Af(z)|^p$$

$$\leq \left[\int_{\Omega} |\mathcal{A}(\zeta,z)| |f(\zeta)|^p |r(\zeta)|^{\frac{\varepsilon p}{s}} dV(\zeta) \right] \left[\int_{\Omega} |\mathcal{A}(\zeta,z)| |r(\zeta)|^{-\varepsilon} dV(\zeta) \right]^{\frac{p}{s}}$$

$$\leq C |r(z)|^{-\frac{\varepsilon p}{s}} \int_{\Omega} |\mathcal{A}(\zeta,z)| |f(\zeta)|^p |r(\zeta)|^{\frac{\varepsilon p}{s}} dV(\zeta)$$

in view of the first inequality (5.36). Consequently

$$\int_{\Omega} |Af(z)|^p dV(z)$$

$$\leq C \int_{\Omega} \left[\int_{\Omega} |\mathcal{A}(\zeta,z)| |r(z)|^{-\frac{\varepsilon p}{s}} dV(z) \right] r(\zeta)^{\frac{\varepsilon p}{s}} |f(\zeta)|^p dV(\zeta)$$

$$\leq C^2 \int_{\Omega} |f(\zeta)|^p dV(\zeta)$$

by the second inequality (5.36). $\square$

4.   We can now combine the information in propositions 5.19, 5.32, 5.35 and 5.36:

Theorem 5.37   i. *An admissible operator A of type $\lambda > 0$ is continuous from*

$$L^p \longrightarrow L^r$$

for $1 \leq p \leq \infty$ and any r with $\frac{1}{r} > \frac{1}{p} - \frac{\lambda}{2n+2}$.

    ii. *A special admissible operator of type ≥ 0 is continuous from*

$$L^p \longrightarrow L^p, \quad 1 < p < \infty.$$

Proposition 5.13 and known properties of isotropic kernels now give our final result.

Theorem 5.38 *The operators P_q, T_q and S_q of the basic integral formula 4.22 are continuous between the following spaces:*

$$P_0 : L^p \longrightarrow L^p, \quad 1 < p < \infty$$
$$P_q : L^1 \longrightarrow C^{\infty}(\overline{\Omega})$$
$$T_q, S_q : L^p \longrightarrow L^r, \quad 1 \leq p \leq \infty, \frac{1}{r} > \frac{1}{p} - \frac{1}{2n+2}$$

(We have not indicated the types of differential forms).

§6 Levi's Problem and Vanishing of Cohomology

1. Let Ω be an arbitrary domain in $\mathbb{C}^n$. We introduce the following vector spaces:

$L_{pq}^s(\Omega)$ *the space of* (p,q)*-forms with coefficients in* L^s.

$Z_s^{pq}(\Omega) = \{f \in L_{pq}^s : \bar\partial f = 0\}$ *the space of* $\bar\partial$*-closed forms in* L_{pq}^s *(cocycles).*

$B_s^{pq}(\Omega) = \{f \in L_{pq}^s : f = \bar\partial u$ *with* $u \in L_{p,q-1}^s\}$ *the space of* $\bar\partial$*-exact forms in* L_{pq}^s

    (coboundaries).

$H_s^{pq}(\Omega) = Z_s^{pq}(\Omega)/B_s^{pq}(\Omega)$ *the cohomology space of* L^s*-bounded cohomology classes* .

Here $1 \leq s \leq \infty$. The inconsistency in the position of the indices (as lower or upper indices) cannot be avoided if we want to stay in accordance with the usual notations. – We can now state our first main result:

Theorem 6.1 (Finiteness theorem) *If* $\Omega \subset\subset \mathbb{C}^n$ *is strictly pseudoconvex then the vector spaces* $H_s^{pq}(\Omega)$ *are finite dimensional for* $q \geq 1$ *and all* s *with* $1 \leq s \leq \infty$. *The spaces* $B_s^{pq}(\Omega)$ *are closed.*

Proof   Since a (p,q)-form f is a $\binom{n}{p}$-tuple of $(0,q)$-forms which are closed resp. exact if and only if f is closed resp. exact we may assume $p = 0$. So let $f \in Z_s^{0q}(\Omega)$, $q \geq 1$. By our main integral formula we have

$$f - P_q f = \bar\partial S_{q-1} f \tag{6.2}$$

where $S_{q-1}f \in L_{0,q-1}^s(\Omega)$, and the operator P_q maps L^s-bounded sets into uniformly equicontinuous sets and is therefore compact from Z_s^{0q} into itself. This implies that the operator $id - P_q$ has closed image of finite codimension. Because of (6.2),

$$B_s^{0q}(\Omega) \supset \operatorname{im}(id - P_q),$$

so B_s^{0q} is also closed and has finite codimension in Z_s^{0q} – which proves the theorem.   □

The finiteness theorem 6.1 already suffices to give a solution of the famous Levi problem (in its easiest version).

Theorem 6.3 *A strictly pseudoconvex domain in* $\mathbb{C}^n$ *is holomorphically convex.*

Let us recall that *a domain (or a complex manifold* X*) is* defined to be *holomorphically convex if for any sequence* z_j *without accumulation point in* X *there is a holomorphic function on* X *which is unbounded on that sequence.*

Proof of 6.3.   Let z_0 be a boundary point of Ω. The Levi polynomial $F(z_0, z)$ is holomorphic, vanishes at z_0, and there is a neighbourhood U of z_0 such that $F(z_0, z) \neq 0$ on $U \cap \overline\Omega - \{z_0\}$. By suitably patching with a C^∞-function we obtain a smooth function

f on $\overline{\Omega}$ which is holomorphic in a neighbourhood of z_0 and vanishes exactly at z_0. The differential forms

$$\overline{\partial}\left(\frac{1}{f^\varkappa}\right), \; \varkappa = 1, 2, 3, \ldots$$

belong to $L_{01}^\infty(\Omega)$ – in fact, they are smooth on $\overline{\Omega}$ –, they are $\overline{\partial}$-closed, and so, by the finiteness theorem, are linear dependent mod $B_\infty^{01}(\Omega)$. In other words, there are constants $a_1, a_2, \ldots, a_k$, with $a_k \neq 0$, and a bounded function u on Ω such that

$$\sum_{\varkappa=1}^k a_\varkappa \overline{\partial}\left(\frac{1}{f^\varkappa}\right) = \overline{\partial}u.$$

Then the function

$$h = \sum_{\varkappa=1}^k a_\varkappa \frac{1}{f^\varkappa} - u$$

is holomorphic on Ω and tends to infinity for $z \to z_0$, which means that Ω is holomorphically convex. $\square$

Remark. Classical arguments show that the union of an ascending sequence of holomorphically convex domains in $\mathbb{C}^n$ is holomorphically convex (Behnke-Stein theorem) [BeS 39]. So we can also claim

Theorem 6.3′ *A pseudoconvex domain in $\mathbb{C}^n$ is holomorphically convex.*

2. The proof of 6.3 would have been slightly easier if $H_\infty^{pq} = 0$. This is in fact the case but requires a fairly delicate reasoning. We start with an extension theorem.

Proposition 6.4 *Let $\Omega = \{\zeta : r(\zeta) < 0\}$, r a strictly plurisubharmonic defining function. Let $\Omega_\delta = \{\zeta : r(\zeta) < \delta\}$. Then there is a positive δ_0 such that for each pair δ_1, δ_2 with $0 \leq \delta_1 \leq \delta_2 \leq \delta_0$ the restriction map*

$$\rho : H_s^{pq}(\Omega_{\delta_2}) \longrightarrow H_s^{pq}(\Omega_{\delta_1})$$

is surjective for $q \geq 1$ and $1 \leq s \leq \infty$.

Proof   (We again assume $p = 0$.)

Consider, for a cocycle $f \in L_{0q}^s(\Omega)$, the form

$$g(z) = P_q f(z) = (-1)^q \int_\Omega f(\zeta) \wedge \overline{\partial}_\zeta K_q(\zeta, z). \tag{6.5}$$

We have on Ω the relation

$$f = g + \overline{\partial}S_{q-1}f;$$

so, since g is smooth on $\overline{\Omega}$ and defines the same cohomology class as f, we assume from the very beginning that f is smooth on $\overline{\Omega}$. But (6.5) defines g as a smooth form even on Ω_0, where $\Omega_0 = \{\zeta : r(\zeta) < \delta_0\}$ is a sufficiently small strictly pseudoconvex neighbourhood of $\overline{\Omega}$, because, for all z in a neighbourhood of $\overline{\Omega}$, the numerator of $K_q(\zeta, z)$ vanishes in a neighbourhood of the set $\Phi(\zeta, z) = 0$. We have to show that $\overline{\partial}g = 0$ on Ω_0. This is true, anyway, on $\overline{\Omega}$. Choose a point $z \in \Omega_0 - \overline{\Omega}$. Then the kernel $K_q(\zeta, z)$ is smooth in $\zeta \in \overline{\Omega}$, and Stokes' theorem gives

$$g(z) = (-1)^q \int_{\Omega} f(\zeta) \wedge \overline{\partial}_\zeta K_q(\zeta, z) \tag{6.6}$$

$$= -\int_{\Omega} \overline{\partial}f(\zeta) \wedge K_q(\zeta, z) + \int_{b\Omega} f(\zeta) \wedge K_q(\zeta, z)$$

$$= \int_{b\Omega} f(\zeta) \wedge K_q(\zeta, z)$$

because $\overline{\partial}f(\zeta) = 0$. But on $b\Omega$ the kernel K_q is defined by a Leray form and therefore satisfies (see II.1.39)

$$\overline{\partial}_z K_q(\zeta, z) = (-1)^{q+1} \overline{\partial}_\zeta K_{q+1}(\zeta, z). \tag{6.7}$$

From (6.6) and (6.7) we get

$$\overline{\partial}_z g(z) = \int_{b\Omega} f(\zeta) \wedge \overline{\partial}_z K_q(\zeta, z)$$

$$= (-1)^{q+1} \int_{b\Omega} f(\zeta) \wedge \overline{\partial}_\zeta K_{q+1}(\zeta, z)$$

$$= -\int_{b\Omega} \overline{\partial}(f(\zeta) \wedge K_{q+1}(\zeta, z))$$

$$= 0.$$

This settles the claim for $\delta_1 = 0$. If $\delta_1 = \delta$ is arbitrary between 0 and δ_0 one considers the kernels K_q^δ constructed from the boundary function $r^\delta(\zeta) = r(\zeta) - \delta$ and repeats the above arguments. $\square$

We need one more fact about the operators in our fundamental integral formula, which immediately follows from their construction:

Lemma 6.8 *Suppose $r \in C^\infty$ and let $f \in L^s_{0q}(\Omega) \cap C^\infty(\Omega)$. Then $\boldsymbol{T}_{q-1}f$ and $\boldsymbol{S}_{q-1}f$ are again in $C^\infty(\Omega)$.*

If r is of class C^k, $k \geq 4$, then $\boldsymbol{T}_q f$ and $\boldsymbol{S}_{q-1}f$ are still in C^2 which would suffice for the following arguments.

3.   After these preparations we can now turn to the proof of the cohomology vanishing theorems. The idea of the following construction is due to Laufer – see [Lau 75]. In the next chapter (Ch.IV, §7) we will choose a different approach to solve the same problem in a more general context.

We assume the situation of proposition 6.4, set $\Omega = \{\zeta : r(\zeta) < 0\}, \Omega_0 = \{\zeta : r(\zeta) < \delta_0\}$ and only consider domains $\Omega_1, \Omega_2, \ldots$ selected in the family $\Omega_\delta = \{\zeta : r(\zeta) < \delta\}, 0 \leq \delta \leq \delta_0$.

Let $f_1, \ldots, f_k \in Z_s^{0q}(\Omega_0)$ be a finite system of generators of that space modulo $B_s^{0q}(\Omega_0)$. We may assume that $f_\varkappa$ are even smooth on $\overline{\Omega}_0$. Their restrictions, to any strictly pseudoconvex domain Ω_1 with

$$\Omega \subset \Omega_1 \subset\subset \Omega_0$$

then generate $Z_s^{0q}(\Omega_1) \mod B_s^{0q}(\Omega_1)$. Let $z = z^1$ be a coordinate function. The forms

$$f_\varkappa, z f_\varkappa, \ldots, z^h f_\varkappa, \ldots$$

are linearly dependent mod $B_s^{0q}(\Omega_0)$; in other words there is a polynomial $p_\varkappa(z)$ of degree ≥ 0 such that

$$p_\varkappa(z) f_\varkappa = \overline{\partial} u_\varkappa, \tag{6.9}$$

with $u_\varkappa \in L_{0,q-1}^s(\Omega_0)$. Multiplying all the $p_\varkappa(z)$ gives a polynomial $p(z)$ with $p(z) f_\varkappa \in B_s^{0q}(\Omega_0)$. Since $f_\varkappa$ generate $Z_s^{0q}(\Omega_0) \mod B_s^{0q}(\Omega_0)$, we have

$$p(z) f = \overline{\partial} u \tag{6.10}$$

for any $f \in Z_s^{0q}$. The form u, which is in $L_{0,q-1}^s(\Omega_0)$, need not be smooth on Ω_0 even if f is. But if f is smooth we can replace u in (6.10) by a smooth form! In fact, applying 4.22' to Ω_0 we have

$$u = \boldsymbol{P}_{q-1} u + \boldsymbol{T}_{q-1} \overline{\partial} u + \overline{\partial} \boldsymbol{S}_{q-2} u, \tag{6.11}$$

and by (6.10):

$$p(z) f = \overline{\partial} [\boldsymbol{P}_{q-1} u + \boldsymbol{T}_{q-1} p(z) f] = \overline{\partial} v, \tag{6.12}$$

where now v is smooth because of Lemma 6.8. To sum up: *there is a polynomial $p(z)$ of degree $d \geq 0$ such that*

$$p(z) f = \overline{\partial} u \tag{6.10}$$

for all $f \in Z_s^{0q}(\Omega_0)$; u can be chosen smooth on Ω_0 if f is smooth.

Now let Ω_1 be strictly pseudoconvex with $\Omega \subset\subset \Omega_1 \subset\subset \Omega_0$. A cohomology class in $H_s^{0q}(\Omega_1)$ can be represented by a smooth (on $\overline{\Omega}_0$) cocycle $f \in Z_s^{0q}(\Omega_0)$. Differentiating (6.10) with respect to z gives

$$\frac{\partial p}{\partial z}(z) f + p(z) \frac{\partial f}{\partial z} = \overline{\partial} \frac{\partial u}{\partial z}. \tag{6.13}$$

But $\frac{\partial f}{\partial z}$ is a smooth cocycle on $\overline{\Omega}_0$ and is therefore annihilated by $p(z) \mod B_s^{0q}(\Omega_0)$. So

$$\frac{\partial p}{\partial z}(z) f = \overline{\partial} \left[\frac{\partial u}{\partial z} + v \right] = \overline{\partial} w$$

where now $w \in C^{\infty}_{0,q-1}(\Omega_0)$ (not necessarily integrable over Ω_0); it is of course integrable over Ω_1. So we see: *there is a polynomial of degree $\leq d - 1$, namely $\frac{\partial p}{\partial z}(z)$, such that*

$$\frac{\partial p}{\partial z}(z)f = \bar{\partial}u \tag{6.14}$$

for all $f \in Z^{0q}_s(\Omega_1)$, with $u \in L^s_{0q-1}(\Omega_1)$. If f is smooth on Ω_1 then u can be chosen smooth on Ω_1.

We can repeat this procedure d times until we arrive at a polynomial p of degree 0, that is a constant $\neq 0$, such that

$$pf = \bar{\partial}u \tag{6.15}$$

for all $f \in Z^{0q}_s(\Omega_d)$, with $u \in L^s_{0q-1}(\Omega_d)$, where $\Omega \subset \Omega_d \subset\subset \Omega_0$. But this means

$$H^{0q}_s(\Omega_d) = 0,$$

therefore, because of the surjectivity of the restriction,

$$H^{0q}_s(\Omega) = 0, \quad q \geq 1.$$

To sum up we have the vanishing theorem

Theorem 6.16 *If $\Omega \subset\subset \mathbb{C}^n$ is strictly pseudoconvex, then the cohomology groups*

$$H^{pq}_s(\Omega) = 0, \quad q \geq 1, 1 \leq s \leq \infty.$$

Moreover, if f is a $\bar{\partial}$-closed smooth (p,q)-form with coefficients in L^s, then there is a solution $u \in L^s_{p,q-1}(\Omega) \cap C^{\infty}$ such that

$$\bar{\partial}u = f.$$

We can get more than just an L^p-bounded solution to the CR equations:

Theorem 6.17 *There is (for Ω strictly pseudoconvex) a linear operator*

$$F = F_{q-1} : Z^{0q}_1(\Omega) \longrightarrow L^1_{0,q-1}(\Omega)$$

with the following properties:

  i. *$\bar{\partial}Ff = f$ for all f.*

  ii. *$F|Z^{0q}_s(\Omega)$ is, for each s, a continuous map from $Z^{0q}_s(\Omega)$ into $L^s_{0,q-1}(\Omega)$.*

So F is an L^s-bounded solution operator to the Cauchy-Riemann equations on Ω. – Such an operator had been constructed quite explicitly on the ball – here, in our more general situation, we can only prove its existence. We shall, however, pass to more explicit constructions somewhat later.

Proof   Denote, for each s with $1 \leq s \leq \infty$, the Banach space Z_s^{0q} simply by Z_s, and let D_s be the domain of definition of $\overline{\partial}$ in $L_{0,q-1}^s$, i.e.

$$D_s = \{f \in L_{0,q-1}^s : \overline{\partial} f \in Z_s\}.$$

We have the inclusions

$$D_\infty \subset D_s \subset D_1; \quad Z_\infty \subset Z_s \subset Z_1.$$

By the previous theorems, the linear map

$$\overline{\partial} : D_s \longrightarrow Z_s$$

is surjective for each s. Moreover, there are linear integral operators

$$\boldsymbol{P} : Z_1 \longrightarrow Z_1; \quad \boldsymbol{S} : Z_1 \longrightarrow D_1$$

such that, for each $f \in Z_s$,

$$f = \boldsymbol{P} f + \overline{\partial} \boldsymbol{S} f. \tag{6.18}$$

By the preceding estimates, $\boldsymbol{P}$ maps each Z_s into itself (in fact: into Z_∞) and is a compact operator, and $\boldsymbol{S}$ maps Z_s into D_s, continuously for all s.

Set

$$B_s = \mathrm{Im}\,\{\overline{\partial} \boldsymbol{S} : Z_s \longrightarrow Z_s\};$$

again we have

$$B_\infty \subset B_s \subset B_1.$$

Let us fix s and write (6.18) as

$$\overline{\partial} \boldsymbol{S} = id - \boldsymbol{P} : Z_s \longrightarrow Z_s. \tag{6.18'}$$

By the theory of compact operators, $A_s = \ker \overline{\partial} \boldsymbol{S}$ and B_s are closed subspaces, such that

$$\dim A_s = \mathrm{codim}\, B_s = l < \infty.$$

But $\overline{\partial} \boldsymbol{S} f = 0$ implies, in view of 5.38, that $f \in C_{0,q}^\infty(\overline{\Omega})$. This means that *the spaces A_s coincide*, we call them simply A. Let us now choose topological complements in Z_s to A and B_s:

$$Z_s = A \oplus A_s^\perp = B_s \oplus B_s^\perp.$$

We want to choose $B_s^\perp \subset Z_\infty$ independent of s. To see that this is possible we take l forms $g_1, \ldots, g_l \in Z_1$ which are linearly independent modulo B_1. Applying (6.18) yields

$$g_\lambda = \boldsymbol{P} g_\lambda + \overline{\partial} \boldsymbol{S} g_\lambda,$$

so we can replace the g_λ by their projections $\boldsymbol{P} g_\lambda$ to obtain another such system which now consists of forms smooth up to the boundary. Thus assume the $g_\lambda \in C_{0,q}^\infty(\overline{\Omega})$ linearly independent modulo B_1, a fortiori modulo B_s, and choose $B^\perp$ as the space generated by the g_λ. Since codim $B_s = l$, independently of s, we have the desired decomposition

$$Z_s = A \oplus A_s^\perp = B_s \oplus B^\perp.$$

This setup can be described in more detail. Namely, obviously,

$$Z_s = Z_1 \cap L^s.$$

Because of (6.18) we also have

$$B_s = B_1 \cap L^s.$$

Finally, having chosen $A_1^\perp$ we can obtain a topological complement to A in Z_s by setting

$$A_s^\perp = A_1^\perp \cap L^s.$$

So the above decomposition reads as

$$Z_s = Z_1 \cap L^s = A \oplus (A_1^\perp \cap L^s) = (B_1 \cap L^s) \oplus B^\perp. \tag{6.19}$$

We now choose a base $f_1, \ldots, f_l$ in A and, *using the surjectivity of $\bar\partial$*, forms

$$h_1, \ldots, h_l \in D_\infty$$

with

$$\bar\partial h_\lambda = g_\lambda.$$

The correspondence

$$R f_\lambda = h_\lambda$$

defines a linear injection

$$R : A \longrightarrow D_\infty.$$

Combining R with S yields a linear map

$$R \oplus S : Z_1 = A \oplus A_1^\perp \longrightarrow D_1$$

such that

$$V = \bar\partial \circ (R \oplus S) : Z_1 = A \oplus A_1^\perp \longrightarrow B_1 \oplus B^\perp$$

is a topological isomorphism. We now set

$$F = (R \oplus S) \circ V^{-1};$$

to obtain the required solution operator $F : Z_1 \to L^1$; the decomposition (6.19) now shows that $V|Z_s$ is a topological isomorphism of Z_s and, consequently, $F|Z_s$ a continuous map into L^s. □

The operators F_q do not necessarily respect differentiability in the interior. This is one justification for more explicit constructions, to be dealt with later. It is, however, important to note that F_0 has the desired regularity:

Theorem 6.20   *i. For $f \in Z_1^{01} \cap C^\infty(\Omega)$ we have $F_0 f \in L^1 \cap C^\infty(\Omega)$.*

   ii. F_0 is continuous in the Fréchet topologies from $C_{01}^\infty(\overline{\Omega}) \cap \ker \overline{\partial} \longrightarrow C^\infty(\Omega)$.

Proof   The first statement follows from $\overline{\partial} F_0 f = f$ and the interior regularity of the $\overline{\partial}$-operator on functions (I.9.14).

To prove *ii.* we consider a sequence $(f_j, F_0 f_j)$ which converges in the product topology to (f, g). Then $f_j \to f$ in L^1, $F_0 f_j \longrightarrow F_0 f$ in L^1, and $F_0 f \in C^\infty(\Omega) \cap L^1$ because of *i.*. But, since $F_0 f_j \longrightarrow g$ in $C^\infty(\Omega)$, we must have $g = F_0 f$ on Ω. The continuity of F_0 now follows from the closed graph theorem. $\square$

§7   The Henkin-Ramírez Formula

The possibility of solving the $\overline{\partial}$-equation with good estimates has a wealth of applications. We discuss one of them, namely global integral representations for functions with holomorphic kernel.

Constructing such a kernel has been easy on the ball because the Levi polynomial

$$F(\zeta, z) = \sum_j F_j(\zeta, z)(\zeta^j - z^j) = \sum_j r_j(\zeta)(\zeta^j - z^j)$$

of the defining function $r(\zeta) = \|\zeta\|^2 - 1$ could immediately be used to define a holomorphic Leray form

$$\frac{1}{F(\zeta, z)} \sum_j F_j(\zeta, z) \, d\zeta^j = \frac{1}{F(\zeta, z)} \sum_j r_j(\zeta) \, d\zeta^j.$$

On a general strictly pseudoconvex domain the corresponding Levi polynomial had to be modified, and this modification destroyed the holomorphy off the boundary diagonal and ultimately led to the – fortunately compact – error terms P_q in the formulae of §4. The results of §6 now allow to remedy this defect and – with considerably more effort – to construct an integral formula on strictly pseudoconvex domains with essentially the same features as on the ball. In order to avoid complicated cohomological reasonings we will stay as close as possible to our integral formula approach.

1.   Let us again collect our data:

$$F(\zeta, z) = \sum_j \left(r_j(\zeta) - \frac{1}{2} \sum_k r_{jk}(\zeta)(\zeta^k - z^k) \right)(\zeta^j - z^j) \tag{7.1}$$

$$= \sum_j F_j(\zeta, z)(\zeta^j - z^j)$$

is the Levi polynomial of a strictly plurisubharmonic defining function r for the strictly pseudoconvex domain Ω.

We can choose δ, ε and $\gamma_0 > 0$ so that

$$2 \operatorname{Re} F(\zeta, z) \geq r(\zeta) - r(z) + \gamma_0 \|\zeta - z\|^2 \tag{7.2}$$

on the set

$$\overline{V}_\delta \times \overline{\Omega}_\delta \cap \{(\zeta, z) : \|\zeta - z\| \leq \varepsilon\},$$

where $\Omega_\delta = \{\zeta : r(\zeta) < \delta\}$ and $V_\delta = \{\zeta : |r(\zeta)| < \delta\}$. Let now $\varphi(\zeta, z)$ be a smooth function on $\mathbb{C}^n \times \mathbb{C}^n$ satisfying $0 \leq \varphi \leq 1$ and

$$\varphi(\zeta, z) = \begin{cases} 0 & \text{for } \|\zeta - z\| \leq \frac{\varepsilon}{2} \\ 1 & \text{for } \|\zeta - z\| \geq \frac{3}{4}\varepsilon. \end{cases} \tag{7.3}$$

Now, if $\delta > 0$ is sufficiently small and if the point $(\zeta, z) \in \overline{V}_\delta \times \overline{\Omega}_\delta$ satisfies

$$\frac{\varepsilon}{2} \leq \|\zeta - z\| \leq \varepsilon,$$

we deduce from (7.2)

$$2 \operatorname{Re} F(\zeta, z) > 0. \tag{7.4}$$

So $\log F(\zeta, z)$ is well defined on the above set, and we can define, on $\overline{V}_\delta \times \overline{\Omega}_\delta$, the form

$$f(\zeta, z) = \begin{cases} \overline{\partial}_z \left[\varphi(\zeta, z) \frac{\log F(\zeta, z)}{F(\zeta, z)^2} \right], & \frac{\varepsilon}{2} \leq \|\zeta - z\| \leq \varepsilon \\ 0 & \text{otherwise} \end{cases} \tag{7.5}$$

f is a C^{k-2}-map

$$f : \overline{\Omega}_\delta \longrightarrow Z^{01}(\overline{\Omega}_\delta) \cap C^\infty(\Omega_\delta),$$

in other words, a $\overline{\partial}_z$-closed double form on $\overline{V}_\delta \times \overline{\Omega}_\delta$ of double type $(0, 0; 0, 1)$ depending C^∞ on z, C^{k-2} on ζ and z, with all z-derivatives again C^{k-2} in ζ and z. We can now apply the solution operator F_0 of Theorem 6.20 to f and obtain a function

$$v(\zeta, z) = F_0 f(\zeta, z) \tag{7.6}$$

satisfying

$$\overline{\partial}_z v(\zeta, z) = f(\zeta, z). \tag{7.7}$$

Because of the continuity properties of F_0 this function v is, when restricted to $z \in \overline{\Omega}_{\delta/2}$, a C^{k-2}-map from

$$\overline{V}_\delta \longrightarrow C^\infty(\overline{\Omega}_{\delta/2}).$$

We now introduce on $\overline{V}_\delta \times \overline{\Omega}_{\delta/2}$ the functions

$$u(\zeta, z) = v(\zeta, z) - \varphi(\zeta, z) \frac{\log F(\zeta, z)}{F(\zeta, z)^2} \quad \text{for } \|\zeta - z\| \leq \varepsilon \tag{7.8}$$

and

$$h(\zeta, z) = \begin{cases} F(\zeta, z) \exp\left(F^2(\zeta, z)u(\zeta, z)\right), & \|\zeta - z\| \leq \varepsilon \\ \exp\left(F^2(\zeta, z)v(\zeta, z)\right), & \|\zeta - z\| \geq \frac{3}{4}\varepsilon. \end{cases} \qquad (7.9)$$

If $\frac{3}{4}\varepsilon \leq \|\zeta - z\| \leq \varepsilon$ we have $u = v - \frac{\log F}{F^2}$ and therefore

$$Fe^{F^2 u} = Fe^{F^2 v}e^{-\log F} = e^{F^2 v}.$$

Moreover, $\overline{\partial}_z v(\zeta, z) = f(\zeta, z) = 0$ for $\|\zeta - z\| \geq \varepsilon$, and u is holomorphic in z; consequently h is holomorphic in z and a C^{k-2}-map

$$h : \overline{V}_\delta \longrightarrow \mathcal{O}(\overline{\Omega}_{\delta/2}). \qquad (7.10)$$

For $\|\zeta - z\| \leq \varepsilon$ we have

$$h - F = F\left(e^{F^2 u} - 1\right) = O(\|\zeta - z\|^3), \qquad (7.11)$$

which, together with (7.2), gives

$$2\,\mathrm{Re}\, h \geq r(\zeta) - r(z) + \gamma\|\zeta - z\|^2 - C\|\zeta - z\|^3. \qquad (7.12)$$

We can now easily choose first ε and then δ so small that on

$$\overline{V}_\delta \times \overline{\Omega}_\delta$$

the function h satisfies the estimates

$$|h(\zeta, z)| \geq c > 0 \quad \text{for } \|\zeta - z\| \geq \varepsilon \qquad (7.13)$$

$$\mathrm{Re}\, h(\zeta, z) \geq r(\zeta) - r(z) + \gamma\|\zeta - z\|^2, \quad \text{for } \|\zeta - z\| \leq \varepsilon. \qquad (7.14)$$

Summing up we have

Theorem 7.15 (Ramírez-Henkin) *Let the strictly pseudoconvex domain Ω be given as*

$$\Omega = \{\zeta : r(\zeta) < 0\}$$

with strictly plurisubharmonic defining C^k-function r. Then there are positive constants δ, ε, c and γ and a C^{k-2}-function $h(\zeta, z)$ on

$$\overline{V}_\delta \times \overline{\Omega}_\delta$$

with the following properties:

  i. h is holomorphic in z, all z-derivatives are C^{k-2} in ζ and z.

  ii. For $\|\zeta - z\| \leq \varepsilon$,

$$h(\zeta, z) = a(\zeta, z)F(\zeta, z),$$

  where $a(\zeta, z)$ is again holomorphic in z and C^{k-2} in (ζ, z), with $a(\zeta, \zeta) = 1$, and where F is the Levi polynomial of r.

   iii. In particular, $h(\zeta, \zeta) \equiv 0$.

   iv. $|h(\zeta, z)| \geq c$ for $\|\zeta - z\| \geq \varepsilon$.

   v. Re $h(\zeta, z) \geq r(\zeta) - r(z) + \gamma \|\zeta - z\|^2$ for $\|\zeta - z\| \leq \varepsilon$.

Definition 7.16 *A function h defined on a neighbourhood of $b\Omega \times \Omega$ with properties i to iii above is called a Ramírez-Henkin function.*

Such a function can be built up in many ways. A different construction (see [Ran 86]) leads to a Ramírez-Henkin function h satisfying, instead of iv and v, the estimate

$$iv'. \quad \mathrm{Re}\; h(\zeta, z) \;\geq\; 0 \text{ for}$$
$$(\zeta, z) \;\in\; \overline{V}_\delta \times \overline{\Omega}_\delta \text{ with } r(\zeta) - r(z) + \gamma|\zeta - z|^2 > 0.$$

The existence of h is intimately tied up with the holomorphic convexity of Ω. In fact,

$$u(\zeta, z) = \frac{1}{h(\zeta, z)} \tag{7.17}$$

has, for $\zeta \in b\Omega$ fixed, the following properties: it is holomorphic in z on $\overline{\Omega} - \{\zeta\}$ and tends to infinity for $z \to \zeta$. The holomorphic convexity of Ω thus follows from the existence of h, and $u(\zeta, \cdot)$ is a family of holomorphic functions tending to ∞ at ζ and depending continuously (in fact: C^{k-2}) on the parameter ζ. This idea can be more neatly expressed as follows

Theorem 7.18 (Peak function theorem) *There is a C^{k-2}-function $e(\zeta, z)$ on $b\Omega \times \Omega_\eta$ such that*

   i. e is holomorphic in z.

   ii. $e(\zeta, \zeta) = 1$.

   iii. $|e(\zeta, z)| < 1$ on $\overline{\Omega} - \{\zeta\}$.

The function

$$e(\zeta, z) = \exp\{-h(\zeta, z)\},$$

where h is a Ramírez-Henkin function with iv', does the trick. The family $e(\zeta, \cdot)$ is a C^{k-2}-family of "holomorphic peak functions".

2. "The" Ramírez-Henkin function $h(\zeta, z)$ from (7.15) will provide us with the right denominator for a global integral kernel; in order to build up a correct numerator we have to decompose

$$h(\zeta, z) = \sum_{j=1}^{n} h_j(\zeta, z)(\zeta^j - z^j), \tag{7.19}$$

where the functions h_j have to be, like h, C^{k-2}-maps from $\overline{V}_\delta$ into the space $\mathcal{O}(\overline{\Omega}_\delta)$. This requires, beside some nice geometric constructions, once again solution of a $\overline{\partial}$-problem "with parameters". Here is the key:

Lemma 7.20 *Let U be a neighbourhood of $\overline{\Omega}$, where $\Omega \subset\subset \mathbb{C}^n$ is strictly pseudoconvex, H a complex hyperplane in $\mathbb{C}^n$ and X a real C^l-manifold. Suppose $f : X \to \mathcal{O}(U \cap H)$ is a C^l-map. Then there is a neighbourhood $V \subset\subset U$ of $\overline{\Omega}$ and a C^l-map $\widehat{f} : X \to \mathcal{O}(V)$ with $\rho \circ \widehat{f} = f$, where $\rho : \mathcal{O}(V) \to \mathcal{O}(V \cap H)$ is the restriction map for holomorphic functions.*

In other words: $f(x, z)$, for $x \in X$ and $z \in U \cap H$, is a function on $X \times (U \cap H)$ which is holomorphic in z and of class C^l in x and z (including all z-derivatives). Then there is on a slightly smaller neighbourhood V of $\overline{\Omega}$, a function $\widehat{f}(x, z)$ on $X \times V$ with $\widehat{f}|X \times (V \cap H) = f$, and $\widehat{f}$ is again holomorphic in z and C^l in (x, z).

Proof   We may assume that H is the hyperplane $z_1 = 0$. If U_1 is a sufficiently small neighbourhood of $H \cap \overline{\Omega}$ in $\mathbb{C}^n$, the function $f(x, \cdot)$ can be considered as a holomorphic function on U_1, i. e.

$$f : X \longrightarrow \mathcal{O}(U_1)$$

is a C^l-map. Let now χ be a smooth patching function,

$$\chi(z) = \begin{cases} 1 \text{ in } U_3(H \cap \overline{\Omega}) \\ 0 \text{ outside } U_2(H \cap \overline{\Omega}) \\ 0 \le \chi \le 1, \end{cases}$$

where U_2 and U_3 are open neighbourhoods of $H \cap \overline{\Omega}$ in $\mathbb{C}^n$ with $U_3 \subset\subset U_2 \subset\subset U_1$, and consider

$$g(x, z) = \begin{cases} \overline{\partial}_z \left(\frac{1}{z_1} \chi(z) f(x, z) \right), & z \in U_1 \\ 0, & z \notin U_1 \end{cases}$$

Since

$$\overline{\partial} \left(\frac{1}{z_1} f(x, z) \chi(z) \right) = \frac{1}{z_1} f(x, z) \overline{\partial} \chi(z)$$

and $\overline{\partial}\chi \equiv 0$ on U_3, g is a smooth $\overline{\partial}$-closed form on a neighbourhood V of $\overline{\Omega}$ which depends C^l-smoothly on x. Theorem 6.20 provides us with a function of class C^l

$$u : X \longrightarrow C^\infty(V),$$

where V is a neighbourhood of $\overline{\Omega}$, such that

$$\overline{\partial} u(x, z) = g(x, z).$$

We can now set

$$\widehat{f}(x, z) = \chi(z) f(x, z) - z_1 u(x, z).$$

$\square$

Lemma 7.21 *Let U, Ω and X be as above, and set, for $1 \leq k \leq n$,*

$$H_k = \{z \in U : z^1 = \ldots = z^k = 0\}.$$

Suppose

$$f : X \longrightarrow \mathcal{O}(U)$$

is a C^l-map satisfying

$$f(x, z) \equiv 0$$

for all $z \in H_k$. Then there is a neighbourhood $V \subset\subset U$ of $\overline{\Omega}$ and C^l-maps

$$f_j : X \longrightarrow \mathcal{O}(V), \quad j = 1, \ldots, k,$$

such that

$$f(x, z) = \sum_{j=1}^{k} f_j(x, z) z^j. \tag{7.22}$$

Proof   If $k = 1$ we can set

$$f_1(x, z) = \frac{f(x, z)}{z^1}$$

to obtain (7.22) for all n. (Cauchy's integral formula tells us that f_1 is as regular as f). Now assume the lemma true for $k - 1$ (and all $n \geq k$). Set

$$\Omega^1 = \Omega \cap \{z : z^1 = 0\}.$$

Then Ω^1 is a (not necessarily connected) strictly pseudoconvex region in

$$\mathbb{C}^{n-1} = \{z : z^1 = 0\},$$

and composition of f with the restriction map ρ to $z^1 = 0$ gives a C^l-map

$$\rho \circ f : X \longrightarrow \mathcal{O}(V(\overline{\Omega}^1)),$$

where V is a neighbourhood of $\overline{\Omega}^1$ in $\mathbb{C}^{n-1}$. If $\delta > 0$ is sufficiently small, the region $\{z^1 = 0\} \cap \{r(z) < \delta\} = \Omega_1'$ is a neighbourhood of $\overline{\Omega}_1$, contained in V which is strictly pseudoconvex with a smooth boundary ($b\Omega^1$ need not be smooth!), and so $\rho \circ f$ satisfies the induction hypotheses on Ω_1': consequently

$$\rho \circ f(x, z) = \sum_{j=2}^{k} g_j(x, z') z^j,$$

where the g_j are C^l-maps from X into $\mathcal{O}(W)$, $W \subset\subset V$ being a suitable neighbourhood of $\overline{\Omega}^1$ in $\mathbb{C}^{n-1}$. We have written $z = (z^1, z')$. Now use the preceding lemma: there are C^l-maps

$$f_j : X \longrightarrow \mathcal{O}(Z), \quad j = 2, \ldots, k,$$

where $\overline{\Omega} \subset Z \subset U$, Z an open set in $\mathbb{C}^n$, such that the restrictions to $Z \cap \mathbb{C}^{n-1}$ satisfy

$$\rho \circ f_j = g_j.$$

If we now define

$$F(x,z) = f(x,z) - \sum_{j=2}^{k} f_j(x,z)z^j, \tag{7.23}$$

then F satisfies

$$F(x,0,z') \equiv 0.$$

The case $k = 1$ yields

$$F(x,z) = z^1 f_1(x,z); \tag{7.24}$$

combining (7.23) and (7.24) yields the claim. $\square$

We can now easily deduce an important decomposition theorem whose non-parametrized version goes back a long way [HefH 50]:

Theorem 7.25 (Hefer's theorem with parameters) *Let again $\Omega \subset\subset \mathbb{C}^n$ be a strictly pseudoconvex domain, X a C^l-manifold and*

$$f : X \longrightarrow \mathcal{O}(U)$$

be a C^l-map, where U is a pseudoconvex neighbourhood of $\overline{\Omega}$. Then there is a neighbourhood V of $\overline{\Omega}$ and there are C^l-maps

$$q_j : X \longrightarrow \mathcal{O}(V \times V), \quad j = 1, \ldots, n,$$

such that for $(\zeta, z) \in V \times V$,

$$f(x,\zeta) - f(x,z) = \sum_{j=1}^{n} q_j(x,\zeta,z)(\zeta^j - z^j). \tag{7.26}$$

Proof   Since $U \times U$ is pseudoconvex it can be exhausted by strictly pseudoconvex sets, therefore we can find a strictly pseudoconvex domain $\Theta \subset \mathbb{C}^n \times \mathbb{C}^n$ with

$$\Omega \times \Omega \subset\subset \Theta \subset\subset U \times U.$$

Consider the function

$$F(x,\zeta,z) = f(x,\zeta) - f(x,z);$$

it satisfies $F(x,\zeta,\zeta) = 0$. This means that after the linear coordinate change

$$\left. \begin{array}{l} u^i = \zeta^i - z^i \\ u^{n+i} = z^i \end{array} \right\} \quad i = 1 \ldots n$$

in $\mathbb{C}^n \times \mathbb{C}^n$ we have $F = 0$ on $\{u_1 = \ldots = u_n = 0\}$. The preceding lemma gives the decomposition

$$F(x,u) = \sum_{j=1}^{n} F_j(x,u)u^j. \tag{7.27}$$

Going back to the ζ, z-coordinates transforms (7.27) into (7.26). $\square$

We now apply Theorem 7.25 to the Ramírez-Henkin function h of Theorem 7.15:

Theorem 7.28 *In the situation of 7.15 there are a positive constant δ and C^{k-2}-maps*

$$h_j : \overline{V}_\delta \longrightarrow \mathcal{O}(\overline{\Omega}_\delta)$$

such that

$$h(\zeta, z) = \sum_{j=1}^{n} h_j(\zeta, z)(\zeta^j - z^j)$$

satisfies the properties and estimates of Theorem 7.15.

Proof   Recall that $\overline{V}_\delta = \{\zeta : |r(\zeta)| \leq \delta\}$, $\overline{\Omega}_\delta = \{z : r(z) \leq \delta\}$. We have, by the preceding theorem, the following decomposition of the Ramírez-Henkin function h:

$$h(x, \zeta) - h(x, z) = \sum_{j=1}^{n} q_j(x, \zeta, z)(\zeta^j - z^j)$$

for $x \in \overline{V}_\delta$, $\zeta, z \in \overline{\Omega}_\delta$ (after shrinking δ). Now set $x = \zeta$. Then

$$-h(\zeta, z) = \sum q_j(\zeta, \zeta, z)(\zeta^j - z^j),$$

which is, up to notations, the required formula.   □

3.   Let us now use the functions $h_j(\zeta, z)$ to define the Leray form

$$k(\zeta, z) = \frac{1}{h(\zeta, z)} \sum_{j=1}^{n} h_j(\zeta, z) \, d\zeta^j. \tag{7.29}$$

The estimates of Theorem 7.15 tell us that $k(\zeta, z)$ is defined on

$$\{(\zeta, z) : \zeta \in b\Omega_\eta, z \in \Omega_\eta\}$$

for $|\eta|$ sufficiently small. The associated Cauchy-Fantappiè kernels

$$\Omega_q(k), \quad q = 0, 1, 2, \ldots \tag{7.30}$$

have the following properties:

    i. $K(\zeta, z) \overset{def}{=} \Omega_0(k)$ *is holomorphic in z and of class C^{k-3} in ζ and z; all its z-derivatives are of class C^{k-3}.*

    ii. $\Omega_q(k) = 0$ *for $q \geq 1$.*

Writing again

$$b(\zeta, z) = \frac{1}{\|\zeta - z\|^2} \sum (\bar{\zeta}^j - \bar{z}^j)\, d\zeta^j$$

for the generating form of the BMK kernels, we introduce the transition kernels

$$A_q(\zeta, z) \stackrel{\text{def}}{=} A_q(k, b); \tag{7.31}$$

they are of class C^{k-3} in ζ and z; all their z and $\bar{z}$-derivatives are again of class C^{k-3}.

Definition 7.32 *The kernels $K_0(\zeta, z)$ and $A_q(\zeta, z)$ are called Ramírez-Henkin kernels associated with Ω (resp. its boundary function r).*

Koppelman's homotopy formula now reads, on $b\Omega_\delta \times \Omega_\delta$, as follows:

$$\begin{aligned}
B_{n0} &= K - \bar{\partial}_\zeta A_0 \\
B_{nq} &= (-1)^{q+1} \bar{\partial}_\zeta A_q + \bar{\partial}_z A_{q-1}; \; q \geq 1.
\end{aligned} \tag{7.33}$$

We can now execute the familiar programme: replacing B_{nq} in the boundary integral by (7.33) and applying Stokes' theorem we arrive at

Theorem 7.34 (Ramírez-Henkin-Lieb) *Let f be in $C^1_{0q}(\overline{\Omega})$, where Ω is strictly pseudoconvex with C^k-boundary, $k \geq 3$. Let $K(\zeta, z)$ and $A_q(\zeta, z)$ be the Ramírez-Henkin kernels associated with Ω. Then we have, for $z \in \Omega$:*

i. $q = 0$

$$f(z) = \int_{b\Omega} f(\zeta) K(\zeta, z) + \int_{b\Omega} \bar{\partial} f(\zeta) \wedge A_0(\zeta, z) - \int_{\Omega} \bar{\partial} f(\zeta) \wedge B_{n0}(\zeta, z),$$

with holomorphic boundary integral;

ii. $q \geq 1$

$$\begin{aligned}
f(z) = \int_{b\Omega} \bar{\partial} f(\zeta) \wedge A_q(\zeta, z) - \int_{\Omega} \bar{\partial} f(\zeta) \wedge B_{nq}(\zeta, z) \\
+ \bar{\partial}_z \left[\int_{b\Omega} f(\zeta) \wedge A_{q-1}(\zeta, z) - \int_{\Omega} f(\zeta) \wedge B_{n,q-1}(\zeta, z) \right].
\end{aligned}$$

We abbreviate the above integral operators as usual:

$$\boldsymbol{K} f(z) = \int_{b\Omega} f(\zeta) K(\zeta, z)$$

$$\boldsymbol{R}_{q-1} f(z) = \int_{b\Omega} f(\zeta) \wedge A_{q-1}(\zeta, z) - \int_{\Omega} f(\zeta) \wedge B_{n,q-1}(\zeta, z)$$

and obtain (7.34) in the following form:

Theorem 7.34' *In the situation of 7.34, for* $f \in C^1_{0q}(\overline{\Omega})$,

$$f = Kf + R_0\overline{\partial}f, \qquad\qquad q = 0$$
$$f = R_q\overline{\partial}f + \overline{\partial}R_{q-1}f, \qquad\qquad q \geq 1.$$

4.   The consequences of the above homotopy formula are obvious:

Theorem 7.35 *If f is a continuous function on $b\Omega$, then the function*

$$Kf(z) = \int\limits_{b\Omega} f(\zeta)K_0(\zeta, z)$$

is holomorphic in Ω; if f is holomorphic on $\overline{\Omega}$, then, for $z \in \Omega$,

$$f(z) = Kf(z).$$

Theorem 7.36 *Let f be a $\overline{\partial}$-closed $(0, q)$-form on $\overline{\Omega}$ of class C^1, $q \geq 1$. Then*

$$f = \overline{\partial}R_{q-1}f.$$

So we have solved the CR equations on Ω by exhibiting an explicit solution in terms of a linear integral formula.

The basic homotopy formula on the ball was slightly different: there we avoided boundary integrals and obtained operators P, T and S which differ from K resp. R by a $\overline{\partial}$-closed term. This allowed us to easily obtain L^p-estimates. It is naturally also possible to pass to volume integrals in Theorem 7.34 by using a suitable "extended Ramírez-Henkin function" which corresponds to the extended Levi polynomial of §4 and §5, together with the extended norm function P, and then obtain an explicit homotopy formula which simultaneously generalises Theorem 3.14 and Theorem 4.22. We state the results as

Theorem 7.37 *Let $b\Omega$ be of class C^k, $k \geq 4$. There are double differential forms of class C^{k-4} outside the diagonal on $\overline{\Omega} \times \overline{\Omega}$*

$$P \text{ of double type } (n, n; 0, 0)$$
$$T_q \text{ of double type } (n, n - q - 1; 0, q)$$
$$S_{q-1} \text{ of double type } (n, n - q; 0, q - 1),$$

with the following properties:

  i. $P(\zeta, z)$ is holomorphic in z and C^{k-4} on $\overline{\Omega} \times \Omega$.

ii. The integrals

$$\boldsymbol{P}f(z) = \int\limits_{\Omega} f(\zeta) P(\zeta, z)$$

$$\boldsymbol{T}_q f(z) = \int\limits_{\Omega} f(\zeta) \wedge T_q(\zeta, z)$$

$$\boldsymbol{S}_q f(z) = \int\limits_{\Omega} f(\zeta) \wedge S_q(\zeta, z)$$

exist for f in L^1_{0q} resp. $L^1_{0,q+1}$ and $\boldsymbol{T}_q$ and $\boldsymbol{S}_q$ define L^1-bounded linear operators.

iii. The identities

$$f = \boldsymbol{P}f + \boldsymbol{T}_0\overline{\partial}f, \qquad\qquad\qquad q = 0$$
$$f = \boldsymbol{T}_q\overline{\partial}f + \overline{\partial}\boldsymbol{S}_{q-1}f, \qquad\qquad q \geq 1$$

hold on L^1_{0q}.

The proof repeats the arguments of §3, §4 and §5, after replacing the Levi polynomial by the Ramírez-Henkin function. We skip the details because, later on, we shall work with a variant of the basic integral formula of §4 on manifolds where an analogue of the Ramírez-Henkin formula is too difficult to establish. So Theorem 7.37 will never be used.

5.   All of the preceding theorems depend on the basic integral formula 4.22′

$$f = \boldsymbol{P}_q f + \boldsymbol{T}_q\overline{\partial}f + \overline{\partial}\boldsymbol{S}_{q-1}f$$

and the integral estimates of §5, for the case $q > 0$. For $q = 0$ we have

$$f = \boldsymbol{P}_0 f + \boldsymbol{T}_0\overline{\partial}f,$$

where the operator $\boldsymbol{P}_0$ is L^p-bounded for $1 < p < \infty$.

Definition 7.38   *The orthogonal projection*

$$P : L^2(\Omega) \longrightarrow L^2(\Omega) \cap \mathcal{O}(\Omega)$$

is called the Bergman projection of Ω.

In the case of the ball we had $P = \boldsymbol{P}_0$, i.e. we could compute the Bergman projector by our integral formulae. In the more general case of a strictly pseudoconvex domain, $\boldsymbol{P}_0$ is certainly not the Bergman projector because $\boldsymbol{P}_0 f$ is in general not holomorphic.

But Theorem 6.20 and the symmetry statement (4.9) can be used to show that $\boldsymbol{P}_0$ differs from the Bergman projector P by a compact operator. We will pursue this idea in a later chapter in much more detail and so allow ourselves to skip this point here.

§8 Convex Domains of Finite Type

We finally take up the report on convex finite type domains which we gave in the first two paragraphs. On using, instead of the Levi polynomial, the holomorphic support function $F(\zeta, z)$ constructed by Diederich and Fornæss, and globalising it as in §7 (this was carefully done by W. Alexandre [Ale 2001]), we can establish, following the development of the preceding paragraphs, a basic homotopy formula for convex domains:

Theorem 8.1 *Let* $\Omega \subset\subset \mathbb{C}^n$ *be a smoothly bounded finite type convex domain. There are linear integral operators* T_q, $q \geq 1$, *and* S_q, $q \geq 0$, *such that for* $f \in C^1_{0q}(\overline{\Omega})$, $q \geq 1$, *the identity*

$$f = T_q \overline{\partial} f + \overline{\partial} S_{q-1} f$$

holds. The operators T_q, S_q *are* L^p-*bounded.*

(Because of the convexity of Ω, there are no "error" operators P_q.)

This theorem immediately solves the CR equations in Ω. Moreover, the precise estimates of Diederich and Fornæss indicated in §2 yield

Theorem 8.2 *Let* Ω *be as above. Then there is a linear solution operator* S_{q-1} *for the equation*

$$\overline{\partial} u = f, \ \overline{\partial} f = 0, \ f \in L^p_{0q}(\Omega),$$

satisfying

$$\overline{\partial} S_{q-1} f = f, \ \|S_{q-1} f\|_{L^r} \leq \text{const.} \ \|f\|_{L^p},$$

where $r = r(p) > p$.

The proof is in [Fis 2001]. — A different construction of a solution operator leading to the same estimates was given by Cumenge [Cum 97], [Cum 2001]. The "regularity order" r above also depends on q — see Hefer's paper [HefT 2002].

§9 Notes

The first results on integral formulae on strictly pseudoconvex domains are those presented in §7; they are due to Ramírez [Ram 70], Henkin [Hen 69] and Lieb [Lie 70] (around 1970) and use global integral representations. So they draw heavily on the "qualitative" theory of Stein manifolds.

It became only gradually clear that, in order to develop large parts of Stein theory, local representations, which are much more elementary, suffice: they can be glued together to finally yield a basic homotopy formula

$$f = P_q f + T_q \overline{\partial} f + \overline{\partial} S_{q-1} f \tag{9.1}$$

with (for $q > 0$) compact error terms $\boldsymbol{P}_q$ – see §4. From here one deduces the fundamental finiteness and vanishing theorems of the theory by means of functional analysis, provided the operators in (9.1) satisfy some estimates. The work of Henkin [Hen 77], Kerzman [Ker 71], Range [Ran 82] and Kerzman/Stein [KeS 78] reflects this line of argument – cf. also [Lie 84], and the books [Ran 86], [HeL 84] and [LTh 97].

The main elementary properties of strictly pseudoconvex domains as presented in the first two paragraphs were formulated in the work of Grauert [Gra 58], Kohn [Koh 63, Koh 64] and Andreotti/Grauert [AnG 62].

Levi metrics were essentially used by Greiner/Stein [GrS 77] and play a crucial role in Lieb/Range's study of the Neumann problem [LiR 83] (which we shall later present in this book). The normalisation proposition 1.14 was found by S. Kaldorf (Bonn, unpublished). None of the above properties carry over to the more general class of "weakly" pseudoconvex domains – see [KoN 73] and [ForS 87]. – The symmetry property of the Levi polynomial (2.12.iii) was discovered and used by Kerzman/Stein [KeS 78], later by Ligocka [Lig 84] and Lieb/Range [LiR 83]. It is essential in the sequel.

Cauchy type integral formulae for the ball as in §3 were found by Hua [Hua 58] and Leray [Ler 56], among others. Theorem 3.14 contains these formulae as a special case and is, at the same time, a model for our more general homotopy formula (9.1) resp. (4.22).

The first cohomology finiteness theorems for strictly pseudoconvex domains were established by Grauert [Gra 58] and Kohn [Koh 63, Koh 64]; their relation to our results is explained in the last chapter. In our case the finiteness theorem 6.1 is an easy consequence of our basic homotopy formula and the compactness of the operators $\boldsymbol{P}_q$, and it immediately implies a solution of Levi's problem. The actual vanishing of cohomology in $\mathbb{C}^n$, that is the solution of the Cauchy-Riemann system

$$\overline{\partial}u = f, \qquad \overline{\partial}f = 0 \tag{9.2}$$

in $L^s(\Omega)$, requires a far more delicate proof whose idea is due to Laufer [Lau 75]. The functional analytic argument leading to Theorems 6.16 and 6.20 is taken from [LTh 97].

The Ramírez-Henkin functions in §7 were first constructed by Ramírez [Ram 70] and Henkin [Hen 69], independently. Our argument closely follows [Ran 86], who, himself, relies on Henkin's construction; Ramírez' proof is much more requiring.

The solution of a parameter-dependent $\overline{\partial}$-problem given by 6.20 is essential for the construction, and this is also true for the second step in the build-up of the Ramírez-Henkin kernel, namely for Hefer's theorem with parameters. The framework of the proof of 7.25 is already in Hefer's thesis from 1940 [HefH 50], but the parameter dependence could not be treated at that time. We have again followed [Ran 86].

Admissible kernels related to ours (§5) were studied by Folland/Stein [FoS 74] and Kerzman/Stein [KeS 78]. Typical is their product structure of factors exhibiting isotropic and anisotropic singularities. Kernels of this structure already turn up in Lieb's work from 1970. The anisotropic estimate 5.4 which is of course basic for the whole theory was established by Grauert/Lieb [GrL 70] and Henkin [Hen 70], independently. Special admissible kernels have so far only been considered in the purely anisotropic case; we need the general case later. The methods of estimation were worked out by many people; suggestions and results of E. Stein's have been particularly important.

The precise results for the $\overline{\partial}$-equation, which we have established for strictly pseudoconvex domains, are not true for arbitrary pseudoconvex domains. We have mentioned, in §8, that they carry over to convex domains of finite type. Other classes of pseudoconvex domains which have been defined with the aim of generalising the "strictly pseudoconvex theory", or in order to find counter-examples, are: domains with bounded strictly plurisubharmonic exhaustion function [DiF 77₁], domains with global plurisubharmonic defining function [BoS 91], exotic "worm-like" domains [DiF 77₂].

If one drops the smoothness requirement on the boundary, several classes of strictly pseudoconvex domains with non-smooth boundary have been considered by, e. g., Henkin/Leiterer [HeL 84] and Range/Siu [RaS 73].

For all these classes, sharp estimates for the Neumann problem, as proved for strictly pseudoconvex domains in Chapter VIII, are still unknown.

Chapter IV

Strictly Pseudoconvex Manifolds

In this chapter we want to carry over the basic integral formula (4.22) of the previous chapter to strictly pseudoconvex bounded subdomains of arbitrary complex manifolds ("*strictly pseudoconvex manifolds*"). The essential ideas remain unchanged: we start with an integral representation involving the complex Laplacian, which leads to a BMK type formula, and transform it, by a Koppelman homotopy formula, into a representation which depends on the strict pseudoconvexity of the boundary. But as compared to the case of $\mathbb{C}^n$, the absence of global coordinates on a manifold provides for a fair amount of complications: practically all calculations known from $\mathbb{C}^n$ are only true "up to error terms" which have to be controlled, and this accounts for a host of additional compact operators.

We start with developing the parametrix method for the real Laplacian on Riemannian manifolds, leading to the basic results of Hodge theory on compact Riemannian manifolds. These results, with their method of proof, can indeed be considered as a model for our study of strictly pseudoconvex manifolds (where, however, things are more complicated). Theorem 3.11, for instance, can be looked at as a real version of the Bochner-Martinelli-Koppelman formula. We will concentrate on those arguments which are essential for later developments and refer to [Rha 60] for many details.

It is only in §5 that we return to complex analysis. We combine the parametrix method for the complex Laplacian with the Cauchy-Fantappiè calculus (the link is established in 5.19) to arrive at a BMK formula for hermitian manifolds. Proceeding as on strictly pseudoconvex domains in $\mathbb{C}^n$ we obtain a homotopy formula on strictly pseudoconvex manifolds which is the main result of the chapter (Theorem 5.54). This formula has the same applications as in $\mathbb{C}^n$: finiteness theorems, solution of Levi's problem, vanishing theorems.

Along the way we carry over a number of important concepts from $\mathbb{C}^n$ to general manifolds: isotropic kernels and their generalisations, admissible kernels, the Bochner-Martinelli-Koppelman kernel for hermitian manifolds. The relation between hermitian and Riemannian metrics, which is fairly important for our theory, is carefully explained in §5.

As mentioned in the introduction to our book, Hodge theory on compact Riemannian manifolds can also be considered as a model for Dolbeault cohomology an strictly pseudoconcave manifolds. This idea has been pursued by Michel [Mic 92]; his work develops methods and results which are non-trivial analogues to the theory of strictly pseudoconvex manifolds as explained here and in the following chapters. We do not include Michel's results but refer to [Mic 92] and also to [HeL 88] and [HefT 99].

§1   The Real Laplacian

1.   Let X be a real n-dimensional oriented C^∞-manifold equipped with a fixed Riemannian metric, given, in local coordinates, by

$$ds^2 = \sum g_{jk}(x)\, dx^j\, dx^k;\tag{1.1}$$

we require the locally defined functions g_{jk} to be of class C^∞. Metric and orientation define a *Hodge $*$-operator*

$$* = *_x : E_x^p \longrightarrow E_x^{n-p}\tag{1.2}$$

between the spaces of complex covectors of the indicated dimensions p resp. $n - p$ at each point $x \in X$; $*$ carries over to differential forms on a subset $M \subset X$ by pointwise application. Let us recall its definition and main properties:

If $\omega^1, \ldots, \omega^n$ is a positively oriented orthonormal base of real 1-covectors (in $x \in X$), then for any subset $K \subset N$

$$*\omega^K = \varepsilon_{K\,N-K}^N \omega^{N-K}.\tag{1.3}$$

We have again (as in Chapter I) used the following conventions:

$N = \{1, \ldots, n\}$, $K = \{i_1 < \ldots < i_p\} \subset N$ a strictly ordered subset, $N - K = \{j_1 < \ldots < j_{n-p}\}$ the strictly ordered complementary set of K in N, $KN - K = \{i_1, \ldots, i_p, j_1, \ldots, j_{n-p}\}$,

$$\varepsilon_{K\,N-K}^N = \text{\textit{sign of the permutation sending } } KN - K \text{ \textit{into} } N,$$
$$\omega^K = \omega^{i_1} \wedge \ldots \wedge \omega^{i_p}.$$

On arbitrary covectors $*$ is defined by $\mathbb{C}$-linear extension.

Lemma 1.4   *i. $*$ is well-defined (independently of the choice of the positive orthonormal base).*

ii.   (1.3) holds for arbitrary arrangements of the indices (not necessarily monotonic).

iii. $** = (-1)^{p(n+1)}$; $*\overline{f} = \overline{*f}$.

iv.   If $f = \sum'_{|K|=p} f_K\, \omega^K$, $g = \sum'_{|K|=p} g_K\, \omega^K$, then

$$f \wedge *g = g \wedge *f = \sum'_{|K|=p} f_K \overline{g}_K * 1 = \langle f, g \rangle * 1,$$

where $\langle , \rangle$ is the hermitian scalar product induced on the space of complex p-covectors by the metric.

$\sum'_{|K|=p}$ means, as usual, summation over the strictly ordered subsets $K \subset N$ of cardinality $|K| = p$.

Lemma 1.5 *Let f be a p-form on X. Then the formula*

$$(*f)(x) = *_x f(x)$$

defines an $(n-p)$-form $*f$ on X.

Definition 1.6 $*1 = dV$ *is called the volume element of the (orientation and) metric.*

Lemma 1.7   *i. $* : E^p(X) \to E^{n-p}(X)$ is an isomorphism between the above spaces of differential forms satisfying*

$$** = (-1)^{p(n+1)}. \tag{1.8}$$

ii. For $f, g \in E^p(X)$,

$$f \wedge *\overline{g} = \overline{g} \wedge *f = \langle f, g \rangle * 1. \tag{1.9}$$

iii.

$$\langle f, g \rangle = \langle *f, *g \rangle \tag{1.10}$$

To compute $*f$ we will in general work with a positive real orthonormal C^∞-system of 1-forms and then use (1.3). This shows, among other things, that $*$ respects differentiability:

$$* : C_p^k(X) \longrightarrow C_{n-p}^k(X).$$

The required bases exist locally but are almost never induced by local coordinates, i. e. $d\omega^j \neq 0$. This fact will give rise to a first group of "error terms" in our later formulae. – If one works with arbitrary C^∞-base forms the simple formula (1.3) no longer holds and has to be replaced by the corresponding formula in [Rha 60].

Now consider a subdomain $\Omega \subset\subset X$. The *scalar product* of two p-forms f and g on Ω is

$$(f, g)_\Omega = \int_\Omega f \wedge *\overline{g} = \int_\Omega \langle f, g \rangle \, dV. \tag{1.11}$$

This defines the Hilbert space $L_p^2(\Omega)$ of square integrable $\mathbb{C}$-valued p-forms with L^2-norm

$$\|f\|_{L^2} = (f, g)_\Omega^{1/2}. \tag{1.12}$$

The case $\Omega = X$ is permitted. – The $*$-operator is a unitary transformation of L^2 (see 1.10).

2. The following development runs closely parallel to Chapter I, §2.

Lemma 1.13 *There is exactly one linear partial differential operator*

$$\delta : C^\infty_{p+1;c}(X) \longrightarrow C^\infty_{p;c}(X)$$

with

$$(df, g) = (f, \delta g) \tag{1.14}$$

for all $f \in C^\infty_{p;c}$ and $g \in C^\infty_{p+1;c}$.

In fact, an easy computation using Stokes' theorem yields

$$\delta = (-1)^{pn+1} * d* : C^\infty_{p+1;c} \longrightarrow C^\infty_{p;c}. \tag{1.15}$$

It is clear that (1.14) holds whenever one of the two forms has compact support.

Definition 1.16   *i. δ is called the formal adjoint of the total differential d.*

ii. $\Delta = d\delta + \delta d$ is the (real) Laplace operator (Laplacian).

Thus, Δ is a second order linear partial differential operator:

$$\Delta = \Delta_p : C^\infty_p(X) \longrightarrow C^\infty_p(X)$$

which satisfies

$$(\Delta f, g) = (df, dg) + (\delta f, \delta g) = (f, \Delta g), \tag{1.17}$$

if f and g are twice continuously differentiable and one of the two forms has compact support. Like d and δ it sends real forms into real forms.

3. We do not give, at this point, explicit formulae in local coordinates for δ or Δ, but we still state the important

Lemma 1.18 (Green's formulae)   *i. For $f \in C^1_{p-1}(X)$, $g \in C^1_p(X)$,*

$$\langle df, g \rangle \, dV = \langle f, \delta g \rangle \, dV + d(f \wedge *\overline{g}). \tag{1.19}$$

ii. If $f, g \in C^1_{p-1}(\overline{\Omega})$, $C^1_p(\overline{\Omega})$, respectively, where Ω is a relatively compact domain with smooth boundary in X, then

$$(df, g)_\Omega = (f, \delta g)_\Omega + \int_{b\Omega} f \wedge *\overline{g}. \tag{1.20}$$

§2 Generalised Isotropic Operators

1.   Let X be as before an oriented n-dimensional (C^∞-) Riemannian manifold. There is a neighbourhood W of the diagonal $\Delta = \{(x, x) : x \in X\} \subset X \times X$ such that the *geodesic distance* $\rho(x, y)$ is a continuous function on W, C^∞ and positive on $W - \Delta$, with $\rho^2(x, y)$ a C^∞ function on all of W. We extend ρ (from a smaller neighbourhood of Δ) to all of $X \times X$ to obtain a function ρ with the following properties:

   i. $\rho(x, y) = $ *geodesic distance on W*

   ii. $\rho > 0$ *and C^∞ on $X \times X - \Delta$*

   iii. $\rho^2(x, y)$ *is C^∞ on $X \times X$*

   iv. $\rho(x, y) = \rho(y, x)$

Definition 2.1 *A double differential form $\mathcal{K}(x, y)$ on $X \times X - \Delta$ is a generalised isotropic kernel (for short: isotropic) of order $\geq l$ (and type $\geq l + n$) if the following holds:*

   i. *Locally,*
$$\mathcal{K}(x, y) = O(\rho(x, y)^l).$$

   ii. *There is an integer $t \leq l$ such that*
$$\mathcal{K}(x, y) = \rho(x, y)^t \mathcal{L}(x, y)$$

   with a double form $\mathcal{L}(x, y)$ which is everywhere (including Δ) smooth.

The first condition means, of course, that given any point $(x_0, x_0) \in \Delta$ there is a coordinate neighbourhood and a constant c such that each coefficient $k(x, y)$ of $\mathcal{K}(x, y)$ satisfies
$$|k(x, y)| \leq c\rho(x, y)^l.$$

– We write, generically,
$$\mathcal{K} = \mathcal{E}_l$$

to indicate that $\mathcal{K}$ is isotropic of order $\geq l$.   The definition also makes sense for kernels defined on a subset $U \subset X \times X$. The isotropic kernels of the first chapter – when ρ is a geodesic distance function – are special cases of the kernels we have just introduced; we need the more general class of kernels only in the first four paragraphs of this chapter (because we also consider odd-dimensional manifolds).

The most important properties of this notion are summarised in the following proposition.

Proposition 2.2 *Let $f(x, y)$ be a smooth function on $X \times X - \Delta$. Let D_x be a differentiation with respect to a local coordinate function x^i, and D_y the same differentiation with respect to the coordinate y^i.*

i. *If f is everywhere smooth, then $f = \mathcal{E}_l$ with $l > 0$ if and only if all derivatives of f up to order $l - 1$ (with respect to local coordinates) vanish on the diagonal. In particular, $D_x f$ and $D_y f$ are isotropic of order $\geq \max(0, l - 1)$.*

ii. *If $f = \mathcal{E}_l$ and $\mathcal{K}$ an isotropic kernel of order $\geq m$, then*

$$f \cdot \mathcal{K} = \mathcal{E}_{l+m}.$$

iii. *$\mathcal{E}_l + \mathcal{E}_m = \mathcal{E}_{\min(l,m)}.$*

iv. *Let f be isotropic of order l, $l \in \mathbb{Z}$. Then*

$$D_x f = \mathcal{E}_{l-1}, \quad D_y f = \mathcal{E}_{l-1}, \quad (D_x + D_y)f = \mathcal{E}_l. \tag{2.3}$$

Proof *i*) follows from Taylor's formula, *ii*) and *iii*) are trivial. So let us prove *iv*). We have

$$f(x,y) = g(x,y)\rho^t, \quad f(x,y) = O(\rho^l),$$

consequently

$$g(x,y) = O(\rho^{l-t}), \quad l - t \geq 0,$$

where $g(x,y)$ is smooth on $X \times X$. Then for $x \neq y$:

$$D_x f(x,y) = D_x g(x,y)\rho^t + t\rho^{t-1}g D_x \rho.$$

The first term is of order $\geq l - 1$ because of *i*). The second term can be written as

$$t\rho^{t-2}g\rho D_x\rho. \tag{2.4}$$

But $\rho D_x\rho$ is smooth even for $x = y$ (because ρ^2 is smooth) and it is $O(\rho)$. So (2.4) has the required form and satisfies the required estimate. Finally, for a smooth function $g(x,y)$ one has — according to [Rha 60] —

$$D_x g(x,y) + D_y g(x,y) = \mathcal{E}_h \tag{2.5}$$

if $g(x,y) = \mathcal{E}_h$. We use this together with its special case

$$\rho D_x\rho + \rho D_y\rho = \mathcal{E}_2 \tag{2.6}$$

to check the last statement of *iv*). $\square$

Proposition 2.7 *Let $(x^1, \ldots, x^n)$ be a system of local coordinates on a relatively compact open set and*

$$R^2(x,y) = \sum_{i,j} g_{ij}(x)(x^i - y^i)(x^j - y^j).$$

Then

$$\rho^2(x,y) = R^2(x,y) + \mathcal{E}_3(x,y). \tag{2.8}$$

Moreover, there are positive constants a and A such that

$$a\rho(x,y) \leq R(x,y) \leq A\rho(x,y) \tag{2.9}$$

(where $R = +\sqrt{R^2}$).

This is proved in [Rha 60].

From (2.8) we easily deduce

$$d_x d_y \rho^2(x, y) = -2 \sum g_{ij} \, dx^i dy^j + \mathcal{E}_1; \tag{2.10}$$

moreover, if we choose orthonormal base fields, $\omega^1, \ldots, \omega^n$, then

$$d_x d_y \rho^2(x, y) = -2 \sum_j \omega^j \Theta^j + \mathcal{E}_1, \tag{2.11}$$

where Θ^j is the base field ω^j in terms of y. These formulae will be important in the next paragraph.

We now introduce integral operators defined by double differential forms as in Chapter I, §4. So if $\mathcal{K}(x, y)$ is a double form of double degree (p, q) on a product domain $\Omega \times \Theta \subset X \times X$, it defines an *integral operator K with kernel $\mathcal{K}$* by

$$Kf(y) = \int_\Omega f(x) \wedge *\overline{\mathcal{K}(x, y)} = (f, \mathcal{K}) \tag{2.12}$$

from $E_p(\Omega)$ to $E_q(\Theta)$, provided the integrals exist. The formal adjoint integral operator is given by the kernel

$$\mathcal{K}^*(x, y) = \overline{\mathcal{K}(y, x)}. \tag{2.13}$$

K is an isotropic (more precisely: generalised isotropic) operator of order $\geq l$ (and type $\geq l + n$) if its kernel has this property; we write, generically,

$$K = E_l.$$

2.   Our new isotropic operators are slightly more general than the operators of the first chapter but they have the same regularity properties which we now collect; all proofs of chapter I, §§5 and 10, carry over with minor modifications. Cf. also [Rha 60].

The metric provides us with an intrinsic definition of L^p-norms for differential forms on Ω: if f is such a form, then

$$|f|(x) = \langle f(x), f(x) \rangle^{1/2}$$

is a well-defined function on Ω, and we set

$$\|f\|_{L^p(\Omega)}^p = \int_\Omega |f|^p \, dV, \quad 1 \leq p < \infty$$

resp.

$$\|f\|_{L^\infty(\Omega)} = \sup_{x \in \Omega} |f|(x).$$

Proposition 2.14 *Let $\mathcal{K}$ be an isotropic kernel of order $m \geq 1 - n$ on $X \times X$, and $\Omega \subset\subset X$ a relatively compact subdomain of X. Then the operator*

$$Kf(y) = \int\limits_{\Omega} f(x) \wedge *\overline{\mathcal{K}(x,y)}$$

with kernel $\mathcal{K}$ has the following properties:

i. *It is continuous as an operator from*

$$L^r(\Omega) \longrightarrow L^s(\Omega) \quad for\ 1 \leq r \leq \infty \ and\ \frac{1}{s} > \frac{1}{r} - \frac{n+m}{n} \tag{2.15}$$

and from

$$L^{\infty} \longrightarrow C^{\alpha}, \quad \alpha < 1, \tag{2.16}$$

where C^{α} is the Hölder space of exponent α.

ii. *If $f \in C^k(\Omega) \cap L^1(\Omega)$, then so is Kf.*

(We have not indicated degrees of differential forms).

We also need

Proposition 2.17 *K, defined as above, is compact from $L^r \to L^r$, $1 \leq r \leq \infty$.*

Proof   Let $r = 1$. Choose, for $\varepsilon > 0$, a smooth function φ_{ε} such that

$$\varphi_{\varepsilon}(x,y) = \begin{cases} 0 & \text{for } \rho(x,y) \leq \frac{\varepsilon}{2} \\ 1 & \text{for } \rho(x,y) \geq \frac{3}{4}\varepsilon \end{cases}$$

and set $\mathcal{K}_{\varepsilon}(x,y) = \varphi_{\varepsilon}(x,y)\mathcal{K}(x,y)$. The operators

$$K_{\varepsilon}f(y) = \int\limits_{\Omega} f(x) \wedge *\overline{\mathcal{K}_{\varepsilon}(x,y)}$$

have smooth kernels and thus are compact (Use, e.g., Ascoli-Arzelà). It suffices to prove $\lim_{\varepsilon \to 0} K_{\varepsilon} = K$ (in the L^1-operator norm). We can, moreover, reduce the claim, by using local coordinates and partition of unity, to the case of $\Omega \subset\subset \mathbb{R}^n$ and f and $\mathcal{K}$ functions; then

$$K_{\varepsilon}f(y) = \int\limits_{\Omega} f(x)\overline{\mathcal{K}_{\varepsilon}(x,y)}\, dV(x)$$

$$Kf(y) = \int\limits_{\Omega} f(x)\overline{\mathcal{K}(x,y)}\, dV(x).$$

Let $f \in L^1(\Omega)$ with $\|f\|_{L^1} = 1$. Then

$$\|Kf - K_\varepsilon f\|_{L^1} = \int_\Omega |Kf(y) - K_\varepsilon f(y)|\, dV(y)$$

$$= \int_\Omega \left| \int_\Omega f(x)[1 - \varphi_\varepsilon(x,y)]\overline{\mathcal{K}(x,y)}\, dV(x) \right| dV(y)$$

$$\leq \int_\Omega \int_\Omega |f(x)| \frac{|C_\varepsilon(x,y)|}{\rho(x,y)^{n-1}}\, dV(x)dV(y),$$

where $C_\varepsilon(x,y)$ is a continuous function with support in $\rho(x,y) \leq \varepsilon$ which is uniformly bounded in x, y and ε. So we can continue the estimate

$$\leq \mathrm{const} \int_\Omega \int_{\substack{\Omega \\ \rho(x,y)\leq\varepsilon}} |f(x)| \frac{1}{\rho(x,y)^{n-1}}\, dV(x)dV(y)$$

$$= \mathrm{const} \int_\Omega |f(x)| \int_{\substack{\Omega \\ \rho(x,y)\leq\varepsilon}} \frac{dV(y)}{\rho(x,y)^{n-1}}\, dV(x).$$

The inner integral is estimated uniformly in x by a constant C_ε with $C_\varepsilon \to 0$ for $\varepsilon \to 0$. This yields as the next step of our chain of inequalities

$$\leq \mathrm{const} \cdot C_\varepsilon \int_\Omega |f(x)|\, dV(x)$$

$$= \mathrm{const} \cdot C_\varepsilon \to 0.$$

This proves the claim. – The case $1 < r < \infty$ follows in the same way; $r = \infty$ follows from (2.16) $\qquad\square$

§3 The Parametrix

1. Let the dimension n of X be at least 3. We will compute, for the geodesic distance function ρ, the expression

$$\Delta\rho^{2-n} = \delta d\rho^{2-n},$$

where the operators work on the x-variable. It is clear that this is an isotropic kernel which is, in view of (2.3), of order $\geq -n$. But there occurs a cancellation of singularities:

Theorem 3.1 $\Delta\rho^{2-n} = \mathcal{E}_{1-n}.$

(The differentiation takes place pointwise, outside the diagonal.)

Even more is true: a careful calculation in Riemannian normal coordinates would yield

$$\Delta\rho^{2-n} = \mathcal{E}_{2-n}. \tag{3.2}$$

We will not need this more precise result.

To prove 3.1 we first need

Lemma 3.3 *Let $x^1, \ldots, x^n$ be a system of local coordinates, $D_i = \frac{\partial}{\partial x^i}$. Suppose the metric is given as*

$$ds^2 = \sum g_{ij}(x)\, dx^i dx^j,$$

and denote the inverse of the matrix (g_{ij}) by (g^{ij}). Then for a C^2-function f

$$\Delta f = -\sum g^{ij}(x) D_i D_j f + \overline{D} f,$$

where $\overline{D}$ is a first order differential operator.

Proof   We can choose local orthonormal base fields $X_1, \ldots, X_n$ and their dual forms $\omega^1, \ldots, \omega^n$. Then

$$df = \sum_i (X_i f)\omega^i$$

$$*df = \sum_i (X_i f) * \omega^i$$

$$d * df = \sum_{i,j} X_j X_i f\, \omega^j \wedge *\omega^i + \ldots$$

$$\delta df = -\sum_i X_i^2 f + \ldots$$

where the dots stand for differential operators of lower order (which arise because $d\omega^i \neq 0$). Expressing

$$X_i = \sum a_i^k D_k$$

we arrive at

$$\delta df = -\sum_{k,l} \sum_i a_i^k a_i^l D_k D_l f + \ldots,$$

and from

$$\langle X_i, X_j \rangle = \delta_{ij}, \qquad \langle D_k, D_l \rangle = g_{kl}$$

we deduce

$$\sum_i a_i^k a_i^l = g^{kl}.$$

$\square$

Proof of Theorem 3.1. a) It remains to show that

$$\Delta\rho^{2-n} = O(\rho^{1-n}).$$

This is a local question, and so we work in a local coordinate chart $x^1, \ldots, x^n$ and decompose

$$\Delta = -\sum g^{ij}(x)D_iD_j + \overline{D}$$
$$= \Delta_0 + \overline{D}$$

as in the Lemma. Then

$$\Delta\rho^{2-n} = \Delta_0\rho^{2-n} + \overline{D}\rho^{2-n} = \Delta_0\rho^{2-n} + O(\rho^{1-n}).$$

We next introduce R with

$$R^2 = \sum g_{ij}(x^i - y^j)(x^j - y^j).$$

Suppose that $n = 2k$, an even number. Then

$$\Delta_0\left(\frac{1}{\rho^{n-2}} - \frac{1}{R^{n-2}}\right) = \Delta_0\left(\frac{(R^2)^{k-1} - (\rho^2)^{k-1}}{\rho^{2k-2}R^{2k-2}}\right);$$

(2.8) can be applied to the numerator, and we easily get

$$\Delta_0\left(\frac{1}{\rho^{n-2}} - \frac{1}{R^{n-2}}\right) = O\left(\frac{1}{\rho^{n-1}}\right).$$

b) The general case is more difficult (and unnecessary for complex analysis). Here we write

$$\Delta_0\rho^{2-n} = \Delta_0\left(\frac{1}{R^{n-2}} \cdot \frac{R^{n-2}}{\rho^{n-2}}\right).$$

Now an easy calculation shows

$$\Delta_0\left(\frac{1}{R^{n-2}} \cdot \frac{R^{n-2}}{\rho^{n-2}}\right) = \frac{R^{n-2}}{\rho^{n-2}}\Delta_0\frac{1}{R^{n-2}} + \frac{1}{R^{n-2}}\Delta_0\frac{R^{n-2}}{\rho^{n-2}}$$
$$- 2\sum g^{ij}(x)D_i\left(\frac{1}{R^{n-2}}\right)D_j\left(\frac{R^{n-2}}{\rho^{n-2}}\right). \quad (3.4)$$

We note

$$\frac{R^{n-2}}{\rho^{n-2}} = O(1).$$

$$D_i\left(\frac{R^{n-2}}{\rho^{n-2}}\right) = (n-2)\left(\frac{R}{\rho}\right)^{n-3}D_i\frac{R}{\rho}$$

$$D_iD_j\left(\frac{R}{\rho}\right)^{n-2} = (n-2)(n-3)\left(\frac{R}{\rho}\right)^{n-4}D_j\frac{R}{\rho}D_i\frac{R}{\rho} + (n-2)\left(\frac{R}{\rho}\right)^{n-3}D_iD_j\frac{R}{\rho}.$$

From (2.8) we deduce

$$\rho D_i \rho = R D_i R + \mathcal{E}_2$$

and

$$R D_i R = O(\rho)$$

and so

$$D_i \frac{R}{\rho} = \frac{\rho D_i R - R D_i \rho}{\rho^2} = \frac{\rho^2 D_i R - R \rho D_i \rho}{\rho^3} = \frac{\rho^2 D_i R - R^2 D_i R + \mathcal{E}_3}{\rho^3}$$

$$= \frac{\mathcal{E}_3 D_i R}{\rho^3} + \frac{\mathcal{E}_3}{\rho^3} = \frac{\mathcal{E}_4}{R \rho^3} + \frac{\mathcal{E}_3}{\rho^3} = O(1).$$

One more differentiation leads to

$$D_i D_j \frac{R}{\rho} = O(\rho^{-1}).$$

We see that

$$D_i \left(\frac{R}{\rho} \right)^{n-2} = O(1)$$

$$D_i D_j \left(\frac{R}{\rho} \right)^{n-2} = O\left(\frac{1}{\rho} \right).$$

This implies that the last two terms in (3.4) are $O(\rho^{1-n})$.

c) We are left with $\Delta_0 R^{2-n}$. Plugging in the definition of Δ_0 we first obtain

$$\Delta_0 \frac{1}{R^{n-2}} = (n-2) R^{-n} \sum_{i,j} g^{ij}(x) [(1-n) D_i R \cdot D_j R + R D_i D_j R]. \qquad (3.5)$$

On the other hand,

$$2 R D_i R = D_i R^2 = 2 \sum_l g_{il}(x)(x^l - y^l) + \mathcal{E}_1$$

$$D_i D_j R^2 = 2 g_{ij}(x) + \mathcal{E}_1,$$

and so

$$(D_i R)(D_j R) + R D_i D_j R = \frac{1}{2} D_i D_j R^2 = g_{ij}(x) + \mathcal{E}_1.$$

Thus, (3.5) becomes

$$\Delta_0 \frac{1}{R^{n-2}} = (n-2)R^{-n}\sum_{i,j} g^{ij}(x)[(1-n)D_iRD_jR - D_iRD_jR + g_{ij}(x) + \mathcal{E}_1]$$

$$= (n-2)R^{-n}\sum g^{ij}(x)[g_{ij}(x) - nD_iRD_jR] + O(\rho^{1-n})$$

$$= (n-2)R^{-n}[n - n\sum g^{ij}(x)D_iRD_jR] + O(\rho^{1-n})$$

$$= (n-2)R^{-n}\left[n - \frac{n}{R^2}\sum g^{ij}(x)RD_iR \cdot RD_jR\right] + O(\rho^{1-n})$$

$$= (n-2)R^{-n}\left[n - \frac{n}{R^2}\sum g^{ij}(x)g_{il}(x)(x^l - y^l)g_{jk}(x)(x^k - y^k)\right]$$
$$+ O(\rho^{1-n})$$

$$= (n-2)R^{-n}n\left[1 - \frac{1}{R^2}\cdot R^2\right] + O(\rho^{1-n}) = O(\rho^{1-n}).$$

$\square$

2. **Definition 3.6** *Set for* $n > 2$ *and* $0 \le k \le n$:

$$\Gamma_k = c_{nk}\frac{1}{\rho^{n-2}}(d_x d_y \rho^2)^k$$

with

$$c_{nk} = \frac{(-1)^k}{2^k k!}\frac{\Gamma(n/2 + 1)}{(n-2)n\pi^{n/2}} = \frac{(-1)^k}{2^k k!}c_n.$$

Γ_k *is called a parametrix for the Laplacian on k-forms.*

The justification for introducing Γ_k is

Theorem 3.7 *There is an isotropic kernel* $\mathcal{E}_{1-n}$ *of order* $\ge 1 - n$, *such that, for each* $f \in C^2_{k,c}(X)$ *the equality*

$$f = (\Delta f, \Gamma_k) + (f, \mathcal{E}_{1-n})$$
$$= (df, d\Gamma_k) + (\delta f, \delta\Gamma_k) + (f, \mathcal{E}_{1-n})$$

holds.

So Γ_k is an "approximate fundamental solution"; in the case of $\mathbb{R}^n$ and the Euclidean metric we would have been able to show that $\mathcal{E}_{1-n} \equiv 0$, i.e. that Γ_k is a precise fundamental solution for Δ. In fact, we have done this in Chapter I for $\mathbb{C}^n$ and the complex Laplacian; the proof for the real Laplacian is identical. (The relation between Γ_q in Chapter I and the above parametrix will be explained in §5 when we return to complex manifolds.)

3. **Proof** of Theorem 3.7   a) Since f has compact support and Γ_k and $d\Gamma_k$, $\delta\Gamma_k$ have integrable singularities, we have

$$(\Delta f, \Gamma_k) = (df, d\Gamma_k) + (\delta f, \delta\Gamma_k).$$

b) Choose a neighbourhood W of the diagonal such that for $y \in X$ there are orthonormal base fields $\omega^1, \ldots, \omega^n$ defined on the neighbourhood

$$W_y = \{x : (x, y) \in W\}.$$

Let $\chi(x, y)$ be a smooth real function with $\operatorname{supp} \chi \subset W$ and $\chi \equiv 1$ on $W' \subset W$ where W' is a smaller neighbourhood of the diagonal Δ. Set

$$f(x) = \chi(x, y) f(x) + (1 - \chi(x, y)) f(x).$$

Then

$$
\begin{aligned}
(\Delta f, \Gamma_k) &= (\Delta\chi f, \Gamma_k) + (\Delta(1 - \chi) f, \Gamma_k) \qquad\qquad (3.8)\\
&= (\Delta\chi f, \Gamma_k) + ((1 - \chi) f, \Delta\Gamma_k)\\
&= (\Delta\chi f, \Gamma_k) + (f, (1 - \chi)\Delta\Gamma_k)\\
&= (\Delta\chi f, \Gamma_k) + (f, \mathcal{E}_{1-n}),
\end{aligned}
$$

because of 3.1

c) Let us fix y and write again f instead of χf. So $\operatorname{supp} f \subset \{x : (x, y) \in W\}$, and

$$f = {\sum_{|K|=k}}' f_K \omega^K.$$

We define a differential operator Δ_d by

$$\Delta_d f = {\sum_{|K|=k}}' \Delta f_K \cdot \omega^K$$

and use from Riemannian geometry – see [Rha 60]:

Lemma 3.9 $\Delta f = \Delta_d f + Lf$, *where L is a differential operator of order 0.*

Thus

$$(\Delta f, \Gamma_k) = (\Delta_d f, \Gamma_k) + (Lf, \Gamma_k) = (\Delta_d f, \Gamma_k) + (f, \mathcal{E}_{2-n}).$$

From (2.11) and (3.5) we obtain

$$\Gamma_k = c_n \frac{1}{\rho^{n-2}} \sum_{|K|=k} \omega^K \Theta^K + \mathcal{E}_{3-n},$$

and consequently

$$(\Delta_d f, \Gamma_k) = \sum_{|K|=k} (\Delta f_K, \frac{c_n}{\rho^{n-2}})\Theta^K + (\Delta_d f, \mathcal{E}_{3-n})$$

$$= \sum_{|K|=k} (\Delta f_K, \Gamma_0)\Theta^K + (f, \mathcal{E}_{1-n})$$

by the definition of Γ_0 and by partial integration in the second term. The case $k = 0$ of functions which we will treat in the next lemma then yields

$$(\Delta_d f, \Gamma_k) = f(y) + (f, \mathcal{E}_{1-n}).$$

Going back to (3.4) and (3.8) we finally prove our claim.   □

So everything is reduced to the case of functions.

Lemma 3.10 *Let f be a function with compact support. Then there is an $\mathcal{E}_{1-n}$ which does not depend on f, such that*

$$(\Delta f, \Gamma_0) = f + (f, \mathcal{E}_{1-n}).$$

Proof   a) We fix y and set

$$B_\varepsilon = \{x : \rho(x, y) \leq \varepsilon\}, \quad S_\varepsilon = \{x : \rho(x, y) = \varepsilon\}, \quad X_\varepsilon = X - B_\varepsilon.$$

Then – by Green's formula –

$$(\Delta f, \Gamma_0) = \lim_{\varepsilon \to 0} (\Delta f, \Gamma_0)_{X_\varepsilon}$$

$$= \lim_{\varepsilon \to 0} (f, \Delta\Gamma_0)_{X_\varepsilon} - \lim_{\varepsilon \to 0} \int_{S_\varepsilon} f \wedge *d\Gamma_0$$

$$= (f, \mathcal{E}_{1-2n}) - \lim_{\varepsilon \to 0} \int_{S_\varepsilon} f * d\Gamma_0.$$

b) To compute the limit, we note

$$\lim_{\varepsilon \to 0} \int_{S_\varepsilon} f(x) * d\Gamma_0(x, y) = f(y) \lim_{\varepsilon \to 0} \int_{S_\varepsilon} *d\Gamma_0(x, y).$$

Now

$$d\Gamma_0 = (2 - n)c_n \frac{\rho d\rho}{\rho^n},$$

and the limit becomes

$$f(y) \frac{(2 - n)c_n}{\varepsilon^n} \lim_{\varepsilon \to 0} \int_{S_\varepsilon} *\rho d\rho.$$

We introduce local coordinates x^j and orthonormal base fields ω^j such that

$$x^j(y) = 0, \quad dx^j(y) = \omega^j(y).$$

Then up to $\mathcal{E}_1$-error terms,

$$
\begin{aligned}
*\rho d\rho &= * \sum g_{ij}(x) x^i \, dx^j \\
&= * \sum g_{ij}(y) x^i \, dx^j \\
&= * \sum_j x^j \omega^j \\
&= \sum x^j (-1)^j \omega^1 \wedge \ldots \wedge \widehat{\omega^j} \wedge \ldots \wedge \omega^n.
\end{aligned}
$$

Stokes' theorem now results in

$$
\begin{aligned}
\int_{S_\varepsilon} *\rho d\rho &= \int_{B_\varepsilon} \sum (-1)^j dx^j \omega^1 \wedge \ldots \wedge \widehat{\omega^j} \wedge \ldots \wedge \omega^n + O(\varepsilon^{n+1}) \\
&= n \int_{B_\varepsilon} dV + O(\varepsilon^{n+1}) \\
&= \frac{n\pi^{n/2}}{\Gamma(n/2+1)} \varepsilon^n + O(\varepsilon^{n+1}).
\end{aligned}
$$

Consequently

$$f(y) \frac{(2-n)c_n}{\varepsilon^n} \lim_{\varepsilon \to 0} \int_{S_\varepsilon} *\rho d\rho = f(y).$$

$\square$

4.   We now use the above theorems to obtain an important integral representation in terms of Γ on a relatively compact subdomain of X.

Theorem 3.11   *Let $\Omega \subset\subset X$ be a domain with piecewise smooth boundary. Then for any form $f \in C^1_k(\overline{\Omega})$,*

$$
\begin{aligned}
f(y) = &-\int_{b\Omega} f(x) \wedge *d\Gamma_k(x,y) + \int_\Omega df(x) \wedge *d\Gamma_k(x,y) \\
&+ \int_\Omega \delta f(x) \wedge *\delta\Gamma_k(x,y) + \int_{b\Omega} \delta\Gamma_k(x,y) \wedge *f + \int_\Omega f(x) \wedge *\mathcal{E}_{1-n}(x,y),
\end{aligned}
$$

where $\mathcal{E}_{1-n}(x,y)$ is an isotropic kernel of the indicated order which does not depend on f.

Proof   Set for $\varepsilon > 0$

$$\Omega_\varepsilon = \{x \in \Omega : \rho(x, b\Omega) > \varepsilon\},$$

where ρ denotes geodesic distance. We choose smooth functions χ_ε with $0 \leq \chi_\varepsilon \leq 1$ and

$$\chi_\varepsilon = \begin{cases} 1 & \text{for } x \in \Omega_{2\varepsilon} \\ 0 & \text{for } x \notin \Omega_\varepsilon. \end{cases}$$

Now let y be an arbitrary point of Ω and $\varepsilon > 0$ such that $y \in \Omega_{3\varepsilon}$. By (3.7) we have

$$f(y) = (\chi_\varepsilon f)(y) = (d(\chi_\varepsilon f), \Gamma_k) + (\delta(\chi_\varepsilon f), \Gamma_q) + (\chi_\varepsilon f, \mathcal{E}_{1-k}). \tag{3.12}$$

For y as above, Green's formula can be applied to the form

$$f_1 = f - \chi_\varepsilon f$$

and yields

$$(df_1, d\Gamma_k) + (\delta f_1, \delta\Gamma_k) = -\int_{b\Omega} f_1 \wedge *d\Gamma_k + \int_{b\Omega} \delta\Gamma_k \wedge *f_1 + (f_1, \Delta\Gamma_k) \tag{3.13}$$

$$= -\int_{b\Omega} f \wedge *d\Gamma_k + \int_{b\Omega} \delta\Gamma_k \wedge *f + (f_1, \widetilde{\mathcal{E}}_{1-n}).$$

Since $\chi_\varepsilon f = f - f_1$, we insert (3.13) in (3.12) and obtain

$$f(y) = (df, d\Gamma_k) + (\delta f, \delta\Gamma_k) - \int_{b\Omega} f \wedge *d\Gamma_k +$$

$$\int_{b\Omega} \delta\Gamma_k \wedge *f + (\chi_\varepsilon f, \mathcal{E}_{1-n}) + (f_1, \widetilde{\mathcal{E}}_{1-n}), \tag{3.14}$$

with two kernels $\mathcal{E}$ and $\widetilde{\mathcal{E}}$ of the indicated order. Neither $\mathcal{E}$ nor $\widetilde{\mathcal{E}}$ depend on f or χ_ε. So

$$(\chi_\varepsilon f, \mathcal{E}_{1-n}) + ((1 - \chi_\varepsilon)f, \widetilde{\mathcal{E}}_{1-n})$$

does not depend on ε; on the other hand, for $\varepsilon \to 0$, we see

$$(\chi_\varepsilon f, \mathcal{E}_{1-n}) \to (f, \mathcal{E}_{1-n})$$
$$((1 - \chi_\varepsilon)f, \widetilde{\mathcal{E}}_{1-n}) \to 0.$$

Consequently we can replace, in (3.14), the function χ_ε by 1: this yields 3.11. $\qquad\square$

If we use 3.2 instead of 3.1 and compute much more carefully, we can even show

Theorem 3.15 *Under the assumptions of 3.11, the last volume integral has an isotropic kernel of order $2 - n$:*

$$\int_\Omega f(x) \wedge *\mathcal{E}_{2-n}(x, y).$$

For the detailed proof see [Rha 60]. This result makes the use of the above integral formula much easier as we shall see in the next paragraph. – However, our applications to complex analysis will provide for additional error terms $(f, \mathcal{E}_{1-n})$ which are unavoidable in the case of a non-Kähler metric; so we have only established the weaker result 3.11 which is easier to prove though more difficult to work with.

§4   Harmonic Forms and Finiteness Theorems on Compact Manifolds

1.   Let X be a compact oriented n-dimensional Riemannian manifold, $E = E_{1-n}$ an isotropic operator of the indicated order and Ξ a smooth vector field on X. We need – in addition to 2.14 and 2.17 – the following regularity and commutation statements.

Lemma 4.1 *There is an isotropic operator $\widetilde{E} = E_{1-n}$ such that*

$$\Xi E = E\Xi + \widetilde{E}.$$

Lemma 4.2 *For each $\alpha < 1$, E is a continuous map from C^0 into C^α.*

Both lemmas were proved in the first chapter; not so the following

Lemma 4.3 *For $0 \le \beta < \alpha < 1$, ΞE maps C^α continuously into C^β.*

A more general statement comprising 4.3 will be proved in Chapter VII; so we do not give the – fairly delicate – proof here. If one deals with operators of order $\ge 2 - n$ one can dispense with the above lemma – this is the main pay off for establishing the precise information 3.15.

Proposition 4.4 *For each $l = 0, 1, 2, \ldots$ and each α with $0 < \alpha < 1$, $E = E_{1-n}$ is a continuous operator between the following spaces:*

$$E : C^{l+\alpha} \longrightarrow C^{l+1+\beta}, \qquad \text{for } \beta < \alpha, \tag{4.5}$$

and

$$E : C^l \longrightarrow C^{l+\alpha}. \tag{4.6}$$

Proof   (4.6) follows from (4.5) and (4.2). To prove (4.5) we use induction with respect to l. The case $l = 0$ is just Lemma 4.3. Now suppose $f \in C^{l+\alpha}$. Then, because of 4.1,

$$\Xi E f = E\Xi f + \widetilde{E} f.$$

By the induction hypothesis, $E\Xi f \in C^{l+\beta}$, and $\widetilde{E} f \in C^{l+\beta}$ as well. So $\Xi E f \in C^{l+\beta}$ for any Ξ which means that $f \in C^{l+1+\beta}$. $\square$

Remark The C^k and Hölder spaces which we have used have to be defined with respect to local coordinates. The norms that occur are non-intrinsic, but that does not matter.

2. Let us now return to Theorem 3.11. If f is a C^1-form of degree k on X, we have

$$f(y) = \int_X df(x) \wedge *d\Gamma_k(x,y) + \int_X \delta f(x) \wedge *\delta\Gamma_k(x,y)$$

$$+ \int_X f(x) \wedge *\mathcal{R}_k(x,y), \quad \text{with } \mathcal{R}_k = \mathcal{E}_{1-n}, \quad (4.7)$$

because boundary integrals do not come up. The identity of weak and strong extension of differential operators proved in the next chapter allows a transfer of this formula to L^2 or even L^1-forms:

Theorem 4.8 Let $f \in L^1_k(X)$ such that df and δf are also in L^1. Then

$$f = T_k df + T_k^* \delta f + R_k f,$$

where T_k, T_k^* and R_k are isotropic operators of order $\geq 1 - n$ (and type ≥ 1) given as

$$T_k g = (g, d\Gamma_k), \quad T_k^* h = (h, \delta\Gamma_k),$$
$$R_k f = (f, \mathcal{R}_k).$$

Note that all the kernels define continuous operators even on L^1, so the scalar products are well defined.

It is useful to rewrite everything in terms of Δ. So we introduce an E_{2-n}-operator $\Gamma = \Gamma_k$:

$$\Gamma f(y) = \int_X f(y) \wedge *\Gamma_k(x,y)$$

and obtain (after dropping unnecessary indices)

Theorem 4.8' Let $f \in L^1_k(X)$ such that Δf is also in L^1. Then

$$f = \Gamma \Delta f + R f$$

where R is isotropic of order $\geq 1 - n$.

We can – for $f \in L^2_k$ – pass to the transposed operators in 4.8' and then go back to L^1 by approximation. This yields, since $\Gamma = \Gamma^*$,

Theorem 4.8'' For f as above,

$$f = \Delta \Gamma f + R^* f,$$

where R^* is the transposed operator of R with kernel $\mathcal{R}^*(x,y) = \mathcal{R}(y,x)$.

Now suppose that df and δf are in C^l, $l \geq 0$. Then $\boldsymbol{T}df$ and $\boldsymbol{T}^* \delta f$ belong to $C^{l+\alpha}$. If $f \in L^1$, then $\boldsymbol{R}f \in L^{1+\delta}$ for $\delta < \frac{1}{n-1}$. So f itself is in $L^{1+\delta}$ in view of 4.8. Iterating this argument yields $f \in C^\alpha$, $\alpha > 0$, and repeated application of Proposition 4.4 shows $f \in C^{l+\alpha}$. We have thus established the basic

Regularity Theorem 4.9 (for compact manifolds)   *i. Let f be in L^1 with df and δf in C^l. Then $f \in C^{l+\alpha}$ for each $0 \leq \alpha < 1$.*

   ii. If $df, \delta f \in C^{l+\alpha}$ then $f \in C^{l+1+\beta}$ (for $0 < \alpha < 1$, for $\beta < \alpha$).

   iii. If $df, \delta f \in C^\infty$, so is f.

The second part follows by the same argument as $i)$. We can use (4.8′) to obtain the corresponding statement for Δ.

Definition 4.10   *f is harmonic if $f \in L^1$ and the following equivalent conditions hold:*

   i. df and $\delta f = 0$

   ii. $\Delta f = 0$.

By 4.9 harmonic forms are necessarily smooth.

3.   From Theorem 4.8 we can now easily deduce the fundamental

Finiteness-Theorem 4.11   *The space $\mathbb{H}^k$ of harmonic k-forms is finite-dimensional.*

In fact, for $f \in \mathbb{H}^k$,

$$f = \boldsymbol{R}f,$$

and $\boldsymbol{R}$ is a compact operator.

4.   For the remainder of the paragraph we restrict attention to the space L_k^2 of square integrable k-forms; the results could be easily generalised to currents [Rha 60] but this is not our aim.

Let us first solve the equation

$$\Delta u = f, \tag{4.12}$$

where f is in L_k^2. A necessary condition is naturally that f is orthogonal to the harmonic space $\mathbb{H}^k$. We try, inspired by (4.8″), the ansatz

$$u = \Gamma g.$$

This leads to the equation

$$g - \boldsymbol{R}^* g = f$$

which, by the theory of compact operators, is soluble if and only if f is orthogonal to the kernel of the adjoint operator $id - \boldsymbol{R}$. This kernel is finite dimensional, comprises $\mathbb{H}^k$, but may be larger. So set

$$\ker(id - \boldsymbol{R}) = V = \mathbb{H}^k \oplus W.$$

By proposition 4.4 all forms in V are smooth.

Lemma 4.13 *There is exactly one form $f_1 \in W$ such that*

$$(\Delta^2 f_1, h) = (f, h)$$

for all $h \in W$.

Proof   The linear map which associates with each $v \in W$ the linear form on W $h \mapsto (\Delta^2 v, h)$ is injective because $(\Delta^2 v, v) = (\Delta v, \Delta v) \neq 0$.   □

If f is orthogonal to $\mathbb{H}^k$, the form $f - \Delta^2 f_1$ is therefore orthogonal to all of V, and we can find g which solves

$$g - \boldsymbol{R}^* g = f - \Delta^2 f_1. \tag{4.14}$$

By (4.8″) this means

$$\Delta \boldsymbol{\Gamma} g = f - \Delta^2 f_1,$$
$$\Delta[\boldsymbol{\Gamma} g + \Delta f_1] = f,$$

and so $u = \boldsymbol{\Gamma} g + \Delta f_1$ solves (4.12). We have

Theorem 4.15 (Hodge decomposition theorem) *The equation $\Delta u = f$ for $f \in L_k^2$ is soluble if and only if f is orthogonal to the harmonic forms. In particular the image of Δ is closed, and we have the orthogonal decomposition*

$$L_k^2 = \operatorname{im} \Delta \oplus \mathbb{H}^k. \tag{4.16}$$

The decomposition (4.16) allows the definition of two operators:

$$H : L_k^2 \longrightarrow \mathbb{H}^k$$

is the *orthogonal projection* ("harmonic projector");

$$G : L_k^2 \longrightarrow L_k^2$$

is defined as follows:

if $f \in \operatorname{im} \Delta$, then $Gf = u$ is the unique solution of $\Delta u = f$ which is orthogonal to $\mathbb{H}^k$; if $f \in \mathbb{H}^k$ then $Gf = 0$. G is called the Green operator of X. (4.16) can now be rewritten: if $f \in L_k^2$, we have

$$f = \Delta G f + H f. \tag{4.17}$$

It is easily seen that G commutes with Δ (if $\Delta f \in L_k^2$) and also with d, δ and $*$ under the corresponding assumptions ($df \in L_{k+1}^2$ etc.); so, denoting the Green operators in different dimensions by the same symbol, we also have the decompositions

$$f = d\delta G f + \delta d G f + H f \tag{4.18}$$
$$= G \Delta f + H f$$
$$= G d\delta f + G \delta d f + H f$$
$$= d G \delta f + \delta G d f + H f$$

into mutually orthogonal terms. In particular, if $df = 0$, (4.18) yields

$$f = dG\delta f + Hf,$$

which proves the main part of

Theorem 4.19 *Each de Rham cohomology class in $H_2^k(X)$ contains exactly one harmonic form.*

Here

$$H_2^k(X) = \frac{\ker\{d : L_k^2 \to L_{k+1}^2\}}{\operatorname{im}\{d : L_{k-1}^2 \to L_k^2\}}.$$

The same theorem holds, of course, if one works with smooth forms or, at the other extreme of the scale, with currents to define the cohomology – because $\mathbb{H}^k$ contains only smooth forms.

Theorems 4.11 to 4.19 are the main results of Hodge theory on compact manifolds.

§5   Basic Integral Representation on Hermitian Manifolds

We return to complex analysis. The methods of Riemannian geometry which we have descibed in the previous paragraphs will now be combined with the ideas of the third chapter in order to study *strictly pseudoconvex manifolds*, i.e. strictly pseudoconvex subdomains of arbitraty complex manifolds. We establish a basic homotopy formula

$$f = Pf + T\overline{\partial}f + \overline{\partial}Sf$$

with compact operators P, T and S which will be used to solve (a variant of) Levi's problem, and will give, moreover, a rough solution to the $\overline{\partial}$-Neumann problem (to be introduced later). It is, furthermore, the first step towards a detailed investigation of the regularity of that problem.

1.   Let X be a n-dimensional complex manifold equipped with a C^∞ hermitian metric H. First we want to determine the relationship between H and the induced hermitian form on the complexified tangent bundle.

Let $z^j = x^j + iy^j$, $j = 1, \ldots, n$ be local holomorphic coordinates in a neighbourhood of a given point. The (real) tangent space T_pX is then spanned over $\mathbb{R}$ by the basis

$$\frac{\partial}{\partial x^1}, \frac{\partial}{\partial y^1}, \ldots \frac{\partial}{\partial x^n}, \frac{\partial}{\partial y^n}.$$

(In the notations we usually drop the dependence on the base point.) The $\mathbb{R}$-linear automorphism $J : T_pX \to T_pX$ defined by

$$J\left(\frac{\partial}{\partial x^j}\right) = \frac{\partial}{\partial y^j}, \quad J\left(\frac{\partial}{\partial y^j}\right) = -\frac{\partial}{\partial x^j}, \quad j = 1, \ldots, n, \tag{5.1}$$

induces a complex structure on T_pX. J is independent of the choice of coordinates and fulfils

$$J^2 = -\operatorname{id} . \tag{5.2}$$

J turns T_pX into a complex vector space of dimension n and

$$\frac{\partial}{\partial x^j}, \quad j = 1, \ldots, n$$

is a basis over $\mathbb{C}$.

Now H is a hermitian strictly positive definite quadratic form

$$H : T_pX \times T_pX \longrightarrow \mathbb{C}$$

with respect to this structure such that the functions

$$g_{jk} := H\left(\left(\frac{\partial}{\partial x^j}\right), \left(\frac{\partial}{\partial x^k}\right)\right), \quad j, k = 1, \ldots, n$$

are C^∞ with respect to the base point.

Remark 5.3 *Let $(T_pX)^*$ be the $\mathbb{R}$-dual space of T_pX. By*

$$J(dx^j) = dy^j, \quad J(dy^j) = -dx^j$$

J induces a complex structure on $(T_pX)^*$. But very often one considers the complex dual $(T_pX)^{*,\mathbb{C}}$ which is spanned by the $\mathbb{C}$-linear forms $\lambda : T_pX \to \mathbb{C}$. Here the $\mathbb{C}$-dual base of $\{\partial/\partial x^j, j = 1, \ldots, n\}$ is denoted by

$$\{dz^j, j = 1, \ldots, n\}$$

and therefore the hermitian metric H is written in the following ambiguous way

$$H := \sum_{j,k=1}^n g_{jk}(p)\, dz^j \otimes d\bar{z}^k.$$

We take this writing only as a notation and the dz^j should by no means be confused with $dz^j = dx^j + i\,dy^j$ which are elements of the complexified cotangent bundle.

Concerning $\bar{\partial}$-problems, Levi forms, Neumann problems and so on we will be concerned with the complexified tangent bundle

$$\mathbb{C}TX := \mathbb{C} \otimes_{\mathbb{R}} TX$$

its dual bundle and the induced Grassmann algebra.

H induces on X the Riemannian structure given by the Riemannian metric G with

$$G(t, s) := \operatorname{Re} H(t, s), \quad t, s \in T_pX. \tag{5.4}$$

The geodesic distance on X will be calculated with respect to G.

Now G extends uniquely to a hermitian form on $\mathbb{C}TX$, also denoted by G, by the following formula:

Let $t, \tilde{t}, u, \tilde{u} \in T_pX$. Then

$$G(t + iu, \tilde{t} + i\tilde{u}) = G(t, \tilde{t}) + iG(u, \tilde{t}) - iG(t, \tilde{u}) + G(u, \tilde{u}). \tag{5.5}$$

$\mathbb{C}T_pX$ decomposes into the orthogonal spaces $T_p^{1,0}X$ and $T_p^{0,1}X$. A basis of $T_p^{1,0}X$ is given by

$$\frac{\partial}{\partial z^j} = \frac{1}{2}\left(\frac{\partial}{\partial x^j} + \frac{1}{i}\frac{\partial}{\partial y^j}\right), \quad j = 1, \dots, n.$$

We want to express $\tilde{g}_{jk} := G(\frac{\partial}{\partial z^j}, \frac{\partial}{\partial z^k})$ in terms of H.

$$
\begin{aligned}
\tilde{g}_{jk} &= \frac{1}{4}G\left(\frac{\partial}{\partial x^j} - i\frac{\partial}{\partial y^j}, \frac{\partial}{\partial x^k} - i\frac{\partial}{\partial y^k}\right) \\
&= \frac{1}{4}\left[\operatorname{Re} H\left(\frac{\partial}{\partial x^j}, \frac{\partial}{\partial x^k}\right) - i\operatorname{Re} H\left(J\frac{\partial}{\partial x^j}, \frac{\partial}{\partial x^k}\right)\right. \\
&\qquad \left. + i\operatorname{Re} H\left(\frac{\partial}{\partial x^j}, J\frac{\partial}{\partial x^k}\right) + \operatorname{Re} H\left(J\frac{\partial}{\partial x^j}, J\frac{\partial}{\partial x^k}\right)\right] \\
&= \frac{1}{4}[\operatorname{Re} g_{jk} - i\operatorname{Re}(ig_{jk}) + i\operatorname{Re}(-ig_{jk}) + \operatorname{Re} g_{jk}] \\
&= \frac{1}{2}g_{jk}.
\end{aligned}
$$

Remark 5.6 *So when $\{\frac{\partial}{\partial x^j}, j = 1, \dots, n\}$ is an orthonormal basis with respect to H then $\frac{\partial}{\partial z^j}$ is of norm $\frac{1}{\sqrt{2}}$.*

Analogous formulae hold in $T_p^{0,1}X$.

2. Let now $*$ be the Hodge $*$-operator defined with respect to the metric $\operatorname{Re} H = G$ and to the canonical orientation. The properties of §1 carry over with the following modifications.

- *When f is a (p, q) form then $*f$ is of bidegree $(n - q, n - p)$.*

- *Applied to forms of bidegree (p, q) one has*

$$** = (-1)^{p+q}\operatorname{id}. \tag{5.7}$$

- *If $\omega_1, \dots, \omega_n$ is an orthonormal system of $(1, 0)$ forms and if A, B, C are mutually disjoint increasingly ordered multiindices*

$$
\begin{aligned}
A &= (i_1, \dots, i_a), \quad 1 \le i_1 < i_2 < \dots < i_a \le n, \\
B &= (j_1, \dots, j_b), \quad 1 \le j_1 < j_2 < \dots < j_b \le n, \\
C &= (k_1, \dots, k_c), \quad 1 \le k_1 < k_2 < \dots < k_c \le n,
\end{aligned}
$$

$$\omega^A = \omega_{i_1} \wedge \ldots \wedge \omega_{i_a}$$

$$\overline{\omega}^A = \overline{\omega}_{i_1} \wedge \ldots \wedge \overline{\omega}_{i_a},$$

$$(\omega \wedge \overline{\omega})^A = \omega_{i_1} \wedge \overline{\omega}_{i_1} \wedge \ldots \wedge \omega_{i_a} \wedge \overline{\omega}_{i_a}, \quad etc.$$

then

$$*(\omega^A \wedge \overline{\omega}^B \wedge (\omega \wedge \overline{\omega})^C) = \alpha(a,b,c)\omega^A \wedge \overline{\omega}^B \wedge (\omega \wedge \overline{\omega})^{N-(A \cup B \cup C)}, \tag{5.8}$$

with

$$\alpha(a,b,c) = i^n(-1)^{a + \frac{h(h-1)}{2}}, \quad h = a + b + 2c.$$

Here $N - (A \cup B \cup C)$ is the increasingly ordered multiindex with components $\{1, 2, \ldots, n\} \setminus (\{i_1, \ldots, i_a\} \cup \{j_1, \ldots, j_b\} \cup \{k_1, \ldots, k_c\})$.

We shall denote variables on X by x, resp. y, and their representation in local coordinates by ζ, resp. z. So a kernel on a subset of X is for example denoted by $\mathcal{A}(x, y)$ and with respect to local coordinates by $\mathcal{A}(\zeta, z)$.

3. We want to apply the integral formulae of §3 to $(0, q)$-forms. Therefore not all of the components of the parametrix Γ_q play a role. Moreover, as the dimension of the underlying real manifold is even, *all isotropic kernels which arise in the formulae of the preceding chapter are of the special kind defined in Chapter I*; that is they are everywhere smooth if they are of non-negative order, and they are of the form

$$\frac{\mathcal{E}_m(x, y)}{\rho^{2t}}, \quad m - 2t = l,$$

when their order l is negative. So from now on we shall only use the word isotropic in this sense.

Let

$$\Gamma_{0,q}(x, y) := \frac{2c_{2n,q}}{\rho^{2n-2}}(\overline{\partial}_x \partial_y \rho^2)^q. \tag{5.9}$$

$\frac{1}{2}\Gamma_{0,q}$ is the component of Γ_q of double type $(0, q; q, 0)$. For the euclidean metric in $\mathbb{C}^n$, Γ_{0q} appeared already in chapter I where it was simply denoted by Γ_q. One still has selfadjointness

$$\overline{\Gamma_{0,q}(y, x)} = \Gamma_{0,q}(x, y). \tag{5.10}$$

So for $f \in C^2_{0,q;c}(X)$ the component of bidegree $(0, q)$ of the equation (3.7) reads as

$$f(y) = (\Delta f, 2\Gamma_{0,q}(x, y)) + (f, \mathcal{E}_{1-2n}(x, y)).$$

Let $\overline{\partial}$ be the Cauchy-Riemann operator and ϑ its formal adjoint on X:

$$For \ f \in C^1_{0,q+1}(X), g \in C^1_{0,q;c}(X), \quad (\vartheta f, g) = (f, \overline{\partial}g). \tag{5.11}$$

ϑ is a first order partial differential operator with

$$\vartheta = - * \partial * . \tag{5.12}$$

Now the complex Laplacian as a partial differential operator is defined by

$$\Box = \overline{\partial}\vartheta + \vartheta\overline{\partial}$$

acting on C^2 forms.

Like Δ the operator $\Box$ respects the bidegree. Moreover

$$\Box - \frac{1}{2}\Delta$$

is a differential operator of first order with smooth coefficients.

Since $\Gamma_{0,q} = \mathcal{E}_{2-2n}$ (3.7) implies

Theorem 5.13 *There is an isotropic kernel $\mathcal{E}_{1-2n}$ of order $\geq 1 - 2n$, such that for each $f \in C^2_{0,q;c}(X)$ the equality*

$$\begin{aligned}
f &= (\Box f, \Gamma_{0,q}) + (f, \mathcal{E}_{1-2n}) \\
&= (\overline{\partial}f, \overline{\partial}_x\Gamma_{0,q}) + (\vartheta f, \vartheta_x\Gamma_{0,q}) + (f, \mathcal{E}_{1-2n})
\end{aligned}$$

holds.

Of course the second equation extends to $C^1_{0,q;c}(X)$. Thus a formal and completely analogous proof as for (3.11) gives

Theorem 5.14 *Let $\Omega \subset\subset X$ be a domain with piecewise smooth boundary. Then for any form $f \in C^1_{0,q}(\overline{\Omega})$ and $y \in \Omega$*

$$f(y) = (\overline{\partial}f, \overline{\partial}_x\Gamma_{0,q}) + (\vartheta f, \vartheta_x\Gamma_{0,q}) - \int_{b\Omega} f(x) \wedge *\partial_x\overline{\Gamma_{0,q}(x,y)}$$

$$+ \int_{b\Omega} \overline{\vartheta_x\Gamma_{0,q}(x,y)} \wedge *f(x) + (f, \mathcal{E}_{1-2n}).$$

4.   We shall have to apply the calculus of Cauchy-Fantappiè forms to the above theorem. If $\eta, \Theta, \ldots$ are double forms of double type $(1,0;0,0)$ on $W \subset X \times X$, their associated Cauchy-Fantappiè kernels $\Omega_q(\eta), \ldots$ and transition kernels $A_q(\eta, \Theta)$ are defined in Chapter II, 1.36 and 1.37:

$$\Omega_q(\eta) = \frac{(-1)^{q(q-1)/2}}{(2\pi i)^n}\binom{n-1}{q}\eta \wedge (\overline{\partial}\eta_x)^{n-q-1} \wedge (\overline{\partial}\eta)^q$$

$$A_q(\eta, \Theta) = \sum_{\mu,\nu} a_{q\mu\nu} A_{q\mu\nu}(\eta, \Theta)$$

with

$$A_{q\mu\nu} = \left(\frac{1}{2\pi i}\right)^n \eta \wedge \Theta \wedge (\overline{\partial}_x\eta)^\mu \wedge (\overline{\partial}_x\Theta)^{n-q-\mu-2} \wedge (\overline{\partial}_y\eta)^\nu \wedge (\overline{\partial}_y\Theta)^{q-\nu},$$

where the coefficients $a_{q\mu\nu}$ can be read off definition II.1.37. Since η, Θ are not Leray forms – in fact, Leray forms cannot even be defined on an arbitrary manifold – we do not have Koppelman's homotopy formula. We will, nevertheless, establish, for the special forms which we will consider, an analogue of the homotopy formula valid "up to error terms".

5. We start with a reformulation of Theorem 5.14 as a homotopy formula involving Cauchy-Fantappiè kernels; this formula can be considered as the natural generalisation of the Bochner-Martinelli-Koppelman formulas to hermitian manifolds.

Lemma 5.15 $\vartheta_x \Gamma_{0,q} - \partial_y \Gamma_{0,q-1} = \mathcal{E}_{2-2n}$.

Proof (5.7) and (5.12) imply that (5.15) is equivalent to

$$(-1)^q \partial_x \Lambda_q - \partial_y \Lambda_{q-1} = \mathcal{E}_{2-2n}$$

with

$$\Lambda_q = *\Gamma_{0,q}.$$

Choose local holomorphic coordinates and set

$$R^2(\zeta, z) = \sum_{i,j=1}^n g_{ij}(\zeta)(\zeta^i - z^i)(\overline{\zeta}^j - \overline{z}^j).$$

Let $L_1(\zeta), \ldots, L_n(\zeta)$ be a local orthonormal system of $(1,0)$ vector fields with respect to Re H and $\omega_1(\zeta), \ldots, \omega_n(\zeta)$ its dual base. With respect to z these objects are denoted by L_i^z, resp. Θ_j. Set

$$a_j^i = L_i[\zeta^j].$$

Then

$$\frac{1}{2} \sum_{i,j=1}^n g_{ij}(\zeta)a_i^k(\zeta)\overline{a_j^l}(\zeta) = \delta_{kl}, \qquad k,l = 1,\ldots,n.$$

This implies

$$\overline{\partial}_x R^2 = \sum_{s=1}^n (\overline{L}_s R^2)\overline{\omega}_s = \sum_{s=1}^n \left(\sum_{i,j=1}^n g_{ij}(\zeta^j - z^j)\overline{a_j^s} \right) \overline{\omega}_s + \mathcal{E}_2$$

and

$$\partial_y \overline{\partial}_x R^2 = \sum_{t=1}^n L_t^z(\overline{\partial}_x R^2)\Theta_t$$

$$= -\sum_{s,t=1}^n \left(\sum_{i,j=1}^n g_{ij} a_i^t \overline{a_j^s} \right) \overline{\omega}_s \wedge \Theta_t + \mathcal{E}_1$$

$$= -2\sum_{j=1}^n \overline{\omega}_j \wedge \Theta_j + \mathcal{E}_1.$$

From (2.8) it follows

$$\bar{\partial}_x \partial_y \rho^2 = -2 \sum_{j=1}^n \bar{\omega}_j \wedge \Theta_j + \mathcal{E}_1. \tag{5.16}$$

Therefore

$$\Gamma_{0,q} = \frac{(n-2)!}{2\pi^n} \frac{1}{\rho^{2n-2}} \sum_{|J|=q}' \bar{\omega}^J \wedge \Theta^J + \mathcal{E}_{3-2n} \tag{5.17}$$

where $\sum'$ indicates that the summation is only over increasing multi-indices J.

By (5.8) we obtain

$$\Lambda_q = (-1)^{q(q-1)/2} \frac{i^n (n-2)!}{2\pi^n \rho^{2n-2}} \sum_{|J|=q}' \bar{\omega}^J \wedge (\omega \wedge \bar{\omega})^{N-J} \wedge \Theta^J + \mathcal{E}_{3-2n}.$$

So the lemma is equivalent to the following formula

$$\left(\sum_{j=1}^n L_j \rho^2 \omega_j \right) \wedge \sum_{|J|=q}' \bar{\omega}^J \wedge (\omega \wedge \bar{\omega})^{N-J} \wedge \Theta^J$$

$$= \left(\sum_{k=1}^n L_k \rho^2 \Theta_k \right) \wedge \sum_{|K|=q-1}' \bar{\omega}^K \wedge (\omega \wedge \bar{\omega})^{N-K} \wedge \Theta^K + \mathcal{E}_2.$$

(Here we have used the obvious fact $(L_k + L_k^z)\rho^2 = \mathcal{E}_2$.) Now the above equation follows from a simple combinatorial argument. $\qquad\square$

The preceding lemma allows to rewrite (5.14) as a homotopy formula:

Proposition 5.18 *Let $\Omega \subset\subset X$ be a domain with piecewise smooth boundary and $f \in C^1_{0,q}(\overline{\Omega})$. Then there exists a kernel of type $\mathcal{E}_{1-2n}$ which does not depend on f, such that for $y \in \Omega$*

$$f(y) = -\int_{b\Omega} f(x) \wedge *\partial_x \overline{\Gamma_{0,q}(x,y)} + \int_\Omega \bar{\partial} f(x) \wedge *\partial_x \overline{\Gamma_{0,q}(x,y)}$$

$$+ \bar{\partial}_y \int_\Omega f(x) \wedge *\partial_x \overline{\Gamma_{0,q-1}(x,y)} + \int_\Omega f(x) \wedge \mathcal{E}_{1-2n}(x,y).$$

Proof   Set $A = \mathcal{E}_{2-2n}$ with $\mathcal{E}_{2-2n}$ the form of Lemma 5.15 and

$$J(f) := \int_{b\Omega} \overline{\vartheta_x \Gamma_{0,q}} \wedge *f + (\vartheta f, \vartheta \Gamma_{0,q}).$$

Now

$$\vartheta_x \Gamma_{0,q} = \partial_y \Gamma_{0,q-1} + A.$$

Therefore

$$J(f) = \bar{\partial}_y \left\{ \int_{b\Omega} \overline{\Gamma_{0,q-1}} \wedge *f + (\vartheta f, \Gamma_{0,q-1}) \right\} + \left\{ \int_{b\Omega} \overline{A} \wedge *f + (\vartheta f, A) \right\}.$$

We apply Stokes theorem in both brackets (this is possible because of the mildness of the singularity) and obtain

$$J(f) = \bar{\partial}_y \int_\Omega \partial_x \overline{\Gamma_{0,q-1}} \wedge *f + \int_\Omega \partial_x \overline{A} \wedge *f$$

$$= \bar{\partial}_y \int_\Omega f \wedge *\partial_x \overline{\Gamma_{0,q-1}} + (f, \mathcal{E}_{1-2n})$$

If we plug this into (5.14) we get the theorem. □

In the remaining part of this section we want to prove that $- * \partial_x \overline{\Gamma_{0,q}}(x,y)$ is an approximate Cauchy-Fantappiè form. Set

$$\beta(x,y) = \frac{\partial_x \rho^2(x,y)}{\rho^2(x,y)}.$$

Then one can show the following

Proposition 5.19

$$- * \partial_x \overline{\Gamma_{0,q}} = (-1)^{q(q-1)/2} \binom{n-1}{q} \frac{1}{(2\pi i)^n} \beta \wedge (\partial_x \beta)^{n-q-1} \wedge (\bar{\partial}_y \beta)^q + \mathcal{E}_{2-2n}$$

$$= \Omega_q(\beta) + \mathcal{E}_{2-2n}.$$

Proof   (5.17) implies

$$\partial_x \overline{\Gamma_{0,q}} = -\frac{(n-1)!}{2\pi^n} \frac{1}{\rho^{2n}} \sum_{j=1}^n \sideset{}{'}\sum_{|J|=q} (L_j \rho^2) \omega_j \wedge \omega^J \wedge \Theta^J + \mathcal{E}_{2-2n}$$

$$- \frac{(n-1)!}{2\pi^n} \frac{1}{\rho^{2n}} \sideset{}{'}\sum_{|K|=q+1} \sum_{k_\nu \in K} \varepsilon_{k_\nu K_\nu}^K (L_{k_\nu} \rho^2) \omega^K \wedge \overline{\Theta}^{K_\nu} + \mathcal{E}_{2-2n}$$

with the following notations.

For $K = (k_1, \ldots, k_{q+1}), 1 \le \nu \le q+1$, we set

$$K_\nu = (k_1, \ldots, k_{\nu-1}, k_{\nu+1}, \ldots, k_{q+1}).$$

$\varepsilon_{k_\nu K_\nu}^K$ is the sign of the permutation $k_\nu K_\nu \to K$. Therefore by (5.8)

$$- * \partial_x \overline{\Gamma_{0,q}} = \frac{(n-1)! i^n}{2\pi^n \rho^{2n}} (-1)^{1+q(q-1)/2} \cdot$$

$$\cdot \sideset{}{'}\sum_{|K|=q+1} \sum_{k_\nu \in K} \varepsilon_{k_\nu K_\nu}^K (L_{k_\nu} \rho^2) \omega^K \wedge (\omega \wedge \bar{\omega})^{N-K} \wedge \overline{\Theta}^{K_\nu} + \mathcal{E}_{2-2n}.$$

Now we calculate $\Omega_q(\beta)$. Since

$$\overline{\partial}_x \partial_x \rho^2 = (-2) \sum_{j=1}^{n} \omega_j \wedge \overline{\omega_j} + \mathcal{E}_1,$$

$$\overline{\partial}_y \partial_x \rho^2 = (-2) \sum_{j=1}^{n} \omega_j \wedge \overline{\Theta_j} + \mathcal{E}_1,$$

we obtain

$$\Omega_q(\beta) = \frac{(-1)^{q(q-1)/2}}{(2\pi i)^n \rho^{2n}} \binom{n-1}{q} \left(\sum_{j=1}^{n} (L_j \rho^2) \omega_j \right) \wedge \left((-2) \sum_{j=1}^{n} \omega_j \wedge \overline{\omega_j} \right)^{n-q-1}$$

$$\wedge \left((-2) \sum_{j=1}^{n} \omega_j \wedge \overline{\Theta_j} \right)^q + \mathcal{E}_{2-2n}$$

$$= (-1)^{q(q-1)/2} \frac{(-1)^i n}{2\pi^n \rho^{2n}} \sideset{}{'}\sum_{\substack{j=1 \\ |A|=n-q-1 \\ |B|=q}}^{n} (L_j \rho^2) \omega_j \wedge (\omega \wedge \overline{\omega})^A \wedge (\omega \wedge \overline{\Theta})^B$$

$$= (-1)^{q(q-1)/2} \frac{(-1)^i n}{2\pi^n \rho^{2n}} \sum_{\substack{j=1 \\ |B|=q}}^{n} (L_j \rho^2) \omega^{jB} \wedge (\omega \wedge \overline{\omega})^{N-jB} \wedge \overline{\Theta}^B$$

$$= (-1)^{q(q-1)/2} \frac{(-1)^i n}{2\pi^n \rho^{2n}} \sum_{\substack{k_\nu \in K \\ |K|=q+1}} \varepsilon_{k_\nu K_\nu}^K (L_{k_\nu} \rho^2) \omega^K \wedge (\omega \wedge \overline{\omega})^{N-K} \wedge \overline{\Theta}^{K_\nu}.$$

So

$$- * \partial_x \overline{\Gamma_{0,q}} - \Omega_q(\beta) = \mathcal{E}_{2-2n},$$

as desired. $\square$

Definition 5.20   *The kernel*

$$\Omega_q(\beta) = (-1)^{q(q-1)/2} \binom{n-1}{q} \frac{1}{(2\pi i)^n} \beta \wedge (\overline{\partial}_x \beta)^{n-q-1} \wedge (\overline{\partial}_y \beta)^q$$

is called the Bochner-Martinelli-Koppelman kernel for hermitian manifolds and denoted by B_q. – One sets $B_{-1} = 0$.

Combining Propositions 5.18 and 5.19 we get

Theorem 5.21 (BMK formula for hermitian manifolds)   *Let $\Omega \subset\subset X$ be a domain with piecewise smooth boundary and $f \in C^1_{0,q}(\overline{\Omega})$. Then there are isotropic kernels $\mathcal{E}_{1-2n}$ and*

$\mathcal{E}_{2-2n}$, which do not depend on f, such that for $y \in \Omega$

$$f(y) = \int_{b\Omega} f(x) \wedge B_q(x, y) - \int_{\Omega} \overline{\partial} f(x) \wedge B_q(x, y) - \overline{\partial}_y \int_{\Omega} f(x) \wedge B_{q-1}(x, y)$$
$$+ (f(x), \mathcal{E}_{1-2n}(x, y)) + (\overline{\partial} f(x), \mathcal{E}_{2-2n}(x, y)).$$

For the proof one uses

$$\int_{b\Omega} f \wedge \mathcal{E}_{2-2n} = (f, \mathcal{E}_{1-2n}) + (\overline{\partial} f, \mathcal{E}_{2-2n}),$$

$$\overline{\partial}_y (f, \mathcal{E}_{2-2n}) = (f, \mathcal{E}_{1-2n}).$$

6.   In order to apply integral formulae to function theory on strictly pseudoconvex domains on X we have to construct a generating form which is holomorphic with respect to y.

Definition 5.22 A domain $\Omega \subset\subset X$ is called strictly pseudoconvex with C^k smooth boundary $(k \geq 2)$ if there is a neighbourhood U of the boundary and a strictly plurisubharmonic C^k function $r : U \to \mathbb{R}$ with

i. $\Omega \cap U = \{x \in U | r(x) < 0\}$,

ii. $dr \neq 0$ on $b\Omega$.

In particular the Levi form

$$L_r(z; t) = \sum_{i,j=1}^{n} \frac{\partial^2 r(z)}{\partial z^i \partial \overline{z}^j} t_i \overline{t}_j$$

is strictly positive definite on U with respect to local holomorphic coordinates.

From now on we suppose that $r \in C^\infty$. Let $\{V^i\}_i$, $\{U^i\}_i$ be two finite open coverings of $b\Omega$ with $V^i \subset\subset U^i \subset\subset U$ and $z_{(i)}$ be a holomorphic coordinate map on $\overline{U}^i$.

On each U^i the Levi polynomial of r with respect to $z_{(i)}$ is

$$F^i(\zeta_{(i)}, z_{(i)}) = \sum_{k=1}^{n} \frac{\partial r(\zeta_{(i)})}{\partial \zeta_{(i)}^k} (\zeta_{(i)}^k - z_{(i)}^k) - \frac{1}{2} \sum_{k,l=1}^{n} \frac{\partial^2 r(\zeta_{(i)})}{\partial \zeta_{(i)}^k \partial \zeta_{(i)}^l} (\zeta_{(i)}^k - z_{(i)}^k)(\zeta_{(i)}^l - z_{(i)} l).$$

By using Taylor expansion one obtains

Lemma 5.23 $F^i - F^j = \mathcal{E}_3 \in C^\infty$ on $(U^i \cap U^j) \times (U^i \cap U^j)$.

Proof   Taylor expansion gives

$$r(z_{(i)}) - r(\zeta_{(i)}) = -2\,\mathrm{Re}\; F^i(\zeta_{(i)}, z_{(i)}) + L_r(\zeta_{(i)}; \zeta_{(i)} - z_{(i)}) + \mathcal{E}_3.$$

Now

$$L_r(\zeta_{(i)}; \zeta_{(i)} - z_{(i)}) - L_r(\xi_{(j)}; \zeta_{(j)} - z_{(j)}) = \mathcal{E}_3.$$

Thus

$$\mathrm{Re}\,(F^i(\zeta_{(i)}, z_{(i)}) - F^j(\zeta_{(j)}, z_{(j)})) = \mathcal{E}_3.$$

Fix a point P and expand $g = F^i - F^j$ with respect to a holomorphic coordinate z (with $\zeta(P) = z(P) = 0$) in a neighbourhood of P. Since $g + \bar{g} = \mathcal{E}_3$ we obtain

$$\frac{\partial g}{\partial z_i}(0) = \frac{\partial^2 g}{\partial z_i \partial z_j}(0) = 0$$

for all $1 \leq i, j \leq n$. Together with $g(0) = 0$ this gives the conclusion. $\square$

Remark   We can only use $F^i - F^j = \mathcal{E}_2$ in this book.

Also from Taylor expansion and from the plurisubharmonicity of r one has

Lemma 5.24   *There are positive reals $\varepsilon \leq \delta$ with:*

   i. $\bigcup_i V^i \supset \{x \in U \mid |r(x)| \leq 3\delta\} \subset\subset U$,

   ii. for all i: $\rho(x,y) \leq \varepsilon$ and $x \in V^i$ implies $y \in U^i$,

   iii. $|r(x)| \leq \delta$ and $\rho(x,y) \leq \varepsilon$ implies $|r(y)| < 2\delta$,

   iv. $|r(x)| \leq 2\delta$ and $\rho(x,y) \leq \varepsilon$ implies $|r(y)| < 3\delta$,

   v. $\exists c > 0 \,\forall i\, \forall (x,y) \in \overline{U^i} \times \overline{U^i}:$

$$2\,\mathrm{Re}\; F^i(x,y) \geq r(x) - r(y) + c\rho(x,y)^2$$

if $|r(x)| \leq 3\delta, |r(y)| \leq 3\delta, \rho(x,y) \leq \varepsilon.$

In order to patch the Levi polynomials together we choose a partition of unity $\{\xi_i\}_i$ with supp $\xi_i \subset\subset V^i$ and

$$\sum_i \xi_i \equiv 1 \quad \text{on } \{x \in U \mid |r(x)| \leq 3\delta\}.$$

We set

$$\xi_i(x)F^i(x,y) = \begin{cases} \xi_i(x)F^i(x,y) & \text{if } F^i(x,y) \text{ is defined} \\ 0 & \text{if } F^i(x,y) \text{ is not defined but } \xi_i(x) = 0 \end{cases}$$

Then

$$F(x,y) = \sum_i \xi_i(x)F^i(x,y)$$

is well defined in particular on

$$\{(x,y) \in U \times U \mid |r(x)| \leq 2\delta, \rho(x,y) \leq \varepsilon\}. \tag{5.25}$$

If we set

$$\widehat{F}(x,y) := F(x,y) - r(x) \tag{5.26}$$

we have

$$2\operatorname{Re}\widehat{F}(x,y) \geq -r(x) - r(y) + c\rho(x,y)^2 \tag{5.27}$$

on the sets of 5.25 and 5.24iv, v.

For us it is not sufficient that $\widehat{F}$ is only defined near the boundary diagonal of Ω. In the following extension of $\widehat{F}$ we have to take care that $\widehat{F}$ does not vanish if $y \in \Omega$.

So let $\varphi \in C^\infty$ be a function on $X \times X$ with

$$0 \leq \varphi \leq 1, \quad \varphi(x,y) = \varphi(y,x),$$
$$\varphi(x,y) \equiv 1 \quad \text{if } \rho(x,y) \leq \frac{\varepsilon}{2},$$
$$\varphi(x,y) \equiv 0 \quad \text{if } \rho(x,y) \geq \varepsilon.$$

Let $A > 1$ be a constant. We set

$$\Phi_A(x,y) = \varphi(x,y)\widehat{F}(x,y) + A(1 - \varphi(x,y))\rho^2(x,y),$$

and

$$\begin{aligned}
\Phi(x,y) &= \varphi(x,y)\widehat{F}(x,y) + (1 - \varphi(x,y))\rho^2(x,y) \\
&= \frac{1}{A}((A-1)\varphi\widehat{F} + \Phi_A)
\end{aligned}$$

on

$$\{x \in X \mid |r(x)| < 2\delta\} \times X. \tag{5.28}$$

This is possible since $\varphi \equiv 0$ if $\rho \geq \varepsilon$.

If

$$|r(x)| < 2\delta, \quad \rho(x,y) < \frac{\varepsilon}{2}$$

then

$$\Phi = \widehat{F}. \tag{5.29}$$

This implies the holomorphy in y on this set.

If A is sufficiently large one has (with a constant $c_A > 0$)

$$2\operatorname{Re}\Phi_A(x,y) \geq c_A(-r(x) - r(y) + \rho^2(x,y)),$$

and therefore, with $c_0 = c_A/A$,

$$2\operatorname{Re}\Phi(x,y) \geq c_0(-r(x) - r(y) + \rho^2(x,y))$$

if

$$|r(x)| < 2\delta, \quad y \in \overline{\Omega}. \tag{5.30}$$

We remind the reader that as in the case of $\mathbb{C}^n$ we can choose for y fixed near $b\Omega$ the functions $r(x)$ and Im $\Phi(x,y)$ as part of a coordinate system in a neighbourhood of y.

One of the basic properties of Φ will be given in the following lemma. It is one of the crucial facts which will allow the cancellation of singularities in the Neumann problem later on.

Lemma 5.31 *Set $\Phi^*(x,y) = \overline{\Phi(y,x)}$ on $\{x \in X \mid |r(x)| < 2\delta|\}^2$. Then*

$$\Phi - \Phi^* = \mathcal{E}_3 \in C^\infty.$$

Proof

$$\Phi(x,y) - \Phi^*(x,y) = \varphi(x,y)(\widehat{F}(x,y) - \overline{\widehat{F}(y,x)}).$$

We show

$$\widehat{F}(x,y) - \overline{\widehat{F}(x,y)} = \mathcal{E}_3 \in C^\infty$$

where $\varphi(x,y) \neq 0$.

Since $|r(x)| < 2\delta$ and $\rho(x,y) \leq \varepsilon$ we have

$$\sum_i \xi_i(x) \equiv 1. \tag{$*$}$$

Thus

$$\widehat{F}(x,y) - \overline{\widehat{F}(y,x)} = \sum_i \xi_i(x)(F^i(x,y) - r(x)) - \sum_i \xi_i(y)(\overline{F^i(y,x)} - r(y))$$

$$= \sum_i \xi_i(y)[F^i(x,y) - r(x) - \overline{F^i(y,x)} + r(y)] + \sum_i (\xi_i(x) - \xi_i(y))F^i(x,y).$$

The last sum is $\mathcal{E}_3$ because of $(*)$ and (5.23). To conclude the proof we show the following result of Kerzman/Stein [KeS 78]. $\square$

Lemma 5.32

$$F^i(x,y) - r(x) - \overline{F^i(y,x)} + r(y) = \mathcal{E}_3 \in C^\infty.$$

Proof   (Compare also Chapter III.) We drop the index i in F^i and calculate in local coordinates ζ, resp. z.

On the one hand Taylor expansion of r implies

$$\text{Re}\,(F(\zeta,z) - r(\zeta)) = -\frac{1}{2}(r(\zeta) + r(z)) + \frac{1}{2}L_r(\zeta; \zeta - z) + \mathcal{E}_3(\zeta,z).$$

On the other hand Taylor expansion of

$$\frac{\partial r}{\partial \zeta^i}(\zeta) - \frac{\partial r}{\partial \zeta^i}(z)$$

gives

$$
\text{Im}\,(F(\zeta,z) + F(z,\zeta)) = \text{Im}\,\left(\sum_{i=1}^{n} \left(\frac{\partial r}{\partial \zeta^i}(\zeta) - \frac{\partial r}{\partial \zeta^i}(z) \right)(\zeta_i - z_i) \right.
$$

$$
\left. - \sum_{i,j=1}^{n} \frac{\partial^2 r}{\partial \zeta^i \partial \zeta^j}(z)(\zeta^i - z^i)(\zeta^j - z^j) + \mathcal{E}_3(\zeta,z) \right)
$$

$$
= \text{Im}\,(L_r(z;\zeta - z) + \mathcal{E}_3(\zeta,z)) = \mathcal{E}_3(\zeta,z).
$$

If we put both formulae together we get (5.32). $\qquad\square$

Definition 5.33 *The function* $\Phi(x,y)$ *constructed in (5.22) to (5.29) is called the extended Levi polynomial of* Ω *(more precisely, of* r *).*

7. Now we want to modify the integral formulae (5.14) and (5.18) by introducing Φ.

Let $\xi \in C^\infty(X)$ with $0 \le \xi \le 1$, $\xi(x) = 1$ if $|r(x)| \le \delta$, $\xi(x) = 0$ if $|r(x)| \ge \frac{3\delta}{2}$. On $\overline{\Omega} \times \overline{\Omega}$ we introduce the approximate Leray form

$$
\alpha(x,y) = \frac{\xi(x)\partial r(x)}{\Phi(x,y)} = \frac{a(x,y)}{\Phi(x,y)}.
$$

α is smooth off the boundary diagonal.

Definition 5.34 *The kernels* $K_q(x,y) = \Omega_q(\alpha)$ *generated by* α *are called boundary kernels for the strictly pseudoconvex domain* Ω.

We can now proceed to replace the boundary integral

$$
\int_{b\Omega} f \wedge B_q
$$

of Theorem 5.21 by the boundary term

$$
\int_{b\Omega} f \wedge K_q
$$

whose kernel is holomorphic in y (near the diagonal). Since neither β nor α are Leray forms the following constructions are fairly tedious.

We have to first switch to local analogues of the kernels. Let $x_0 \in b\Omega$ be a fixed point and U a sufficiently small Stein neighbourhood of x_0 and ζ a local coordinate map on U. Since $F(\zeta,z)$ vanishes on the diagonal of $\overline{U} \times \overline{U}$ we can find a decomposition according to Hefer's theorem

$$
F(\zeta,z) = \sum_{j=1}^{n} h_j(\zeta,z)(\zeta_j - z_j) \tag{5.35}
$$

(or from the construction of F) with $h_j(\zeta, \cdot)$ holomorphic in z. Set

$$a^0(\zeta, z) = \sum_{j=1}^{n} h_j(\zeta, z)\, d\zeta_j \tag{5.36}$$

and

$$\alpha^0(\zeta, z) = \frac{a^0(\zeta, z)}{F(\zeta, z)}. \tag{5.37}$$

With respect to the same coordinates we set

$$b^0(\zeta, z) = \sum_{j=1}^{n} \sum_{k=1}^{n} g_{jk}(\zeta)(\overline{\zeta}_k - \overline{z}_k)\, d\zeta_j, \tag{5.38}$$

$$R^2(\zeta, z) = b^0(\zeta, z)[\zeta - z] = \sum_{j,k=1}^{n} g_{jk}(\zeta)(\zeta_j - z_j)(\overline{\zeta}_k - \overline{z}_k), \tag{5.39}$$

and

$$\beta^0(\zeta, z) = \frac{b^0(\zeta, z)}{R^2(\zeta, z)}. \tag{5.40}$$

The following useful relations are direct consequences of the definition of F and of the differential geometric equation

$$\rho^2 = R^2 + \mathcal{E}_3.$$

Lemma 5.41

$$\partial_\zeta r(\zeta) = a^0(\zeta, z) + \mathcal{E}_1(\zeta, z),$$
$$\overline{\partial}_\zeta \partial_\zeta r(\zeta) = \overline{\partial}_\zeta a^0(\zeta, z) + \mathcal{E}_1(\zeta, z),$$
$$\overline{\partial}_\zeta F(\zeta, z) = \mathcal{E}_1(\zeta, z),$$
$$b^0(\zeta, z) = \mathcal{E}_1(\zeta, z),$$
$$\partial_\zeta \rho^2(\zeta, z) = b^0(\zeta, z) + \mathcal{E}_2(\zeta, z),$$
$$\overline{\partial}_\zeta \partial_\zeta \rho^2(\zeta, z) = \overline{\partial}_\zeta b^0(\zeta, z) + \mathcal{E}_1(\zeta, z),$$
$$\overline{\partial}_z \partial_\zeta \rho^2(\zeta, z) = \overline{\partial}_z b^0(\zeta, z) + \mathcal{E}_1(\zeta, z).$$

We now introduce on $U \times U$ the transition kernels between α^0 and β^0 (see II.1.37):

$$A_q(\alpha^0, \beta^0) = \sum_{\mu=0}^{q} \sum_{\nu=0}^{n-2-q} a_{q\mu\nu} A_{q\mu\nu}(\alpha^0, \beta^0),$$

with

$$A_{q\mu\nu}(\alpha^0, \beta^0) = \alpha^0 \wedge \beta^0 \wedge (\overline{\partial}_\zeta \alpha^0)^\mu \wedge (\overline{\partial}_\zeta \beta^0)^{n-2-q-\mu} \wedge (\overline{\partial}_z \alpha^0)^\nu \wedge (\overline{\partial}_z \beta^0)^{q-\nu},$$

where the coefficients $a_{q\mu\nu}$ are given in II.1.37.

Theorem II.1.38, now implies

$$\Omega_q(\beta^0) = \Omega_q(\alpha^0) + (-1)^{q+1}\bar{\partial}_\zeta A_q(\alpha^0, \beta^0) + \bar{\partial}_z A_{q-1}(\alpha^0, \beta^0). \qquad (5.42)$$

The next step is a comparison of the global kernels $\Omega_q(\alpha)$, $B_q = \Omega_q(\beta)$, $A_q(\alpha, \beta)$ (transition kernels between $\Omega_q(\alpha)$ and $\Omega_q(\beta)$) with their local counterparts.

Lemma 5.43   *i. On $(b\Omega \cap U) \times U$ one has*

$$\Omega_q(\alpha) = \Omega_q(\alpha^0) + \mathcal{E}_\infty(\zeta, z) \quad \text{for } q \geq 1.$$

$$\Omega_0(\alpha) = \Omega_0(\alpha^0) + \frac{\mathcal{E}_1(\zeta, z)}{\Phi(\zeta, z)^n}.$$

($\mathcal{E}_\infty$ indicates a smooth form which is $\mathcal{E}_j$ for all $j \geq 0$.).

ii. On $U \times U$ one has

$$\Omega_q(\beta) = \frac{R^{2n}}{\rho^{2n}}\Omega_q\beta^0 + \mathcal{E}_{2-2n}.$$

Proof   (*i*) follows from $\bar{\partial}_z\alpha = \bar{\partial}_z\alpha^0 = 0$ near the diagonal and from

$$\Omega_0(\alpha) = \text{const}\frac{\partial_\zeta r \wedge (\bar{\partial}_\zeta\partial_\zeta r)^{n-1}}{\Phi(\zeta, z)^n}$$

$$= \text{const}\frac{(a^0 + \mathcal{E}_1) \wedge (\bar{\partial}_\zeta a^0 + \mathcal{E}_1)^{n-1}}{\Phi(\zeta, z)^n}$$

$$= \Omega_0(\alpha^0) + \frac{\mathcal{E}_1(\zeta, z)}{\Phi(\zeta, z)^n}.$$

(*ii*) follows from the fifth equation of 5.41

$$\Omega_q(\beta) = \frac{1}{\rho^{2n}}\Omega_q(\partial_\zeta\rho^2) = \frac{1}{\rho^{2n}}\Omega_q(b^0 + \mathcal{E}_2) = \frac{1}{\rho^{2n}}\Omega_q(b^0) + \mathcal{E}_{2-2n}$$

$$= \frac{R^{2n}}{\rho^{2n}}\Omega_q(\beta^0) + \mathcal{E}_{2-2n}.$$

$\square$

Now we want to decompose $\Omega_q(\beta)$. In the following calculations $\mathcal{E}_j$, with $j \geq 0$, denotes a C^∞ form.

$$\Omega_q(\beta) = \frac{R^{2n}}{\rho^{2n}}\Omega_q(\beta^0) + \mathcal{E}_{2-2n}$$

$$= \frac{R^{2n}}{\rho^{2n}}(\Omega_q(\beta^0) - \Omega_q(\alpha^0)) + \Omega_q(\alpha^0) + \frac{\mathcal{E}_{2n+1}}{\rho^{2n}}\Omega_q(\alpha^0) + \mathcal{E}_{2-2n}.$$

Now

$$\Omega_q(\beta^0) - \Omega_q(\alpha^0) = (-1)^{q+1}\overline{\partial}_\zeta A_q(\alpha^0, \beta^0) + \overline{\partial}_z A_{q-1}(\alpha^0, \beta^0).$$
$$\frac{R^{2n}}{\rho^{2n}}\overline{\partial}_\zeta A_q(\alpha^0, \beta^0) = \overline{\partial}_\zeta\left(\frac{R^{2n}}{\rho^{2n}}A_q(\alpha^0, \beta^0)\right) - \frac{\mathcal{E}_4 R^{2(n-1)}}{\rho^{2n+2}} \wedge A_q(\alpha^0, \beta^0).$$

$$
\begin{aligned}
\frac{R^{2n}}{\rho^{2n}}A_q(\alpha^0, \beta^0) &= \sum_{\mu,\nu} a_{q\mu\nu}\frac{R^{2n}}{\rho^{2n}}\frac{A_{q\mu\nu}(\alpha^0, b^0)}{(R^2)^{n-1-\mu-\nu}} \\
&= \sum_{\mu,\nu} a_{q\mu\nu}\frac{R^{2(1+\mu+\nu)}}{\rho^{2(1+\mu+\nu)}}\frac{A_{q\mu\nu}(\alpha^0, b^0)}{\rho^{2(n-1-\mu-\nu)}} \\
&= \sum_{\mu,\nu} a_{q\mu\nu}\frac{R^{2(1+\mu+\nu)}}{\rho^{2(1+\mu+\nu)}}A_{q\mu\nu}\left(\alpha^0, \frac{\partial_\zeta\rho^2 + \mathcal{E}_2}{\rho^2}\right) \\
&= A_q\left(\alpha^0, \frac{\partial_\zeta\rho^2 + \mathcal{E}_2}{\rho^2}\right) + \sum_{\mu,\nu}\frac{\mathcal{E}_{3+2\mu+2\nu}}{(\rho^2)^{2+\mu+\nu}} \wedge A_{q\mu\nu}\left(\alpha^0, \frac{\partial_\zeta\rho^2 + \mathcal{E}_2}{\rho^2}\right).
\end{aligned}
$$

$$\frac{\mathcal{E}_4 R^{2(n-1)}}{(\rho^2)^{n+1}} \wedge A_q(\alpha^0, \beta^0) = \sum_{\mu,\nu}\frac{\mathcal{E}_{4+2\mu+2\nu}}{(\rho^2)^{2+\mu+\nu}} \wedge A_{q\mu\nu}\left(\alpha^0, \frac{\partial_\zeta\rho^2 + \mathcal{E}_2}{\rho^2}\right)$$

by the same reasoning. Therefore

$$
\begin{aligned}
&\frac{R^{2n}}{\rho^{2n}}\overline{\partial}_\zeta A_q(\alpha^0, \beta^0) \\
&= \overline{\partial}_\zeta\left[A_q\left(\alpha^0, \frac{\partial_\zeta\rho^2 + \mathcal{E}_2}{\rho^2}\right) + \sum_{\mu,\nu}\frac{\mathcal{E}_{3+2\mu+\nu}}{(\rho^2)^{1+\mu+\nu}}A_{q\mu\nu}\left(\alpha^0, \frac{\partial_\zeta\rho^2 + \mathcal{E}_2}{\rho^2}\right)\right] \\
&\qquad\qquad\qquad + \sum_{\mu,\nu}\frac{\mathcal{E}_{4+2\mu+2\nu}}{(\rho^2)^{2+\mu+\nu}} \wedge A_{q\mu\nu}\left(\alpha^0, \frac{\partial_\zeta\rho^2 + \mathcal{E}_2}{\rho^2}\right)
\end{aligned}
$$

Analogously one gets

$$
\begin{aligned}
\frac{R^{2n}}{\rho^{2n}}\overline{\partial}_z A_{q-1}(\alpha^0, \beta^0) &= \overline{\partial}_z\left[A_{q-1}\left(\alpha^0, \frac{\partial_\zeta\rho^2 + \mathcal{E}_2}{\rho^2}\right)\right. \\
&\qquad \left. + \sum_{\mu,\nu}\frac{\mathcal{E}_{3+2\mu+2\nu}}{(\rho^2)^{1+\mu+\nu}}A_{q-1,\mu,\nu}\left(\alpha^0, \frac{\partial_\zeta\rho^2 + \mathcal{E}_2}{\rho^2}\right)\right] \\
&\qquad + \sum_{\mu,\nu}\frac{\mathcal{E}_{4+2\mu+2\nu}}{(\rho^2)^{2+\mu+\nu}}A_{q-1,\mu,\nu}\left(\alpha^0, \frac{\partial_\zeta\rho^2 + \mathcal{E}_2}{\rho^2}\right)
\end{aligned}
$$

Putting all the pieces together we obtain

$$\Omega_q(\beta) = \Omega_q(\alpha) + (-1)^{q+1}\overline{\partial}_\zeta A_q(\alpha,\beta) + \overline{\partial}_z A_{q-1}(\alpha,\beta) + \left(\Omega_q(\alpha) - \Omega_q(\alpha^0)\right)$$

$$+ \frac{\mathcal{E}_{2n+1}}{\rho^{2n}}\Omega_q(\alpha^0) + \mathcal{E}_{2-2n} + \overline{\partial}_\zeta\left[\left(A_q\left(\alpha^0,\frac{\partial_\zeta\rho^2 + \mathcal{E}_2}{\rho^2}\right) - A_q(\alpha,\beta)\right)\right.$$

$$\left. + \sum_{\mu,\nu}\frac{\mathcal{E}_{3+2\mu+2\nu}}{(\rho^2)^{1+\mu+\nu}}A_{q,\mu,\nu}\left(\alpha^0,\frac{\partial_\zeta\rho^2 + \mathcal{E}_2}{\rho^2}\right)\right]$$

$$+ \sum_{\mu,\nu}\frac{\mathcal{E}_{4+2\mu+2\nu}}{(\rho^2)^{2+\mu+\nu}} \wedge A_{q\mu\nu}\left(\alpha^0,\frac{\partial_\zeta\rho^2 + \mathcal{E}_2}{\rho^2}\right) + \sum_{\mu,\nu}\frac{\mathcal{E}_{4+2\mu+2\nu}}{(\rho^2)^{2+\mu+\nu}}A_{q-1,\mu,\nu}\left(\alpha^0,\frac{\partial_\zeta\rho^2 + \mathcal{E}_2}{\rho^2}\right)$$

$$+ \overline{\partial}_z\left[\left(A_{q-1}\left(\alpha^0,\frac{\partial_\zeta\rho^2 + \mathcal{E}_2}{\rho^2}\right) - A_{q-1}(\alpha,\beta)\right)\right.$$

$$\left. + \sum_{\mu,\nu}\frac{\mathcal{E}_{3+2\mu+2\nu}}{(\rho^2)^{1+\mu+\nu}}A_{q-1,\mu,\nu}\left(\alpha^0,\frac{\partial_\zeta\rho^2 + \mathcal{E}_2}{\rho^2}\right)\right]$$

Later on we shall treat all the terms on the right hand side except the three first one as error terms.

The following calculations are meant for $\zeta \in b\Omega$ such that $F = \Phi$. We have $\Omega_q(\alpha) = \Omega_q(\alpha^0) \equiv 0$ near the boundary diagonal if $q > 0$. Now

$$\frac{\mathcal{E}_{2n+1}}{\rho^{2n}}\Omega_0(\alpha^0) = \frac{\mathcal{E}_{2n+1}}{\Phi^n\rho^{2n}}.$$

(5.43), (i) gives

$$\Omega_0(\alpha) - \Omega_0(\alpha^0) = \frac{\mathcal{E}_1}{\Phi^n}.$$

Moreover

$$A_{q\mu\nu}\left(\alpha^0,\frac{\partial_\zeta\rho^2 + \mathcal{E}_2}{\rho^2}\right) = \frac{\mathcal{E}_1}{\Phi^{1+\mu+\nu}(\rho^2)^{n-1-\mu-\nu}}.$$

Thus

$$\overline{\partial}_\zeta\left(\frac{\mathcal{E}_{3+2\mu+2\nu}}{(\rho^2)^{1+\mu+\nu}}A_{q\mu\nu}\left(\alpha^0,\frac{\partial_\zeta\rho^2 + \mathcal{E}_2}{\rho^2}\right)\right) = \overline{\partial}_\zeta\left(\frac{\mathcal{E}_{4+2\mu+2\nu}}{\Phi^{1+\mu+\nu}(\rho^2)^n}\right)$$

$$= \frac{\mathcal{E}_{3+2\mu+2\nu}}{\Phi^{1+\mu+\nu}(\rho^2)^n} + \frac{\mathcal{E}_{5+2\mu+2\nu}}{\Phi^{1+\mu+\nu}(\rho^2)^{n+1}}$$

$$+ \frac{\mathcal{E}_{5+2\mu+2\nu} + \mathcal{E}_{4+2\mu+2\nu}\wedge\overline{\partial}_\zeta r}{\Phi^{2+\mu+\nu}(\rho^2)^n}$$

An analogous formula holds for

$$\overline{\partial}_z\left(\frac{\mathcal{E}_{3+2\mu+2\nu}}{(\rho^2)^{1+\mu+\nu}}A_{q-1,\mu,\nu}\left(\alpha^0,\frac{\partial_\zeta\rho^2 + \mathcal{E}_2}{\rho^2}\right)\right).$$

$$A_{q\mu\nu}\left(\alpha^0, \frac{\partial_\zeta \rho^2 + \mathcal{E}_2}{\rho^2}\right) = A_{q\mu\nu}(\alpha, \beta) \equiv 0 \text{ near } \Lambda \text{ if } \nu > 0.$$

$$A_{q\mu0}\left(\alpha^0, \frac{\partial_\zeta \rho^2 + \mathcal{E}_2}{\rho^2}\right) - A_{q\mu0}(\alpha, \beta) = \frac{\mathcal{E}_2}{\Phi^{1+\mu+\nu}(\rho^2)^{n-1-\mu-\nu}}$$

by (5.41). Therefore

$$\overline{\partial}_\zeta \left(A_{q\mu0}\left(\alpha^0, \frac{\partial_\zeta \rho^2 + \mathcal{E}_2}{\rho^2}\right) - A_{q\mu0}(\alpha, \beta)\right)$$
$$= \frac{\mathcal{E}_1}{\Phi^{1+\mu}(\rho^2)^{n-1-\mu}} + \frac{\mathcal{E}_3}{\Phi^{1+\mu}(\rho^2)^{n-\mu}} + \frac{\mathcal{E}_3 + \mathcal{E}_2 \wedge \overline{\partial}_\zeta r}{\Phi^{2+\mu}(\rho^2)^{n-1-\mu}}.$$

An analogous formula, but without a term $\ldots \wedge \overline{\partial}_\zeta r$, holds for

$$\overline{\partial}_z \left(A_{q-1,\mu0}\left(\alpha^0, \frac{\partial_\zeta \rho^2 + \mathcal{E}_2}{\rho^2}\right) - A_{q-1,\mu0}(\alpha, \beta)\right).$$

Collecting all the terms and using a partition of unity we obtain

$$\Omega_q(\beta) = \Omega_q(\alpha) + (-1)^{q+1}\overline{\partial}_x A_q(\alpha, \beta) + \overline{\partial}_y A_{q-1}(\alpha, \beta) + \mathcal{E}_{2-2n}$$
$$+ \sum_{\varkappa \geq 0} \frac{\mathcal{E}_{3+2\varkappa}}{\Phi^{1+\varkappa}\rho^{2n}} + \frac{\mathcal{E}_{5+2\varkappa}}{\Phi^{1+\varkappa}(\rho^2)^{n+1}} + \frac{\mathcal{E}_{5+2\varkappa} + \mathcal{E}_{4+2\varkappa} \wedge \overline{\partial}_\zeta r}{\Phi^{2+\varkappa}(\rho^2)^n} \quad (5.44)$$

for $x \in b\Omega, y \in \Omega$.

Consequently we have after an integration by parts for $f \in C^1_{0,q}(\overline{\Omega})$

$$\int_{b\Omega} f \wedge B_q = \int_{b\Omega} f \wedge \Omega_q(\alpha) + \int_{b\Omega} \overline{\partial}f \wedge A_q(\alpha, \beta) + \overline{\partial}_z \int_{b\Omega} f \wedge A_{q-1}(\alpha, \beta)$$
$$+ \int_{b\Omega} f \wedge \mathcal{E}_{2-2n} + \sum_{\varkappa=0}^n \int_{b\Omega} f \wedge \left\{\frac{\mathcal{E}_{3+2\varkappa}}{\Phi^{1+\varkappa}(\rho^2)^n} + \frac{\mathcal{E}_{5+2\varkappa}}{\Phi^{1+\varkappa}(\rho^2)^{n+1}}.\right\} \quad (5.45)$$

In order to transform all the boundary integrals into volume integrals we replace ρ^2 in all the denominators by

$$P(x, y) = \rho^2(x, y) + 2r(x)r(y). \quad (5.46)$$

Set for example

$$A_q := A_q\left(\alpha, \frac{\overline{\partial}_\zeta \rho^2}{P}\right). \quad (5.47)$$

Thus

$$\int_{b\Omega} f \wedge B_q = \int_{\Omega} \overline{\partial} f \wedge \Omega_q(\alpha) + (-1)^q \int_{\Omega} f \wedge \overline{\partial}_x \Omega_q(\alpha) + (-1)^{q+1} \int_{\Omega} \overline{\partial} f \wedge \overline{\partial}_x A_q$$

$$+ \int_{\Omega} \overline{\partial} f \wedge \overline{\partial}_y A_{q-1} + (-1)^q \overline{\partial}_y \int_{\Omega} f \wedge \overline{\partial}_x A_{q-1} + \int_{\Omega} \overline{\partial} f \wedge \mathcal{E}_{2-2n}$$

$$+ \int_{\Omega} f \wedge \mathcal{E}_{1-2n} + \sum_{\varkappa=0}^{n} \int_{\Omega} \overline{\partial} f \wedge \left\{ \frac{\mathcal{E}_{3+2\varkappa}}{\Phi^{1+\varkappa} P^n} + \frac{\mathcal{E}_{5+2\varkappa}}{\Phi^{1+\varkappa} P^{n+1}} \right\}$$

$$+ \sum_{\varkappa=0}^{n} \int_{\Omega} f \wedge \left\{ \frac{\mathcal{E}_{2+2\varkappa}}{\Phi^{1+\varkappa} P^n} + \frac{\mathcal{E}_{3+2\varkappa}}{\Phi^{2+\varkappa} P^n} + \frac{\mathcal{E}_{4+2\varkappa} + r(y)\mathcal{E}_{3+2\varkappa}}{\Phi^{1+\varkappa} P^{n+1}} \right.$$

$$\left. + \frac{\mathcal{E}_{4+2\varkappa}}{\Phi^{1+\varkappa} P^{n+1}} + \frac{\mathcal{E}_{5+2\varkappa}}{\Phi^{2+\varkappa} P^{n+1}} + \frac{\mathcal{E}_{6+2\varkappa} + r(y)\mathcal{E}_{5+2\varkappa}}{\Phi^{1+\varkappa} P^{n+2}} \right\}. \quad (5.48)$$

To get rid of the clumsy formalism we introduce the following notion which is analogous to definitions (5.9) and (5.11) of chapter III.

Definition 5.49 *A double differential form $\mathcal{A}(x,y)$ on $\overline{\Omega} \times \overline{\Omega}$ is an admissible kernel, if it has the following properties:*

i. *$\mathcal{A}$ is smooth on $\overline{\Omega} \times \overline{\Omega} \setminus \Lambda$ (Λ denotes the boundary diagonal.).*

ii. *For any point $(x_0, y_0) \in \Lambda$ there is a neighbourhood $U \times U$ and a representation*

$$\mathcal{A} = \mathcal{E}_j P^{-t_0} \Phi^{t_1} \overline{\Phi}^{t_2} \Phi^{*t_3} \overline{\Phi^*}^{t_4} r^k r^{*l}$$

(here $\Phi^*(x,y) = \overline{\Phi(y,x)}$, $r^*(x) = r(y)$), with j, t_0, t_1, t_2, t_3, t_4, k, l integers and j, t_0, k, $l \geq 0$, $t = -(t_1 + t_2 + t_3 + t_4) \geq 0$.

$\mathcal{E}_j$ is smooth and isotropic of order $\geq j$.

The type of the above representation is the integer λ with

$$\lambda = 2n + j + \min(2, t - k - l) - 2(t_0 + t - k - l).$$

$\mathcal{A}$ is called of type $\geq \lambda$, if $\mathcal{A}$ has everywhere a local representation of type $\geq \lambda$.

$\mathcal{A}$ is called of commutator type $\geq \lambda$ if additionally $t_1 t_3 \geq 0$, $t_2 t_4 \geq 0$, $(t_1 + t_3)(t_2 + t_4) \leq 0$. $\mathcal{A}$ is called special admissible of type ≥ 0 if $\mathcal{A}$ is of type ≥ 0 and $\mathcal{A}$ can be estimated by

$$|\mathcal{A}| \lesssim \frac{1}{P^a |\Phi|^b}, \quad \text{with } b \neq 2,$$

and $2n + \min(2, b) - 2(a + b) \geq 0$.

An operator with admissible kernel is called admissible (of the corresponding type),

$$A_\lambda f(y) = \int_{\Omega} f(x) \wedge \overline{*A_\lambda(x,y)} = (f(\cdot), A_\lambda(\cdot, y))_\Omega .$$

Remarks   The definition carries the notion of an admissible operator from $\mathbb{C}^n$ over to the present more general context. As in $\mathbb{C}^n$, *special admissible operators* have better regularity behaviour than general type 0 operators. The new concept of *commutator type* refers to the following fact (which will be made precise and proved in Chapter VII): *operators of commutator type essentially commute with vectorfields* and therefore respect norms involving derivatives. — To finally explain the restriction $t > 0$ in *ii.*, we use the notion of a *normalised Levi metric*, which is defined in VI.2.16 and VI.2.34, and show that, if we have such a metric, we can transform kernels with $t < 0$ in such with $t \geq 0$ whithout changing the type or the property of being of commutator type. So we can always assume that $t \geq 0$. To do this one proceeds as follows. Let $t_1 + t_2 + t_3 + t_4 > 0$. We assume $t_1 > 0$. The other cases are analogous. Then

$$\mathcal{A} = |\Phi|^2 \mathcal{E}_j \mathrm{P}^{-t_0} \Phi^{t_1-1} \overline{\Phi}^{t_2-1} \Phi^{*\,t_3} \overline{\Phi}^{*\,t_4} r^k r^{*\,l}.$$

By using (2.19) of Chapter VI (2.19*ii* is fulfilled if the metric is a normalised Levi metric) we set

$$|\Phi|^2 = \frac{1}{2}\mathrm{P} + \mathcal{E}_2.$$

Therefore $\mathcal{A}$ can be decomposed in $\mathcal{A}^I + \mathcal{A}^{II}$ with

$$\mathcal{A}^I = \mathcal{E}_j \mathrm{P}^{1-t_0} \Phi^{t_1-1} \overline{\Phi}^{t_2-1} \Phi^{*\,t_3} \overline{\Phi}^{*\,t_4} r^k r^{*\,l},$$
$$\mathcal{A}^{II} = \mathcal{E}_{j+2} \mathrm{P}^{-t_0} \Phi^{t_1-1} \overline{\Phi}^{t_2-1} \Phi^{*\,t_3} \overline{\Phi}^{*\,t_4} r^k r^{*\,l}.$$

This drops $(-t)$ by 2. The type does not change since $t < 0$. Let now $\mathcal{A}$ be of commutator type. So additionally to $t_1 > 0$ we have $(t_1 + t_3)(t_2 + t_4) \leq 0, t_1 t_3 \geq 0, t_2 t_4 \geq 0$. Then

$$(t_1 - 1 + t_3)(t_2 - 1 + t_4) = (t_1 + t_3)(t_2 + t_4) + t + 1 \leq t + 1 \leq 0$$

since $t < 0$ implies $t \leq -1$. $t_1 > 0, t_1 t_3 \geq 0 \Rightarrow t_1 \geq 1, t_3 \geq 0$. Therefore

$$(t_1 - 1)t_3 = t_1 t_3 - t_3 \geq t_3 - t_3 \geq 0.$$

The proof of the last inequality to be shown is slightly more complicated. If $t_2 > 0$ as in the previous case it follows $(t_2 - 1)t_4 \geq 0$. Let $t_2 \leq 0$.

Case $t_2 < 0$. $t_2 t_4 \geq 0 \Rightarrow t_4 \leq 0$. Thus

$$(t_2 - 1)t_4 = t_2 t_4 - t_4 \geq t_2 t_4 \geq 0.$$

Case $t_2 = 0$.   If $t_4 \leq 0$ then $(t_2 - 1)t_4 = -t_4 \geq 0$. So let $t_4 > 0$. But $(t_1 + t_3)t_4 \leq 0$ implies $t_1 + t_3 \leq 0$.

Since $t_1 > 0$ and $t_1 t_3 \geq 0$ we have $t_3 \geq 0$. So $t_1 + t_3 > 0$. This case cannot happen.

By repeating the above method if necessary we can lift t up to $t \geq 0$. At last if $\mathcal{A}$ is special admissible of type ≥ 0 then

$$|\mathcal{A}| \lesssim \frac{1}{\mathrm{P}^{t_0}|\Phi|^m}$$

with $2n + \min 2, m - 2(t_0 + m) \geq 0$. Since then

$$\left| \frac{\mathcal{A}}{|\Phi|^2} \right| \lesssim \frac{1}{P^{t_0} |\Phi|^{m+2}}$$

and $|\mathcal{E}_2| \lesssim P$ we obtain

$$|\mathcal{A}^I| + |\mathcal{A}^{II}| \lesssim \frac{1}{P^{t_0-1} |\Phi|^{m+2}} \lesssim \frac{1}{P^{t_0} |\Phi|^m}$$

since (2.19), Chapter VI, also implies

$$P \lesssim |\Phi|^2 + \rho^2 \lesssim |\Phi|^2.$$

Consequently the transformation described above also preserves special admissibility.

A quick check now yields some examples all of commutator type:

$$\Omega_q(\alpha) \text{ is of type} \geq 2,$$
$$\overline{\partial}_x \Omega_q(\alpha) \text{ is of type} \geq 0,$$
$$A_q \text{ is of type} \geq 2,$$
$$\overline{\partial}_x A_q \text{ is of type} \geq 1,$$
$$\overline{\partial}_y A_q \text{ is of type} \geq 2.$$

$$\frac{\mathcal{E}_{3+2\varkappa}}{\Phi^{1+\varkappa} P^n}, 0 \leq \varkappa \leq n, \text{ is of type} \geq 2.$$

$$\frac{\mathcal{E}_{2+2\varkappa}}{\Phi^{1+\varkappa} P^n}, \frac{\mathcal{E}_{3+2\varkappa}}{\Phi^{2+\varkappa} P^n}, \frac{r(y)\mathcal{E}_{3+2\varkappa}}{\Phi^{1+\varkappa} P^{n+1}}, \text{ etc. are of type} \geq 1.$$

We use again the generic notion

$$\mathcal{K} = \mathcal{E}_j \quad \text{resp.} \quad \mathcal{K} = \mathcal{A}_\lambda, \qquad (5.50)$$
$$K = E_j \quad \text{resp.} \quad K = A_\lambda,$$

to abbreviate the statement: $\mathcal{K}$ is an isotropic kernel of order $\geq j$, etc. With this convention we can state, for instance,

$$*\mathcal{E}_j(x,y) = \mathcal{E}_j(x,y) \qquad (5.51)$$
$$\mathcal{E}_j(y,x) = \mathcal{E}_j(x,y),$$

etc.

We obtain the main result of this paragraph:

Theorem 5.52 (Basic integral representation) *Let* $\Omega \subset\subset X$ *be a strictly pseudoconvex domain with smooth boundary. Then for* $f \in C^1_{0,q}(\overline{\Omega})$ *and* $y \in \Omega$ *the following integral* *representation holds.*

$$f(y) = (-1)^q \int\limits_\Omega f \wedge \overline{\partial}_x \Omega_q(\alpha) + \int\limits_\Omega \overline{\partial} f \wedge \Omega_q(\alpha) + (-1)^{q+1} \int\limits_\Omega \overline{\partial} f \wedge \overline{\partial}_x A_q$$

$$- \int\limits_\Omega \overline{\partial} f \wedge B_q + \overline{\partial}_y \left[(-1)^q \int\limits_\Omega f \wedge \overline{\partial}_x A_{q-1} - \int\limits_\Omega f \wedge B_{q-1} \right]$$

$$+ (f, \mathcal{E}_{1-2n}) + (f, \mathcal{A}_1) + (\overline{\partial} f, \mathcal{E}_{2-2n}) + (\overline{\partial} f, \mathcal{A}_2).$$

(We have written, e. g.

$$\int\limits_\Omega f \wedge \mathcal{E}_{1-2n} = \pm \int\limits_\Omega f \wedge \overline{*(*\overline{\mathcal{E}}_{1-2n})} = (f, \mathcal{E}_{1-2n}),$$

where we used the convention 5.51.)

An easy check gives

Theorem 5.53 $\overline{\partial}_x \Omega_q(\alpha)$ *is $\mathcal{E}_0$, that is everywhere smooth, if $q > 0$. The kernels $\overline{\partial}_x A_q$, $\Omega_0(\alpha)$, $\overline{\partial}_x A_{q-1}$ are of type ≥ 1, and $\overline{\partial}_x \Omega_0(\alpha)$ is special admissible of type ≥ 0.*

We finally rewrite and extend our main result: The regularity theorems for $\mathcal{A}_1$ and $\mathcal{E}_{1-2n}$ kernels as proved in the preceding chapters and Chapter VII show that all the kernels which appear in 5.52 define bounded operators from L^s into itself for $1 < s < \infty$; with the exception of $\overline{\partial}_x \Omega_0(\alpha)$ they yield even compact operators from L^s to L^s for all s, $1 \leq s \leq \infty$. (This latter fact will be proved in Chapter VII.) Now by the density theorem V.2.6 the smooth forms are dense in L^1 for the graph norm $\|f\|_{L^1} + \|\overline{\partial} f\|_{L^1}$. So Theorem 5.52 extends to these spaces – cf. the proof of III. 4.22'. Regrouping the terms and introducing shorter notations we obtain

Theorem 5.54 *(Basic homotopy formula).*   Let $f \in L^1_{0q}$ and $\overline{\partial} f \in L^1_{0q+1}$, where $\overline{\partial} f$ is defined as a distribution. Then

$$f = P_q f + T_q \overline{\partial} f + \overline{\partial} S_{q-1} f.$$

The operators P_q, T_q, S_q are bounded between the following spaces

   i) $P_0 : L^s_{00} \to L^s_{00}, \quad 1 < s < \infty$

   ii) $P_q : L^s_{0q} \to L^s_{0q}, \quad 1 \leq s \leq \infty, q \geq 1.$

   iii) $T_q, S_q : L^s_{0q+1} \to L^s_{0q}, \quad 1 \leq s \leq \infty$

The operators from ii and iii are, moreover, compact.

§6 The Levi Problem on Strictly Pseudoconvex Manifolds

Let $\Omega \subset\subset X$ be a strictly pseudoconvex domain with smooth boundary in a complex manifold. Consider, for $q \geq 1$, a $\bar\partial$-closed $(0,q)$-form $f \in L^2_{0q}(\Omega)$. We obtain, from the basic homotopy formula of the preceding paragraph,

$$f = \bar\partial S_{q-1} f + P_q f \overset{def}{=} \tilde S_q f + P_q f, \tag{6.1}$$

where the operators S_{q-1} and P_q are compact, and $\tilde S_q = \bar\partial S_{q-1}$.

We introduce the spaces

$$Z^{0q}_s(\Omega) = L^s_{0q}(\Omega) \cap \ker \bar\partial,$$
$$H^{0q}_s = Z^{0q}_s / \bar\partial \{ L^s_{0,q-1}(\Omega) \cap \operatorname{dom} \bar\partial \},$$

(the Dolbeault cohomology spaces of L^s-bounded classes) and we will show that the H^{0q}_s are finite-dimensional.

Now (6.1) reads on Z^{0q}_s

$$\tilde S_q = id - P_q, \tag{6.2}$$

and functional analysis tells us that $\tilde S_q$ has closed range of finite codimension:

$$\dim Z^{0q}_s(\Omega) / \mathcal{R}(\tilde S_q) < \infty. \tag{6.3}$$

Since $\bar\partial(L^s_{0,q-1}(\Omega) \cap \operatorname{dom} \bar\partial) \supset \mathcal{R}(\tilde S_q)$, this space is closed in $Z^{0q}_s(\Omega)$ and

$$\dim_{\mathbb{C}} H^{0q}_s(\Omega) < \infty, \quad q \geq 1. \tag{6.4}$$

Theorem 6.5 (Finiteness theorem) *Let $\Omega \subset\subset X$ be a strictly pseudoconvex domain in a complex manifold X. Then the Dolbeault cohomology spaces $H^{0q}_s(\Omega)$ are finite dimensional for $q \geq 1$ and $1 \leq s \leq \infty$.*

We can now easily show that strictly pseudoconvex manifolds are holomorphically convex. Let $x_0 \in b\Omega$ be fixed. It follows from the properties of the Levi polynomial (cf. Lemma 5.24) that there exists a function $h \in C^\infty(\overline\Omega)$ with

- h is holomorphic in a neighbourhood of x_0 in $\overline\Omega$,

- $\{x \in \overline\Omega \mid h(x) = 0\} = \{x_0\}$.

Set for $\nu \in \mathbb{N}$

$$f^\nu := \bar\partial\left(\frac{1}{h^\nu}\right).$$

Then $f^\nu \in C^\infty_{01}(\overline\Omega) \cap \ker \bar\partial$.

Since $\dim H^{01}_\infty(\Omega) < \infty$ there exist constants $c_1, c_2, \ldots, c_s$ not all vanishing and $g \in L^\infty(\Omega) \cap \operatorname{dom} \bar\partial$ with

$$\sum_{v=1}^{s} c_v f^v = \bar\partial g.$$

Thus

$$F := \sum_{v=1}^{s} \frac{c_v}{h^v} - g$$

is a holomorphic function on Ω with $\lim_{x \to x_0} |F(x)| = \infty$.

Theorem 6.6 (Solution of Levi's problem) *Let $\Omega \subset\subset X$ be a strictly pseudoconvex domain with C^∞ boundary in a complex manifold. Then for any boundary point $x_0 \in b\Omega$ there exists a holomorphic function F_{x_0} on Ω such that*

$$\lim_{x \to x_0} |F_{x_0}(x)| = \infty.$$

Corollary 6.7 *Let $\Omega \subset\subset X$ be as in the theorem. Then Ω is holomorphically convex. That means that for any compact $K \subset\subset \Omega$ the holomorphically convex hull*

$$\widehat{K} := \{x \in \Omega \mid |f(x)| \leq \sup_K |f| \text{ for all } f \in \mathcal{O}(\Omega)\}$$

is compact.

§7 Vanishing of Dolbeault Cohomology Groups

We now prove vanishing theorems of the Dolbeault cohomology for L^p forms on strictly pseudoconvex subdomains of Stein manifolds.

Lemma 7.1 *Let D_1 and D_2 be strictly pseudoconvex relatively compact domains in a complex manifold, $U_1 \supset \overline{D_1}$, $U_2 \supset \overline{D_2}$ be open sets and $\rho_1 : U_1 \to \mathbb{C}$, $\rho_2 : U_2 \to \mathbb{C}$ be smooth strictly plurisubharmonic defining functions for D_1, D_2 respectively, with*

$$d\rho_1 \wedge d\rho_2 \neq 0 \text{ where } \rho_1 = \rho_2 = 0.$$

Let V be a neighbourhood of the edge $bD_1 \cap bD_2$ of the boundary. Then there exists a strictly pseudoconvex domain Ω with smooth boundary such that $\Omega \subset D_1 \cap D_2$ and

$$\Omega \setminus V = (D_1 \cap D_2) \setminus V.$$

Moreover, if ρ_1 and ρ_2 are strictly convex (in $\mathbb{C}^n$), then Ω can be chosen to be strictly convex as well.

Proof   Let C be a convex open subset of the third quadrant $\{x < 0, y < 0\}$ of $\mathbb{R}^2$ and $\varepsilon > 0$ such that the boundary of C is a smooth curve which coincides with $\{(x, y) \in \mathbb{R}^2 \mid y = 0, x \leq -\varepsilon\}$ if $x \leq -\varepsilon$ and with $\{(x, y) \in \mathbb{R}^2 \mid x = 0, y \leq -\varepsilon\}$ if $y \leq -\varepsilon$.

bC is a graph over $\{(x, -x) \mid x \in \mathbb{R}\}$. Therefore there exists a smooth positive valued convex function $h : \mathbb{R} \to \mathbb{R}$ such that bC is given by

$$\left\{ (x, y) \in \mathbb{R}^2 \mid \gamma(x, y) := \frac{x + y}{\sqrt{2}} + h\left(\frac{-x + y}{\sqrt{2}}\right) = 0 \right\}$$

and $h(t) = t$ if $t \geq \varepsilon/\sqrt{2}$ and $h(t) = -t$ if $t \leq -\varepsilon/\sqrt{2}$. We have

$$C = \{(x, y) \in \mathbb{R}^2 \mid \gamma(x, y) < 0\}.$$

Let ε be so small that $|\rho_1(z)| + |\rho_2(z)| < \varepsilon$ if $z \in V$. Set for $z \in U_1 \cap U_2$

$$r(z) = \gamma(\rho_1(z), \rho_2(z))$$

and $\Omega = \{z \in U_1 \cap U_2 \mid r(z) < 0\}$.

Since $\Omega = \{z \in U_1 \cap U_2 \mid (\rho_1(z), \rho_2(z)) \in C\} \subset \{z \in U_1 \cap U_2 \mid \rho_1(z) < 0, \rho_2(z) < 0\}$ we have $\Omega \subset D_1 \cap D_2$.

For $z \in \Omega \setminus V$ we have $\rho_1(z) \leq -\varepsilon$ or $\rho_2(z) \leq -\varepsilon$. Then it is clear that $\Omega \setminus V = (D_1 \cap D_2) \setminus V$.

It is trivial that $d\gamma$ vanishes nowhere. Therefore

$$dr(z) = \frac{\partial \gamma}{\partial x}(\rho_1(z), \rho_2(z)) \cdot d\rho_1(z) + \frac{\partial \gamma}{\partial y}(\rho_1(z), \rho_2(z)) \cdot d\rho_2(z)$$

cannot vanish on $b\Omega$ if ε is small enough such that $d\rho_1(z) \wedge d\rho_2(z) \neq 0$ for $\max\{|\rho_1(z)|, |\rho_2(z)|\} \leq \varepsilon$.

Now we want to calculate the Levi form of r on the boundary of Ω. Modulo a rotation γ is given by

$$h(\xi) - \eta$$

with orthonormal coordinates (ξ, η). Therefore γ is a convex function and consequently

$$\begin{aligned}
L_r(z; t) &= \gamma_{xx}(\rho_1, \rho_2)|\langle\partial\rho_1, t\rangle|^2 + \gamma_{yy}(\rho_1, \rho_2)|\langle\partial\rho_2, t\rangle|^2 \\
&\quad + \gamma_{xy}(\rho_1, \rho_2)(2\operatorname{Re}\langle\partial\rho_1, t\rangle\overline{\langle\partial\rho_2, t\rangle}) \\
&\quad + \gamma_x(\rho_1, \rho_2)L_{\rho_1}(z; t) + \gamma_y(\rho_1, \rho_2)L_{\rho_2}(z; t) \\
&\geq \gamma_x(\rho_1, \rho_2)L_{\rho_1}(z; t) + \gamma_y(\rho_1, \rho_2)L_{\rho_2}(z; t).
\end{aligned}$$

Since $\gamma_x = \frac{1}{\sqrt{2}}(1 - h'(\frac{-x+y}{\sqrt{2}}))$, $\gamma_y = \frac{1}{\sqrt{2}}(1 + h'(\frac{-x+y}{\sqrt{2}}))$ and $|h'| \leq 1$ L_r is strictly positive definite. It is obvious that r is strictly convex if ρ_1 and ρ_2 have this property. $\square$

Let us now apply the 'bump method' of Grauert/Andreotti. Suppose Ω is a strictly pseu-doconvex domain with smooth boundary in a complex manifold X and $\rho : \widetilde{U} \to \mathbb{R}$ a C^∞ strictly plurisubharmonic defining function defined on a neighbourhood $\widetilde{U}$ of $b\Omega$.

We choose a finite open covering $U_1, U_2, \ldots, U_t$ of $b\Omega$ such that for $i = 1, 2, \ldots, t$

- $U_i \subset\subset \widetilde{U}$ and there exists a biholomorphic map $\Phi_i : N_i \to \widetilde{G}_i \subset\subset \mathbb{C}^n$ of an open neighbourhood N_i of $\overline{U}_i$ to an open set $\widetilde{G}_i$ of $\mathbb{C}^n$ such that $G_i = \Phi_i(U_i)$ is a strictly convex domain with smooth boundary;

- the boundaries of U_i and Ω intersect transversally, and $\Phi_i(U_i \cap \Omega)$ is the intersection of $\Phi_i(U_i)$ with a strictly convex domain C_i.

Choose C^∞ functions $\rho_i : \widetilde{U} \to \mathbb{R}$, $i = 1, \ldots, t$, with

- $\rho_i \geq 0$, $\operatorname{supp} \rho_i \subset\subset U_i$;

- $\sum_{i=1}^t \rho_i > 0$ on $b\Omega$.

Let $\varepsilon_1, \ldots, \varepsilon_t > 0$ be sufficiently small positive constants such that for $\varphi_i = \varepsilon_i \rho_i$ we have

- $h_r := \rho - \sum_{i=1}^t \varphi_i$, $r = 0, 1, \ldots, t$, is strictly plurisubharmonic on $\widetilde{U}_i$;

- $B_r := \{z \in \widetilde{U} \mid h_r(z) < 0\}$ is a strictly pseudoconvex domain, $dh_r \neq 0$ on bB_r, and bB_r intersects bU_{r+1} transversally ($U_{t+1} := \emptyset$), $r = 0, \ldots, t$.

It is easily seen that for ε_i small enough the above convexity properties of U_i, Φ_i, Ω re-main true for U_i, Φ_i, B_r due to the stability of convexity and transversality under small pertubations.

We have $B_0 = \Omega$, $\Omega \subset\subset B_t$ and

$$B_r \setminus B_{r-1} \subset\subset U_r \text{ for } r = 1, \ldots, t.$$

Lemma 7.2 *Let $r \in \{1, 2, \ldots, t\}$, $q \geq 1$, $1 \leq p \leq \infty$ and let the form $f \in L^p_{0q}(B_{r-1})$ be $\bar\partial$-closed. Then there exist a $\bar\partial$-closed form $f' \in L^p_{0q}(B_r)$ and a form $g \in L^p_{0,q-1}(B_{r-1})$ with*

$$f = f' + \bar\partial g$$

on B_{r-1}.

The lemma asserts that *the restriction from B_r to B_{r-1} maps the Dolbeault cohomology for L^p-$(0, q)$-forms on B_r surjectively onto that on B_{r-1}.*

Proof of 7.2 Let the neighbourhood V of $bU_r \cap bB_{r-1}$ be sufficiently small such that $\operatorname{supp} \rho_r \cap V = \emptyset$. We apply Lemma 7.1. Let D be a strictly pseudoconvex domain with smooth boundary such that

$$D \subset B_{r-1} \cap U_r \text{ and } D \setminus V = (B_{r-1} \cap U_r) \setminus V$$

such that D is – see Lemma 7.1 – biholomorphically equivalent to a strictly convex domain $\tilde{D}$ with smooth boundary.

Obviously we have

$$\overline{bD} \cap \overline{B_{r-1}} \cap \operatorname{supp} \rho_r = \emptyset.$$

Let $\psi \in C_c^\infty(X)$ with $\operatorname{supp} \psi \subset\subset U_r$, $\psi \equiv 1$ in a neighbourhood of $\operatorname{supp} \rho_r$ and

$$\operatorname{supp} \psi \cap \overline{bD} \cap \overline{B_{r-1}} = \emptyset.$$

Now $f \in L_{0q}^p(D)$. From Theorem III.6.16 it follows that there exists $\tilde{g} \in L_{0,q-1}^p(D)$ with

$$\overline{\partial}\tilde{g} = f \text{ on } D.$$

We set $g := \psi\tilde{g}$ on D and $g \equiv 0$ on $B_{r-1} \setminus D$. Then $g \in L_{0,q-1}^p(B_{r-1}) \cap \operatorname{dom} \overline{\partial}$.

Moreover if we set

$$f' := \begin{cases} f - \overline{\partial}g & \text{on } B_{r-1}, \\ 0 & \text{on } B_r \setminus B_{r-1} \end{cases}$$

we obtain $f' \in L_{0,q}^p(B_r) \cap \operatorname{dom} \overline{\partial}$ and $\overline{\partial}f' = 0$. $\qquad\square$

Remark 7.3   In 7.5.4 we want to apply the same method to two slightly different situations.

a) *If $f \in C_{0q}^\infty(B_{r-1})$ then f restricts to a form in $C_{0q}^\infty(D)$ which we also denote by f (D is chosen as above, in particular biholomorphically equivalent to a strictly convex domain $\tilde{D}$ in $\mathbb{C}^n$). This form gives rise to a form $\tilde{f} \in C_{0q}^\infty(\tilde{D})$, $\overline{\partial}\tilde{f} = 0$. Our methods in Chapter III allow to solve $\overline{\partial}\tilde{u} = \tilde{f}$ on convex subdomains of $\tilde{D}$; classical exhaustion methods of $\tilde{D}$ combined with the polynomial convexity of $\tilde{D}$ then yield a solution $\tilde{u} \in C_{0q-1}^\infty(\tilde{D})$ of $\overline{\partial}\tilde{u} = \tilde{f}$ and so a solution $u \in C_{0q-1}^\infty(D)$ of $\overline{\partial}u = f$.*

b) *Similarly, if $f \in C_{0q}^\infty(\overline{B_{r-1}})$, its restriction f to $\overline{D}$ is in $C_{0q}^\infty(\overline{D})$. A famous result of J. J. Kohn's valid in a much more general situation gives a solution $u \in C_{0q-1}^\infty(\overline{D})$ with $\overline{\partial}u = f$.* In our special situation Kohn's result is easily obtained by integral formula methods (see Lieb/Range [LiR 80], Michel [Mic 91], Chaumat/Chollet [ChC 88]). We skip the details.

Corollary 7.4 *Let $\Omega \subset\subset X$ be a strictly pseudoconvex domain with smooth boundary in a complex manifold. Let ρ be a strictly plurisubharmonic C^∞ defining function for Ω. Then we have for $\Omega_\varepsilon := \{z \mid \rho(z) < \varepsilon\}$ if $\varepsilon > 0$ is sufficiently small:*

Let $f \in L_{0q}^p(\Omega)$, $q \geq 1$, $1 \leq p \leq \infty$, be a $\overline{\partial}$-closed form. Then there exist forms $f' \in L_{0q}^p(\Omega_\varepsilon) \cap \ker \overline{\partial}$ and $g \in L_{0,q-1}^p(\Omega) \cap \operatorname{dom} \overline{\partial}$ with

$$f = f' + \overline{\partial}g \quad \text{on } \Omega.$$

Proof   With the notations of Lemma 7.2 we have $\Omega \subset\subset B_t \subset\subset \tilde{U}$. So if $\varepsilon > 0$ is sufficiently small such that $\Omega_\varepsilon \subset B_t$ the assertion follows. $\qquad\square$

Remark 7.5   1) *If* $0 < \delta_0 < \varepsilon/2$ *is sufficiently small then* the construction can be repeated for Ω_{δ_0} instead of Ω and *we obtain the assertion of the last lemma also for the couples* $(\Omega_{\delta_0}, \Omega_{\varepsilon-\delta_0})$. The following remark of course remains also valid for these couples.

2) By using the special case of Theorem III.6.16, where all the forms are C^∞ in the interior *we obtain the same corollary for* $f \in L_{0q}^p(\Omega_\varepsilon) \cap C_{0q}^\infty(\Omega_\varepsilon)$ *with*

$$f' \in L_{0q}^p(\Omega_\varepsilon) \cap C_{0q}^\infty(\Omega_\varepsilon), \; g \in L_{0q-1}^p(\Omega) \cap C_{0q-1}^\infty(\Omega).$$

3) As in $\mathbb{C}^n$ it follows from the basic integral formulae that *any L^p cohomology class can be represented by C^∞ forms which are smooth only in the interior.* But now *it follows from remark 2 that they can be represented by forms which are C^∞ up to the boundary.*

4) *If, in the proof of 7.2, the form f is only smooth and $\bar\partial$-closed on Ω* (no L^p-bounds required), *then the result of remark 3 is still true, with f' and g smooth but not L^p-bounded.* In fact, we can solve $f = \bar\partial\tilde{g}$ with $\tilde{g} \in C_{0q}^\infty(D)$, so the same procedure works. In the same way we can get the corresponding result for forms smooth up to the boundary (now using 7.3.b).

We need an auxiliary lemma.

Lemma 7.6 *Let $\Omega \subset\subset X$ be a domain in a hermitian manifold and $D \subset\subset \mathbb{C}$ be a disc with center w_0 and radius τ.*

Let $\Gamma = \{(z,w) \mid z \in \Omega, w \in D\}$, $f \in L_{0q}^p(\Omega)$, $g \in L_{0,q-1}^p(\Gamma)$, $q \geq 1$, $1 \leq p \leq \infty$, such that there exists a holomorphic polynomial of one variable $p(w)$ with

$$p(w)f(z) = \bar\partial_z g(z,w)$$

on Γ. We suppose that g does not contain $d\bar{w}$ terms.

Then there exists $G \in L_{0,q-1}^p(\Omega)$ with

$$p(w_0)f = \bar\partial_z G$$

on Ω.

Proof   Set
$$G(z) = \frac{i}{2\pi\tau^2} \int\limits_D g(z,w)\, dw \wedge d\bar{w}.$$

It follows that G exists for almost all $z \in \Omega$ with $G \in L_{0,q-1}^p(\Omega)$. Now let $\varphi_1 \in C_{n,n-q}^\infty(\Omega)$, $\varphi_2 \in C_{1,1}^\infty(D)$ be forms with compact support. Then

$$\int\limits_\Omega f(z) \wedge \varphi_1(z) \wedge \int\limits_D p(w)\varphi_2(w) = \int\limits_\Omega \bar\partial_z\varphi_1(z) \wedge \int\limits_D g(z,w) \wedge \varphi_2(w).$$

Since the forms are integrable on Γ the above equation also holds for $\varphi_2(w) = dw \wedge d\overline{w}$. Since $p(w)$ is holomorphic the mean value property implies

$$p(w_0) \int_\Omega f(z) \wedge \varphi_1(z) = \int_\Omega \overline{\partial}_z \varphi_1(z) \wedge G(z)$$

for all $\varphi_1 \in C_{n,n-q}^\infty(\Omega)$ with compact support. This proves the assertion of the lemma. $\square$

Theorem 7.7 *Let $\Omega \subset\subset X$ be a strictly pseudoconvex domain with C^∞ boundary in a Stein manifold. Then for any $\overline{\partial}$-closed form $f \in L_{0q}^p(\Omega)$, $q \geq 1$, $1 \leq p \leq \infty$, there exists a $g \in L_{0,q-1}^p(\Omega) \cap \operatorname{dom} \overline{\partial}$ with*

$$\overline{\partial} g = f.$$

Proof   Let $\rho : U \to \mathbb{R}$ be a strictly plurisubharmonic defining function on a neighbourhood U of $\overline{\Omega}$. Let $\varepsilon > 0$ be so small that Corollary 7.4 applies.

Set $\widetilde{X} := X \times \mathbb{C} = \{(z,w) \mid z \in X, w \in \mathbb{C}\}$, $\widetilde{U} := U \times \mathbb{C}$ and

$$\widetilde{\Omega}_\varepsilon := \{(z,w) \in \widetilde{U} \mid \rho(z) + |w|^2 < \varepsilon\}.$$

Then $\widetilde{X}$ is Stein and $\widetilde{\Omega}_\varepsilon$ is a strictly pseudoconvex domain in it with smooth boundary.

Let $\pi : \widetilde{X} \to X$ be the holomorphic projection on the z component.

Let $f \in L_{0q}^p(\Omega)$ be as in the theorem and f', g as in Corollary 7.4. We have that $\pi^* f'$ is $\overline{\partial}$-closed on $\widetilde{\Omega}_\varepsilon$ and $\pi^* f' \in L_{0q}^p(\widetilde{\Omega}_\varepsilon)$.

Since the cohomology groups of strictly pseudoconvex domains are finite dimensional for $q > 0$ there exists a polynomial $p(w) \neq 0$ and $\widetilde{g} \in L_{0,q-1}^p(\widetilde{\Omega}_\varepsilon)$ such that on $\widetilde{\Omega}_\varepsilon$

$$p(w)\pi^* f' = \overline{\partial}_{z,w} \widetilde{g}.$$

Now we decompose $\widetilde{g}$ with $\widetilde{g}_1 \in L_{0,q-1}^p(\widetilde{\Omega}_\varepsilon)$, $\widetilde{g}_2 \in L_{0,q-2}^p(\widetilde{\Omega}_\varepsilon)$ such that

$$\widetilde{g} = \widetilde{g}_1 + d\overline{w} \wedge \widetilde{g}_2,$$

where $\widetilde{g}_1, \widetilde{g}_2$ do not contain the factor $d\overline{w}$.

Since $\pi^* f'$ does not contain $d\overline{w}$ either we obtain

$$p(w)\pi^* f' = \overline{\partial}_z \widetilde{g}_1.$$

Now let $w_0 \in \mathbb{C}$ with $|w_0|^2 < \varepsilon$ and $p(w_0) \neq 0$. Let $\delta, \tau > 0$ be so small that for $D_\tau := \{w \mid |w - w_0| < \tau\}$

$$\Gamma := \Omega_\delta \times D_\tau \subset \widetilde{\Omega}_\varepsilon.$$

By Lemma 7.6 there exists $G \in L_{0,q-1}^p(\Omega_\delta)$ with

$$p(w_0)f = \overline{\partial}G$$

on Ω_δ.

Now $\Omega_\delta \supset\supset \Omega$ implies that there exists $h \in L^p_{0,q-1}(\Omega)$ with

$$f = \overline{\partial} h$$

on Ω in the distributional sense. $\square$

Remark 7.8 In the proof we did not pass to forms of class C^∞ in order to show that the method would also work in more general settings.

To sum up let $H^{pq}_s(\Omega)$ denote the Dolbeault cohomology group for (p, q)-forms on Ω with L^s coefficients, $1 \le s \le \infty$. For $p = 0$ we have just proved

Theorem 7.9 (Vanishing theorem) *Let $\Omega \subset\subset X$ be a strictly pseudoconvex domain with C^∞ boundary in a Stein manifold. Then for $q \ge 1$ and $1 \le s \le \infty$ the Dolbeault cohomology groups for Ω vanish:*

$$H^{0q}_s(\Omega) = 0.$$

This vanishing theorem also holds for $p > 0$, but since $H^{pq}_s(\Omega)$ is no longer a direct sum of $H^{0q}_s(\Omega)$ as in $\mathbb{C}^n$, these spaces have to be treated in the framework of forms with coefficients in a vector bundle. In order to not overload this book we do not discuss them.

§8 Notes

For the first four paragraphs compare [Rha 60]. Theorem 3.1 is a weaker result than de Rham's, and so are its consequences in §3. We have resigned ourselves to our version for several reasons: 1° it can be proved by elementary calculations, 2° it suffices for the applications to Hodge theory (provided one uses some deep additional estimates for isotropic operators – essentially Lemma 4.3), and 3° de Rham's stronger version of Theorem 3.1 will be lost, anyway, when applied to the complex Laplacian, as we have to do in §5.

General isotropic kernels as defined in §2 are inevitable in the case of Riemannian manifolds; in the complex case which is dealt with in the whole book except the above paragraphs, more special definitions suffice: see the remarks preceding 5.9. The main results of Hodge theory in Chapter IV are taken from [Rha 60] whose proof we essentially follow; they are of course older, and we do not go into their history.

The relation between Hermitian and Riemannian geometry is explained at many places – e.g. in [Wei 58]. Most of the content of §5 is taken from [LiR 83]. Proposition 5.19 establishes the – crucial – connection between the parametrix representation and Cauchy-Fantappiè forms leading to the BMK formula on hermitian manifolds due to Lieb/Range.

The idea of passing to local counterparts of the global kernels in order to apply Koppelman's homotopy formula is again in [LiR 83]. Admissible kernels similar to our Definition 5.49 were introduced and studied by Folland/Stein [FoS 74], Kerzman/Stein [KeS 78] and Lieb/Range [LiR 83]. Our definition slightly modifies that in [LiR 83]. The difficult case of pseudoconcave manifolds which we have not studied, requires a different concept of admissible kernel which was introduced and studied by Michel [Mic 92], see also [HefT 99].

The homotopy formula

$$f = Pf + T\overline{\partial}f + \overline{\partial}Sf \tag{8.1}$$

is one of the main results of our book – and one of the main tools.

Levi's problem, the characterisation of holomorphically convex domains by local conditions on their boundary, arose out of the fundamental work of F. Hartogs and E. E. Levi 1906–1911 [Har 06] [Lev 11]. It contributed immensely to developing complex analysis until it was solved by Oka et al. (1943–1953) [Oka 84]. Variants and generalisations of the problem were later solved, first by Grauert [Gra 58], then by many other mathematicians; for the history of the problem see again [Ran 86].

The solution of Levi's problem based on a finiteness argument for cohomology spaces (Theorem 6.6) is due to Grauert [Gra 58]. The L^p-cohomology finiteness theorem 6.5 is a direct consequence of the homotopy formula (8.1). It was proved for $p = 2$ by Kohn [Koh 63, Koh 64] and Hörmander [Hör 65]; for domains in $\mathbb{C}^n$ (where the cohomology dimension is 0) by Lieb [Lie 70] in case $p = \infty$, by Kerzman [Ker 71] and Øvrelid [Øvr 71$_1$] in all cases. The case of strictly pseudoconvex manifolds goes back to Henkin [Hen 77]. The finiteness of cohomology without bounds was proved by Grauert [Gra 58] who also considers cohomology with values in arbitrary coherent analytic sheaves. Note that the arguments in §6 run completely parallel to the corresponding arguments in $\mathbb{C}^n$ as expanded in III,§6: the homotopy formula (8.1) together with some estimates on the operators involved allows application of elementary functional analysis. On the other hand, the vanishing theorem of §7 is more difficult than its $\mathbb{C}^n$-counterpart. We exploit the fact that Ω is a sublevel set of a globally defined strictly plurisubharmonic function; this allows to apply a variant of Oka's method of passing to higher dimensions [Oka 84]. Our method seems to be new in this context.

The "bump method" which we use was invented by Grauert [Gra 58], it is one of the most powerful methods of passing from local to global information and has been widely used ever since – see [AnG 62], [Hör 65], [Ker 71], as examples.

Chapter V

The $\overline{\partial}$-Neumann Problem

The homotopy formula which we have so far used allows to solve the Cauchy-Riemann equations with L^2 (or even L^p) estimates. This suggests a study of the minimal solution of the equation, i. e. the solution with minimal L^2-norm. To do this we are led to include in our investigation the Hilbert space adjoint $\overline{\partial}^*$ of $\overline{\partial}$ and the corresponding Laplacian (again denoted by $\square$), and formulate the corresponding boundary value problem, the $\overline{\partial}$-*Neumann problem*. The set-up of the problem and a first, still fairly superficial, solution on strictly pseudoconvex manifolds is given in §6, again as an application of our basic homotopy formula. (We only prove an existence result; the important regularity results, which depend on much more sophisticated methods, will be treated in the following chapters.) The set-up needs a density theorem due to Friedrichs which we prove in §2. The much deeper density theorem of §5 due to Hörmander will only be needed in the following chapters.

It should be possible to read just §6 of this chapter together with the statement of Theorem 2.6 and 5.6 in order to understand the following chapters.

§1 Operators on Hilbert Spaces

Here we give some elementary facts on Hilbert space theory. For more details see [Rud 73]. Let $(H, (\cdot, \cdot))$ be a complex Hilbert space where the inner product (f, g) is conjugate linear in the second variable. We denote the norm of f by $\|f\|$, so $\|f\| = (f, f)^{1/2}$. Let $(H^*, \|\cdot\|_*)$ be the dual Hilbert space of H. That means that

$$H^* := \{\lambda : H \longrightarrow \mathbb{C} \mid \lambda \text{ linear, continuous}\}$$

and

$$\|\lambda\|_* := \sup_{\substack{f \in H \\ \|f\| \leq 1}} |\lambda(f)|.$$

If $f \in H$ then $(\cdot, f) \in H^*$ and the map $\varkappa : H \to H^*$, $\varkappa(f) := (\cdot, f)$ is a conjugate linear isomorphism of H onto H^*. Moreover $\varkappa$ is isometric:

$$\|\varkappa(f)\|_* = \|f\| \quad \text{for } f \in H.$$

$\varkappa$ defines an inner product on H^* in a canonical way by

$$(\lambda, \mu)_* := (\varkappa^{-1}(\mu), \varkappa^{-1}(\lambda)), \quad \lambda, \mu \in H^*$$

and one has

$$\|\lambda\|_* = (\lambda, \lambda)_*^{1/2}.$$

If $\lambda \in H^*$, $\varkappa^{-1}(\lambda)$ is the uniquely defined $f \in H$ which is orthogonal to $\ker(\lambda)$ with $\lambda(f) = (f, f)$.

For Hilbert spaces $(H_1, (\cdot, \cdot)_1)$, $(H_2, (\cdot, \cdot)_2)$ $H_1 \times H_2$ will denote the product Hilbert space equipped with the inner product

$$((f_1, f_2), (g_1, g_2)) = (f_1, g_1)_1 + (f_2, g_2)_2$$

for $(f_1, f_2), (g_1, g_2) \in H_1 \times H_2$.

Let $V \subset H_1$ be a linear subspace and $A : V \to H_2$ be a linear map. The *graph* of A is defined by

$$\text{graph}(A) := \{(f, Af) \in H_1 \times H_2 \mid f \in V\}.$$

Thus graph (A) is a linear subspace of $H_1 \times H_2$. The most interesting case is when V is a dense subspace of H_1. Then A will be called *densely defined*. The closed graph theorem asserts for $V = H_1$ that A is continuous if and only if graph (A) is closed in $H_1 \times H_2$. This gives rise to the following definition. $A : V \to H_2$ is called a *closed* map if graph (A) is a closed linear subspace of $H_1 \times H_2$. Closed densely defined linear maps have many properties in common with continuous maps.

We shall apply these considerations to a *partial differential operator P* (PDO). A priori a PDO is an operator acting on sufficiently regular differential forms. By defining appropriate Hilbert spaces containing these forms one can then extend P, normally in many ways, to a closed densely defined operator on these spaces. We shall soon consider examples.

Let $\overline{\text{graph}(A)}$ be the closure of graph (A) in $H_1 \times H_2$. Then $\overline{\text{graph}(A)}$ is the graph of some linear extension of A, $\widetilde{A} : \widetilde{V} \to H_2$, $\widetilde{V} \supset V$, if and only if

$$\{0\} \times H_2 \cap \overline{\text{graph}(A)} = \{(0, 0)\}.$$

Or equivalently

$$\text{For } h \in H_2 : \quad (0, h) \in \overline{\text{graph}(A)} \Rightarrow h = 0.$$

If $\widetilde{A}$ exists, it is the closed extension of A with the minimal domain of definition.

The more interesting problem of finding a *maximal* closed extension is nontrivial in general. We shall discuss this in more detail in the case of a partial differential operator (PDO) P. Here we only make a few remarks.

In the case of a PDO one obtains easily a maximal extension by passing to distributions. But for practical reasons it is more convenient to work with smooth forms which are naturally related to the minimal extension. These notions will be explained later on. The equivalence of these two approaches is furnished in important cases by the so-called *Friedrichs extension lemma* which we shall prove. Let us first derive some general preliminary notions. Let $A : H_1 \to H_2$ be a linear continuous map between Hilbert spaces and $\varkappa_i : H_i \to H_i^*$ be the related conjugate linear isomorphisms for $i = 1, 2$. Set, for $g \in H_2$,

$$A^* g := \varkappa_1^{-1}(\varkappa_2(g) \circ A).$$

Then $A^* : H_2 \to H_1$ is a continuous linear map and it is the only one with

$$(Af, g)_2 = (f, A^* g)_1$$

for all $f \in H_1$, $g \in H_2$. A^* is called the *Hilbert space adjoint* of A.

Since A is continuous

$$\|A\| := \sup_{\|f\|_1 \leq 1} \|Af\|_2 < \infty.$$

It follows from the definition that

$$\|A^*\| = \|A\| = \sup_{\|g\|_2 \leq 1} \|A^* g\|_1$$

and

$$A^{**} = (A^*)^* = A.$$

The basic observation for the construction of A^* has been that $f \mapsto (Af, g)$ defines for any fixed $g \in H_2$ a continuous linear functional on H_1.

We shall now generalise this to linear maps $A : V \to H_2$, where V is a dense linear subspace of H_1. We set

$$\mathrm{dom}\,(A^*) := \{g \in H_2 \mid \exists C = C_g \geq 0 \;\forall f \in V : |(Af, g)_2| \leq C\|f\|_1\}.$$

Let $g \in \mathrm{dom}\,(A^*)$. Because V is dense the functional $V \ni f \mapsto (Af, g)_2$ can be continued to a uniquely defined continuous linear functional $(\cdot, h)_1 \in H_1^*$, with $h \in H_1$. Set

$$A^* g := h.$$

Therefore $A^* : \mathrm{dom}\,(A^*) \to H_1$ is the uniquely defined map with

$$(Af, g)_2 = (f, A^* g)_1$$

for all $f \in V = \mathrm{dom}\,(A)$, $g \in \mathrm{dom}\,(A^*)$. A^* is called the *adjoint operator* of A. It has the following additional properties.

   i. *A^* is a closed linear map;*

   ii. *If A is closed and densely defined then A^* is also closed and densely defined. In this case one has*

$$A^{**} = (A^*)^* = A.$$

   iii. *If A is closed and densely defined then*

$$\overline{\mathcal{R}(A^*)} = \ker(A)^{\perp}, \quad \overline{\mathcal{R}(A)} = \ker(A^*)^{\perp}.$$

(Here $\mathcal{R}$ denotes the range and $X^{\perp}$ the orthogonal complement of a subspace X.)

Let $H_1 = H_2$. If $\mathrm{dom}\,(A) = \mathrm{dom}\,(A^*)$ and $A = A^*$ then A is called a *self-adjoint operator*. In this case one has

$$(Af, g) = (f, Ag) \quad \text{for } f, g \in \mathrm{dom}\,(A),$$

and there exists no proper self-adjoint extension of A.

§2   Hilbert Spaces of Differential Forms

For the applications we have in mind we need to consider n-dimensional complex manifolds X. We suppose that on X we are given a C^∞-smooth hermitian metric ds^2 and recall
the relevant notions from IV.1 and IV.5.

ds^2 induces a volume element dV and in any point of X a Hodge $*$-operator acting on
(p, q)-forms

$$* : E_{p,q} \longrightarrow E_{n-q,n-p}.$$

As in the Euclidean case one has

$$dV = *1. \tag{2.1}$$

Moreover, let f, g be (p, q) differential forms on X. Then $f \wedge *\bar{g}$ is a (n, n)-form which is
a multiple of dV

$$f \wedge *\bar{g} = \langle f, g \rangle \, dV. \tag{2.2}$$

Here $\langle f, g \rangle$ denotes the pointwise inner product on (p, q)-forms which is induced by the
metric. We say that a form f is L^2 integrable on a relatively compact domain Ω of X, if
$\langle f, f \rangle$ is integrable with respect to dV. We denote the space of L^2 integrable (p, q)-forms
on Ω by $L^2_{p,q}(\Omega)$. The inner product

$$(f, g)_{L^2(\Omega)} := (f, g) := \int_\Omega f \wedge *\bar{g}$$

turns $L^2_{p,q}(\Omega)$ into a Hilbert space with norm $\|f\|_{L^2(\Omega)} = (f, f)^{\frac{1}{2}}_{L^2(\Omega)}$

$L^2_{p,q}(\Omega)$ as a topological vector space does not depend on the metric because of $\Omega \subset\subset X$
and the smoothness of ds^2. C^∞-smooth forms with support in Ω are dense in $L^2_{p,q}(\Omega)$. Let
P be a linear partial differential operator on X or $\overline{\Omega}$ with smooth coefficients. In the sequel
we have to take into account the following two facts.

Fact 2.3 *P can be interpreted as a densely defined closed operator* $P : V \to L^2_{p',q'}(\Omega)$,
where $V = C^\infty_{p,q}(\overline{\Omega}) \subset L^2_{p,q}(\Omega)$.

Then its Hilbert space adjoint P^* coincides, *on its domain of definition*, with a well defined
partial differential operator $\widetilde{P}$ on X given by integration by parts against forms of compact support – see (2.5). But whereas $\widetilde{P}$ is defined on the space of all (p', q')-forms with
distribution coefficients, sending such a form into a corresponding (p, q)-form, *the domain
of definition of P^* is*, as we shall see later, *singled out by a boundary condition* and does
not even necessarily comprise all of $C^\infty_{p',q'}(\overline{\Omega})$. It is precisely the space of those forms
$g \in L^2_{p',q'}(\Omega)$ such that $\widetilde{P}g \in L^2_{p,q}(\Omega)$ and the equality

$$(Pf, g)_{L^2(\Omega)} = (f, \widetilde{P}g)_{L^2(\Omega)} \tag{$*$}$$

holds for all f in V. Whenever we write P^* instead of $\widetilde{P}$ we thereby indicate that its
argument g is taken from the domain of P^* (to be more explicitly described in the next

paragraph). The operator $\widetilde{P}$ is called *the formal adjoint* of P; obviously $\widetilde{\widetilde{P}} = P$. Note that if $g \in L^2_{p',q'}(\Omega)$ and $\widetilde{P}g \in L^2_{p,q}(\Omega)$ this does not mean $g \in \operatorname{dom} P^*$, i.e.(∗) need not be satisfied. Also note that P^* depends on the choice of V (we have chosen $V = C^\infty_{0q}(\overline{\Omega})$ here) – although $\widetilde{P}$ remains unchanged as long as V contains sufficiently many forms (e. g. if $V = C^\infty_{0q,c}(\Omega)$).

Fact 2.4 $L^2_{p,q}(\Omega)$ *is over the reals locally the direct sum of finitely many* $L^2(\Omega, \mathbb{R})$. *Therefore locally we can look at P as a system of differential operators acting on vector-valued real functions.*

P is then a matrix of real partial differential operators. Let $P : C^\infty_{p,q}(X) \to C^\infty_{p',q'}(X)$ be a first order PDO with smooth coefficients. Integration by parts gives as its formal adjoint a first order PDO with smooth coefficients $\widetilde{P} : C^\infty_{p',q'}(X) \to C^\infty_{p,q}(X)$ and

$$(Pf, g)_X = (f, \widetilde{P}g)_X \qquad (2.5)$$

for $f \in C^\infty_{p,q}(X)$, $g \in C^\infty_{p',q'}(X)$ if one of the forms f, g has compact support in X. Let $\Omega \subset\subset X$ be a domain with smooth boundary and $\operatorname{dom}_\Omega(P_{\min}) = \operatorname{dom}(P_{\min})$ be the set of $f \in L^2_{p,q}(\Omega)$ such that there exists a sequence $f_n \in C^\infty_{p,q}(\overline{\Omega})$ and $g \in L^2_{p',q'}(\Omega)$, with $\lim_{n \to \infty} f_n = f$ in $L^2_{p,q}(\Omega)$ and $\lim_{n \to \infty} Pf_n = g$ in $L^2_{p',q'}(\Omega)$. Set $P_{\min}(f) := g$ for $f \in \operatorname{dom}(P_{\min})$.

We denote by $C^\infty_{p,q,c}(\Omega)$ the forms in $C^\infty_{p,q}(X)$ with compact support in Ω. Let

$$h \in C^\infty_{p',q',c}(\Omega) \,.$$

Then

$$(g, h) = \lim_{n \to \infty} (Pf_n, h) = \lim_{n \to \infty} (f_n, \widetilde{P}h) = (f, \widetilde{P}h).$$

This shows that $P_{\min}(f)$ is uniquely defined. Set

$$
\begin{aligned}
\operatorname{dom}_\Omega(P_{\max}) \ &= \ \operatorname{dom}(P_{\max}) \\
&= \ \{f \in L^2_{p,q}(\Omega) \mid Pf \in L^2_{p',q'}(\Omega) \text{ in the distributional sense}\}
\end{aligned}
$$

and

$$P_{\max}(f) = Pf \text{ for } f \in \operatorname{dom}(P_{\max}).$$

Very often we shall denote $P_{\max}$ simply by P. Clearly $P_{\max}$ is an extension of $P_{\min}$.

Theorem 2.6 (Friedrichs' extension lemma) *Let $\Omega \subset\subset X$ be a domain with smooth boundary. Then for any $f \in \operatorname{dom}(P_{\max})$ there exists a sequence $f_n \in C^\infty_{p,q}(\overline{\Omega})$ such that $\lim_{n \to \infty} f_n = f$ and $\lim_{n \to \infty} Pf_n = P_{\max}(f)$ with respect to L^2-norms. Shortly this means that*

$$P_{\min} = P_{\max}.$$

Remark A very slight modification of the following proof shows that it is even simpler to obtain the result: *Let $f \in L^1_{p,q}(\Omega)$ such that $Pf \in L^1_{p',q'}(\Omega)$ (in the distributional sense). Then there exists a sequence $f_n \in C^\infty_{p,q}(\overline{\Omega})$ such that $\lim_{n\to\infty} f_n = f$ and $\lim_{n\to\infty} Pf_n = Pf$ with respect to L^1-norms.*

We leave this verification to the reader.

Proof Let $f \in \mathrm{dom}\,(P_{\max})$. If $\varphi \in C^\infty(\overline{\Omega})$ then $\varphi f \in \mathrm{dom}\,(P_{\max})$. Applying a partition of unity argument we can assume that $\Omega \subset\subset X \subset \mathbb{C}^n$, where X is equipped with a hermitian metric, and either

> Case i) $\mathrm{supp}(f) \subset\subset \Omega$, or
>
> Case ii) $\mathrm{supp}(f) \subset\subset U \cap \overline{\Omega}$, where U is an arbitrarily small neighbourhood of a boundary point.

Let f, resp. Pf, locally be given with respect to some base of constant (p,q)-forms resp. (p',q')-forms. By passing to real and imaginary parts P can be interpreted as a matrix of first order differential operators $\mathcal{P} = P_{ji}$ acting on real vector valued functions.

$$\mathcal{P}f = \sum_{s=1}^{A}\sum_{k=1}^{N} \alpha_{sk}\frac{\partial f_s}{\partial x_k} + \sum_{s=1}^{A} \beta_s f_s, \tag{2.7}$$

where α_{sk}, β_s are smooth functions on X and $x_k = \mathrm{Re}\, z_k$, $x_{k+n} = \mathrm{Im}\, z_k$, $k \le n$, $N = 2n$. The fact that N is even is of no importance in the proof. $f \in \mathrm{dom}\,(P_{\max})$ if and only if $P_{ji}f \in L^2(\Omega)$ in the distributional sense for all P_{ji}. So it suffices to consider operators of type $\mathcal{P}$. We shall soon investigate this matrix representation in more detail.

Let $\varphi \in C^\infty_c(\mathbb{C}^n)$, with $\varphi \ge 0$, $\int \varphi\, dx = 1$, where dx is the Euclidean volume element. Set $\varphi_\varepsilon(z) = \varepsilon^{-2n}\varphi\left(\frac{z}{\varepsilon}\right)$.

Case i) Set $f_\varepsilon := f * \varphi_\varepsilon$. Then $f_\varepsilon \in C^\infty_{p,q,c}(\Omega)$ and $f_\varepsilon \to f$ in L^2 for $\varepsilon \to \pm 0$. We show that $\mathcal{P}f_\varepsilon \to \mathcal{P}f$ in $L^2(\mathbb{R}^N)$. We have with $D_k = \frac{\partial}{\partial x_k}$

$$\alpha_{sk}D_k(f_s * \varphi_\varepsilon) - (\alpha_{sk}D_k f_s) * \varphi_\varepsilon = \alpha_{sk}(f_s * D_k\varphi_\varepsilon) - (\alpha_{sk}f_s) * D_k\varphi_\varepsilon + ((D_k\alpha_{sk})f_s) * \varphi_\varepsilon.$$

Here we have used that in the distributional sense

$$D_k(f_s * \varphi_\varepsilon) = (D_k f_s) * \varphi_\varepsilon = f_s * D_k\varphi_\varepsilon.$$

Because of this decomposition we set

$$A = \alpha_{sk}(y)(f_s * D_k\varphi_\varepsilon)(y) - ((\alpha_{sk}f_s) * D_k\varphi_\varepsilon)(y).$$

Then

$$A = \frac{1}{\varepsilon}\int [\alpha_{sk}(y)f_s(x)(D_k\varphi)_\varepsilon(y-x) - \alpha_{sk}(x)f_s(x)(D_k\varphi)_\varepsilon(y-x)]\, dx$$

$$= \int \left(\frac{\alpha_{sk}(y) - \alpha_{sk}(x)}{|x-y|}\right) f_s(x)\frac{|x-y|}{\varepsilon}\varepsilon^{-2n}(D_k\varphi)\left(\frac{y-x}{\varepsilon}\right)\, dx.$$

Therefore

$$|A| \lesssim \varepsilon^{-2n} \int |f_s(x)| \left| (D_k \varphi) \left(\frac{y-x}{\varepsilon} \right) \right| \left| \frac{y-x}{\varepsilon} \right| dx.$$

Now

$$M = \int |(D_k \varphi)(x)| |x| \, dx < \infty.$$

Therefore the measure

$$d\mu(x) = |D_k \varphi(x)| |x| \, dx$$

has finite mass M. We set

$$d\mu_\varepsilon(x) = \frac{1}{\varepsilon^{2n}} d\mu \left(\frac{x}{\varepsilon} \right).$$

Thus the mass of μ_ε is M and

$$|A| \lesssim |f_s| * \mu_\varepsilon.$$

We show

$$\| |f_s| * \mu_\varepsilon \|_{L^2} \lesssim \|f_s\|_{L^2} \tag{2.8}$$

with a constant independent of ε.

$$\|f_s\|_{L^2}^2 \gtrsim \int_x \int_y |f_s(y)|^2 \, dy \, d\mu_\varepsilon(x)$$

$$= \int_y \left(\int_x |f_s(y-x)|^2 \, d\mu_\varepsilon(x) \right) dy$$

$$\gtrsim \int_y \left(\int_x |f_s(y-x)| \, d\mu_\varepsilon(x) \right)^2 dy = \| |f_s| * \mu_\varepsilon \|_{L^2}^2,$$

because of the Schwarz inequality with respect to $d\mu_\varepsilon$. Now

$$\|(D_k \alpha_{sk}) f_s * \varphi_\varepsilon\|_{L^2} \lesssim \|f_s\|_{L^2}. \tag{2.9}$$

Therefore (2.8) and (2.9) imply

$$\|\alpha_{sk} D_k (f_s * \varphi_\varepsilon) - (\alpha_{sk} D_k f_s) * \varphi_\varepsilon\|_{L^2} \lesssim \|f_s\|_{L^2} \tag{2.10}$$

with a constant independent of ε.

Remark Note that $D_k f_s$ is a distribution.

If we add up, we obtain

$$\|\mathcal{P} f_\varepsilon - (\mathcal{P}f) * \varphi_\varepsilon\|_{L^2} \lesssim \|f\|_{L^2}. \tag{2.11}$$

Let $\delta > 0$ and ψ a smooth form with compact support in Ω such that

$$\|f - \psi\|_{L^2} < \delta.$$

Then
$$\|\mathcal{P}f_\varepsilon - (\mathcal{P}f) * \varphi_\varepsilon\|_{L^2} \lesssim \delta + \|\mathcal{P}\psi_\varepsilon - (\mathcal{P}\psi) * \varphi_\varepsilon\|_{L^2}.$$

The right-hand side can be made arbitrarily small for small δ and ε and we are done.

Case ii)   Let U be so small that there exists an open convex set Γ with $0 \in \overline{\Gamma}$ and

$$x + y \in U \setminus \overline{\Omega} \text{ for } x \in b\Omega \cap \text{supp}(f) \text{ and } y \in \Gamma.$$

Let $\mathcal{P}f = u$ on Ω. We extend f and u to be 0 outside Ω. Then in the distributional sense on $\mathbb{R}^N$ we have

$$\mathcal{P}f = u + g, \tag{2.12}$$

where $g|_\Omega = 0$, $g|_{\mathbb{R}^N \setminus \overline{\Omega}} = 0$ and g is a distribution on $\mathbb{R}^N$ supported on $b\Omega \cap \text{supp}(f)$. Let φ_ε be as in case i), but with $\text{supp}\,\varphi \subset \Gamma$. Set $f_\varepsilon = f * \varphi_\varepsilon$. Then $f_\varepsilon \to f$ in $L^2_{p,q}(\mathbb{R}^N)$. By (2.11) and (2.12) we obtain

$$\|\mathcal{P}f_\varepsilon - u * \varphi_\varepsilon - g * \varphi_\varepsilon\|_{L^2(\Omega)} \lesssim \|f\|_{L^2(\Omega)}.$$

Let $y \in \overline{\Omega}$ and $\varepsilon > 0$. Then by the definition of a convolution with a distribution g we have

$$(g * \varphi_\varepsilon)(y) = g[\varphi_\varepsilon(y - \cdot)] = 0$$

because of

$$\text{supp}\,\varphi_\varepsilon(y - \cdot) \cap \text{supp}\,g = b\Omega \cap \text{supp}(f) \cap \{x \mid y - x \in \varepsilon\Gamma\} = \emptyset,$$

for the points $x + \varepsilon\Gamma$ lie outside $\overline{\Omega}$. Consequently

$$\|\mathcal{P}f_\varepsilon - (\mathcal{P}f)_\varepsilon\|_{L^2(\Omega)} \lesssim \|f\|_{L^2(\Omega)}.$$

By arguing as in the end of case i) the proof is achieved.   □

Remark   Very little of the boundary regularity was used. So the theorem can be generalised to much more general situations.

§3   The Generalised Cauchy Condition

We want to generalise the methods of §2 in §4 to get a much deeper result. This needs some preparations.

We shall now treat P and its formal adjoint always as differential operators. Related Hilbert space operators are denoted by $P_{\min}$, $P^*_{\min}$ (note that $P_{\min} = P_{\max}$ by (2.6)).

Definition 3.1   *Let $P : C^\infty_{p,q}(X) \to C^\infty_{p',q'}(X)$ be a first order differential operator with smooth coefficients and $\widetilde{P} : C^\infty_{p',q'}(X) \to C^\infty_{p,q}(X)$ its formal adjoint differential operator. Set for $f \in L^2_{p,q}(\Omega)$*

$$\widehat{f}|_\Omega = f, \quad \widehat{f}|_{X \setminus \Omega} = 0.$$

Then $\widehat{f} \in L^2_{p,q}(X)$. We say that f satisfies the generalised Cauchy condition (gCc) for P if

$$P\widehat{f} \in L^2_{p',q'}(X)$$

in the distributional sense.

We first treat the local situation where $\Omega \subset\subset X \subset \mathbb{C}^n$. This can always be achieved by a partition of unity argument. Let $\widetilde{v}_1, \ldots, \widetilde{v}_A$ be a basis of constant forms for the vector space of (p,q)-forms in $\mathbb{C}^n$. By the Gram-Schmidt orthogonalization process we can pass over to a basis $v_1(x), \ldots, v_A(x)$ with smooth coefficients such that

$$v_i \wedge *\overline{v}_j \equiv 0 \quad \text{for } i \neq j,$$
$$v_i \wedge *\overline{v}_i = a_i \, dx,$$

for all $x \in X$, $a_i(x) > 0$, $dx = dx_1 \wedge dx_2 \wedge \ldots \wedge dx_{2n}$. We construct an analogous base $w_1, \ldots, w_B$ for (p',q')-forms with $b_i > 0$ and

$$w_i \wedge *\overline{w}_i = b_i \, dx \quad \text{for } i = 1, \ldots, B.$$

If we decompose $f \in C^\infty_{p,q}(X)$ and $Pf \in C^\infty_{p',q'}(X)$ with respect to these bases, we get

$$f = \sum_{i=1}^{A} f_i v_i,$$

$$Pf = \sum_{j=1}^{B} (Pf)_j w_j,$$

(3.2)

with

$$(Pf)_j = \sum_{s=1}^{A} P_{js} f_s = \sum_{s=1}^{A} \left(\sum_{k=1}^{N} \alpha^j_{sk} \frac{\partial f_s}{\partial x_k} + \beta^j_s f_s \right),$$

(3.3)

where α_{sk}, β_s are smooth functions. Let $g = \sum_{j=1}^{B} g_j w_j \in C^\infty_{p',q',c}(X)$. Then

$$Pf \wedge *\overline{g} = \sum_{j=1}^{B} (Pf)_j \overline{g}_j b_j \, dx.$$

So we get

$$(Pf, g)_{L^2(\Omega)} = \sum_{j=1}^{B} \sum_{s=1}^{A} \int_{\Omega} \left(\sum_{k=1}^{N} \alpha^j_{sk} \frac{\partial f_s}{\partial x_k} + \beta^j_s f_s \right) \overline{g}_j b_j \, dx$$

$$= \sum_{j=1}^{B} \sum_{s=1}^{A} \int_{\Omega} f_s \overline{\beta^j_s} g_j b_j \, dx + \sum_{j=1}^{B} \sum_{s=1}^{A} \sum_{k=1}^{N} \rho^j_{sk},$$

with $(dx)_k = dx_1 \wedge \ldots \wedge dx_{k-1} \wedge dx_{k+1} \wedge \ldots \wedge dx_N$,

$$\rho_{sk}^j = \int_\Omega \alpha_{sk}^j \frac{\partial f_s}{\partial x_k} \bar{g}_j b_j \, dx$$

$$= (-1)^{k-1} \int_\Omega df_s \wedge \alpha_{sk}^j \bar{g}_j b_j \, (dx)_k$$

$$= \int_{b\Omega} (-1)^{k-1} f_s \alpha_{sk}^j \bar{g}_j b_j \, (dx)_k - \int_\Omega f_s \frac{\partial(\alpha_{sk}^j \bar{g}_j b_j)}{\partial x_k} \, dx.$$

Therefore we have

$$(Pf, g)_{L^2(\Omega)} = \sum_{j=1}^{B} \sum_{s=1}^{A} \sum_{k=1}^{N} \int_{b\Omega} (-1)^{k-1} \alpha_{sk}^j f_s \bar{g}_j b_j \, (dx)_k$$

$$- \sum_{s=1}^{A} \int_\Omega f_s \overline{\left[\sum_{j=1}^{B} \left(\sum_{k=1}^{N} \frac{1}{b_j} \frac{\partial(\overline{\alpha_{sk}^j} b_j g_j)}{\partial x_k} - \overline{\beta_s^j} g_j \right) \right]} b_j \, dx.$$

Set

$$(\widetilde{P}g)_s := \sum_{j=1}^{B} \left(\overline{\beta_s^j} g_j - \sum_{k=1}^{N} \frac{1}{b_j} \frac{\partial(\overline{\alpha_{sk}^j} b_j g_j)}{\partial x_k} \right) \tag{3.4}$$

and

$$\widetilde{P}g := \sum_{s=1}^{A} (\widetilde{P}g)_s v_s. \tag{3.5}$$

Then $\widetilde{P}$ is the formal adjoint of P on X. Moreover it follows from the meaning of the phrase

"$Pf \in L^2_{p',q'}$ in the distributional sense for $f \in L^2_{p,q}(X)$"

that it is equivalent to the following:

"There exists $h \in L^2_{p',q'}(X)$ with

$$(h, g)_X = (f, \widetilde{P}g)_X \text{ for all } g \in C^\infty_{p',q',c}(X)."$$

This can be easily seen because P is of first order and both criteria therefore use one integration by parts. The only difference comes from the differentiation of smooth terms which stem from the metric.

But the second criterion in turn says nothing more than that the functional

$$C^\infty_{p',q',c}(X) \ni g \longmapsto (f, \widetilde{P}g)_X$$

can be continued to a bounded linear functional on $L^2_{p',q',c}(X)$.

In particular we obtain that $f \in L^2_{p,q}(\Omega)$ satisfies (gCc) for P if and only if there is a $C > 0$ with

$$|(f, \widetilde{P}g)_{L^2(\Omega)}| \le C\|g\|_{L^2(\Omega)} \quad \text{for all } g \in C^\infty_{p',q',c}(X). \tag{3.6}$$

By Theorem (2.6) this is equivalent to saying that

$$f \in \text{dom }_\Omega((\widetilde{P})_{\min})^* = \text{dom }_\Omega((\widetilde{P})_{\max})^*.$$

Remark 3.7   *i. Our main application will be the relationship between condition (gCc) for $\overline{\partial}^*$ as a PDO (denoted by ϑ) and $\overline{\partial}^*$ as a Hilbert space adjoint. In view of this if we apply the above criterion to $Q = \widetilde{P}$ we obtain*

$$f \text{ satisfies (gCc) for } \widetilde{Q} \text{ if and only if } f \in \text{dom }_\Omega((Q_{\min})^*).$$

ii. Another conclusion of the above calculations and (3.4) gives for a first order PDO P that $f \in \text{dom }_\Omega((\widetilde{P}_{\min})^*) \cap C^1_{p,q}(\overline{\Omega})$ if and only if

$$\iota^*_{b\Omega}\left(\sum_{s=1}^A \sum_{k=1}^N (-1)^{k-1} \alpha^j_{sk} f_s\, (dx)_k\right) = 0, \quad j = 1, \ldots, B.$$

Here $\iota_{b\Omega} : b\Omega \hookrightarrow X$ denotes the inclusion map. This looks less strange if we set $P = \overline{\partial}^*$ or ϑ as a PDO. Then $((\widetilde{P})_{\min})^* = \overline{\partial}^*$ and we obtain a boundary condition for $f \in \text{dom }_\Omega(\overline{\partial}^*) \cap C^1_{p,q}(\overline{\Omega})$.

The vanishing of the above forms on $b\Omega$ means that they are multiples of dr, where r is a defining function of Ω. We shall now determine this condition. By changing the enumeration we can assume that locally at $z_0 \in b\Omega$: $\frac{\partial r}{\partial x_N} \ne 0$. Then for $k < N = 2n$

$$(dx)_k = (-1)^k \frac{\frac{\partial r}{\partial x_k}}{\frac{\partial r}{\partial x_N}} (dx)_N + dr \wedge \ldots$$

Therefore the boundary condition reads as

$$\sum_{s=1}^A \left(\sum_{k=1}^N \frac{\partial r}{\partial x_k} \alpha^j_{sk}\right) f_s = 0, \quad j = 1, 2, \ldots, B. \tag{3.8}$$

This can be invariantly written as

$$P(rf) = 0 \quad \text{on } b\Omega, \tag{3.9}$$

where r is a defining function for Ω. Thus $f \in C^1_{pq}(\overline{\Omega}) \cap \text{dom }((\widetilde{P}_{\min})^*)$ if and only if (3.9) holds.

Remark $i.$ Especially important will be the case when the matrix

$$\left(\sum_{k=1}^{N} \frac{\partial r}{\partial x_k} \alpha_{sk}^j\right)_{\substack{s=1,\ldots,A \\ j=1,\ldots,B}} \tag{3.10}$$

has constant rank.

$ii.$ In all the above matrix equations we can pass over to the real and imaginary parts. This will enlarge the system but every assertion can be carried over.

$iii.$ All these considerations are not affected by the partition of unity and the localisation.

§4 The Friedrichs-Hörmander Lemma

Here we follow the approach given in [Hör 65]. In the local situation in $\mathbb{R}^N$ we shall now consider a system of real differential equations

$$\sum_{s=1}^{A}\sum_{k=1}^{N} \alpha_{sk}^j \frac{\partial u_s}{\partial x_k} + \beta_s^j u_s = f_j, \quad j = 1,\ldots,B \tag{4.1}$$

which we write in the form

$$Au + Bu = f. \tag{4.2}$$

Let U be an open set in $\mathbb{R}^N$, $r \in C^\infty(U)$ be real valued with $dr \neq 0$ when $r = 0$. Suppose U is so small that together with some x_i, we can take r as part of a coordinate system. Set

$$U^\pm = \{x \in U \mid \pm r(x) \geq 0\}.$$

Proposition 4.3 *Suppose in the interior of U^- there are given $u, f \in L^2(U^-)$ which solve the system of PDE (4.2) and which vanish outside a compact subset of U^-. Furthermore we suppose that u satisfies (gCc) for the PDO $A + B$ on U. Then there is a sequence $u^\nu \in C_c^\infty(U)$, with $\operatorname{supp} u^\nu$ contained in the interior of U^-, such that*

$$\lim_{\nu\to\infty} u^\nu = u, \quad \lim_{\nu\to\infty} Au^\nu + Bu^\nu = f \quad \text{in } L^2(U^-).$$

Proof Assume in a first step that there is an open convex set Γ with $0 \in \overline{\Gamma}$ and

$$(\operatorname{supp} u \cap \{x \in U \mid r(x) = 0\}) - \Gamma \subset U^-.$$

Extend u and f to be 0 outside U^-. Then

$$Au + Bu = f \quad \text{on } U,$$

because (gCc) holds for u.

Choose $\varphi \in C^\infty(\mathbb{R}^N)$ with $\varphi \geq 0$, $\operatorname{supp}\varphi \subset \Gamma$ and $\int \varphi\, dx = 1$. Then as in the proof of Theorem (2.6) we have with $\varphi_\varepsilon(x) = \frac{1}{\varepsilon^N}\varphi\left(\frac{x}{\varepsilon}\right)$.

$$\lim_{\varepsilon\to 0} A(u * \varphi_\varepsilon) + B(u * \varphi_\varepsilon) = f \quad \text{in } L^2(U).$$

Let $-1 < \varepsilon < 0$.

$$u * \varphi_\varepsilon(y) = \int u(x)\varphi_\varepsilon(y - x)\, dx$$

$$= \int_{x\in\operatorname{supp} u} u(x)\frac{\varphi\left(\frac{x-y}{|\varepsilon|}\right)}{\varepsilon^N}\, dx.$$

Here it suffices to integrate only over x with $y \in x - \varepsilon\Gamma \subset x - \Gamma$. That means that $\operatorname{supp}(u * \varphi_\varepsilon)$ is in the interior of U^-.

By a partition of unity argument we can achieve the theorem with finitely many conveniently chosen Γ_i. $\square$

Remark   Again if everything is formulated in L^1-terms the proof can be carried over. (gCc) for u of course means that

$$A\widehat{u} + B\widehat{u} \in L^1_*(U).$$

Now we suppose that the solution of the system satisfies the (gCc) condition only for a PDO defined by a submatrix of $A + B$.

Let $B_0 \leq B$, $f^0 = (f_1, \ldots, f_{B_0})$. Let the first B_0 lines of (4.2) be written in the form

$$A_0 u + B_0 u = f^0. \tag{4.4}$$

We define the two matrices

$$A(\operatorname{grad} r) = \left(\sum_{k=1}^N \frac{\partial r}{\partial x_k}\alpha_{sk}^j\right)_{\substack{s=1,\ldots,A \\ j=1,\ldots,B}}$$

$$A_0(\operatorname{grad} r) = \left(\sum_{k=1}^N \frac{\partial r}{\partial x_k}\alpha_{sk}^j\right)_{\substack{s=1,\ldots,A \\ j=1,\ldots,B_0}} \tag{4.5}$$

Let $U^\pm$ be as before.

Theorem 4.6 (Friedrichs-Hörmander Lemma)   *Suppose we are given in the interior U_0^- of U^- a solution $u \in L^2(U^-)$ of $Au + Bu = f$, $f \in L^2(U^-)$, which vanishes outside a compact subset of U^-. We assume that u satisfies (gCc) for the PDO $A_0 + B_0$ so that we have*

$$A_0 u + B_0 u = f^0.$$

Moreover we suppose that $A(\operatorname{grad} r)$ and $A_0(\operatorname{grad} r)$ have constant rank in a neighbour-hood of $\{x \in U \mid r(x) = 0\}$. Then there is a sequence $u^\nu \in C^\infty(U^-)$, which vanishes outside a fixed compact subset of U^-, such that

$$\lim_{\nu \to \infty} u^\nu = u, \quad \lim_{\nu \to \infty} Au^\nu + Bu^\nu = f \quad in \ L^2(U^-)$$

and u^ν satisfies (gCc) with respect to $A_0 u + B_0 u$, i.e.one has $A_0(\operatorname{grad} r)u^\nu = 0$ when $r = 0$.

Proof   Let $R = \operatorname{rank} A(\operatorname{grad} r)$, $R_0 = \operatorname{rank} A_0(\operatorname{grad} r)$.

1st step $r(x) = x_N$, $\alpha^j_{sN} = 0$ except when $j = k = 1, \ldots, R_0$ and when $B + 1 - j = A + 1 - s = 1, 2, \ldots, R - R_0$. Moreover we assume that $\alpha^j_{sN} = 1$ in these cases. This means that the matrix $A(\operatorname{grad} r) = (\alpha^j_{sN})$ looks as follows:

$$\underbrace{\left. \begin{pmatrix} E_{R_0} & \cdots \\ \cdots & E_{R-R_0} \end{pmatrix} \right\} B}_{A}$$

where E_T means the $T \times T$ unit matrix, and all the other entries in the above $A \times B$-matrix are 0. The two unit matrices sit in the upper left corner and the lower right corner, respectively.

Later on we shall reduce the general case to this by a transformation. Set $u = 0$, $f = 0$ on $U \setminus U_0^-$. Then

$$A_0 u + B_0 u = f^0 \quad \text{on } U.$$

We shall now regularize u in the $(x_1, \ldots, x_{N-1})$ directions. Choose $\psi \in C_c^\infty(\mathbb{R}^{N-1})$, with $\psi \geq 0$, $\int \psi(x') \, dx' = 1$ $(x' = (x_1, \ldots, x_{N-1}))$. Set for $\psi_\varepsilon(x') = \frac{1}{\varepsilon^{N-1}} \psi \left(\frac{x'}{\varepsilon} \right)$. Then

$$u^\varepsilon(y) := \int_{\mathbb{R}^{N-1}} u(x', y_N) \psi_\varepsilon(y' - x') \, dx'$$

exists almost everywhere with respect to y_N and one has

$$\lim_{\varepsilon \to 0} u^\varepsilon = u \quad \text{in } L^2(U).$$

α^j_{sN} is constant, therefore we get analogously as in the proof of (2.6)

$$\begin{aligned} \lim_{\varepsilon \to 0} Au^\varepsilon + Bu^\varepsilon &= f \quad \text{in } L^2(U^-), \\ \lim_{\varepsilon \to 0} A_0 u^\varepsilon + B_0 u^\varepsilon &= f^0 \quad \text{in } L^2(U). \end{aligned} \tag{4.7}$$

For despite the fact that u^ε is no longer of class C^∞ in the x_N-variable we can repeat the proof of Theorem 2.6. Each term in the proof still makes sense and the intermediate results in the argument remain valid. Whenever the x_N-variable comes up the constancy of the α^j_{sN} makes the corresponding assertion trivial. Of course, some terms may come into a virtual distributional state but only for a very short time before they fall back to the world of honest functions.

Moreover we have $u^\varepsilon|_{U \setminus U_0^-} = 0$. Therefore u^ε satisfies (gCc) for $A_0 + B_0$.

For $j \neq N$ one has

$$\frac{\partial u^\varepsilon}{\partial x_j} \in L^2(U). \tag{4.8}$$

Therefore we conclude because of the special form of our matrix (α^j_{sN}):

$$\begin{aligned}
\frac{\partial u^\varepsilon_s}{\partial x_N} &\in L^2(U) \quad \text{if } s \leq R_0, \\
\frac{\partial u^\varepsilon_s}{\partial x_N} &\in L^2(U^-) \quad \text{if } s \geq A + 1 + R_0 - R.
\end{aligned} \tag{4.9}$$

These are the only $\frac{\partial}{\partial x_N}$-derivatives occurring in (4.2) and (4.7). Let $\varphi \in C_c^\infty(\mathbb{R}^N)$ with $\varphi \geq 0$, $\int \varphi \, dx = 1$, $\operatorname{supp} \varphi \subset \{x \in \mathbb{R}^N \mid x_N > 0\}$. Set for $\delta > 0$

$$\varphi^{+\delta}(x) = \frac{1}{\delta^N} \varphi\left(\frac{x}{\delta}\right),$$

$$\varphi^{-\delta}(x) = \frac{1}{\delta^N} \varphi\left(-\frac{x}{\delta}\right).$$

$$u_s^{\varepsilon\delta} = u_s^\varepsilon * \varphi^{-\delta}, \quad \text{if } s = 1, 2, \ldots, B_0,$$
$$u_s^{\varepsilon\delta} = u_s^\varepsilon * \varphi^{+\delta}, \quad \text{if } s = B_0 + 1, \ldots, A.$$

Then $u_s^{\varepsilon\delta} \in C^\infty(\mathbb{R}^N)$. Let $s \leq B_0$.

$$u_s^{\varepsilon\delta}(y) = \int\limits_{x_N \leq 0} u_s^\varepsilon(x) \frac{1}{\delta^N} \varphi\left(\frac{x - y}{\delta}\right) \, dx = 0$$

if $y_N \geq 0$. So $\operatorname{supp} u_s^{\varepsilon\delta} \subset U^-$ when $s \leq B_0$. (4.8) and (4.9) imply:

For $k < N$ and all s

$$\lim_{\delta \to 0} \frac{\partial u_s^{\varepsilon\delta}}{\partial x_k} = \frac{\partial u_s^\varepsilon}{\partial x_k} \quad \text{in } L^2(U).$$

For $s \leq R_0$

$$\lim_{\delta \to 0} \frac{\partial u_s^{\varepsilon\delta}}{\partial x_N} = \frac{\partial u_s^\varepsilon}{\partial x_N} \quad \text{in } L^2(U).$$

For $s \geq A + 1 + R_0 - R$

$$\lim_{\delta \to 0} \frac{\partial u_s^{\varepsilon\delta}}{\partial x_N} = \frac{\partial u_s^\varepsilon}{\partial x_N} \quad \text{in } L^2(U^-).$$

Therefore if we approximate u at first by u^ε and then u^ε by $u^{\varepsilon\delta}$ with $\delta = \delta(\varepsilon)$ we can achieve that for a sequence $\varepsilon_v \to 0$, $u^v = u^{\varepsilon_v \delta(\varepsilon_v)}$, we have

$$\lim_{v \to \infty} u^v = u, \quad \lim_{v \to \infty} (Au^v + Bu^v) = f$$

in $L^2(U^-)$ and that $u^\vee$ satisfies (gCc) for the PDO $A_0 + B_0$.

2nd step Let $r(x_0) = 0$. By a change of coordinates in a small neighbourhood of x_0 we can assume that $r = x_N$. The rank condition on A and A^0 does not change under diffeomorphisms. We denote the elements again by α_{sN}^j. Then

$$\left(\alpha_{sN}^j\right)_{\substack{s=1,\ldots,A \\ j=1,\ldots,B_0}}$$

has constant rank R_0 in a smaller neighbourhood. We can suppose that the matrix

$$(\alpha_{sN}^j)_{j,s \leq R_0}$$

is nonsingular.

We define

$$u_j' = \sum_{l=1}^{A} \alpha_{lN}^j u_l \qquad\qquad j = 1,\ldots, R_0,$$

$$u_j' = u_j \qquad\qquad R_0 < j \leq A.$$

Then the PDE reads as

$$\sum_{s=1}^{A}\sum_{k=1}^{N} \alpha_{sk}'^{j} \frac{\partial u_s'}{\partial x_k} + \sum_{s=1}^{A} \beta_s'^{j} u_s' = f_j, \quad j = 1,\ldots, B$$

with (gCc) for the first B_0 lines. We have

$$\alpha_{sN}'^{j} = \delta_{js} \qquad \text{for } j = 1, 2, \ldots, R_0, \qquad s = 1,\ldots, A,$$
$$\alpha_{sN}'^{j} = 0 \qquad \text{for } B_0 \geq j > R_0, \qquad s = 1,\ldots, A$$

because of $\operatorname{rank} (\alpha_{sN}'^{j})_{\substack{s=1,\ldots,A \\ j=1,\ldots,B_0}} = R_0$.

By subtracting linear combinations of the first R_0 equations from the others we can obtain

$$\alpha_{sN}'^{j} = 0 \quad \text{if } s \leq R_0, R_0 < j \leq B.$$

Now the first B_0 equations are already in the form of the first step with

$$\operatorname{rank} \left(\alpha_{sN}'^{j}\right)_{\substack{j=R_0+1,\ldots,B \\ s=B_0+1,\ldots,A}} = R - R_0.$$

By switching to a new system of functions $u_{R_0+1}', \ldots, u_A'$ as linear combinations of u_{R_0+1}, $\ldots, u_A$ in the same way we can transform the system into the desired form.

The global assertion of the Theorem can be achieved by a partition of unity argument. $\square$

Remark   If everything is formulated in L^1-terms the proof is analogous.

(gCc) for u of course means again that $A_0\widehat{u} + B_0\widehat{u} \in L^1_*(U)$.

§5   The Self-adjointness of the Complex Laplacian and Hörmander's Density Theorem.

Let X be a hermitian manifold and $\Omega \subset\subset X$ be a domain with smooth boundary. On X the operator $\overline{\partial}$ and its formal adjoint ϑ are defined as PDOs. We likewise denote the maximal extension of $\overline{\partial}$ by $\overline{\partial}$ or $\overline{\partial}_{p,q}$ if applied on (p,q)-forms in $L^2_{p,q}(\Omega)$. By Theorem (2.6) we have

$$\overline{\partial} = \overline{\partial}_{\min}. \tag{5.1}$$

Remark   We can define $\overline{\partial}_{\min}$ and $\overline{\partial}_{\max}$ on L^1_{0q} in the same way as for L^2_{0q}, to wit

$$\text{dom}\,(\overline{\partial}_{\min}) = \{f \in L^1_{0q} : \exists f_j \in C^\infty_{0q}(\overline{\Omega}), f_j \underset{L^1}{\rightarrow} f\}$$

$$\text{dom}\,(\overline{\partial}_{\max}) = \{f \in L^1_{0q} : \overline{\partial} f \in L^1_{0,q+1}\},$$

and obtain, as in L^2, the equalities

$$\overline{\partial}_{\max} = \overline{\partial}_{\min} = \overline{\partial}.$$

The Hilbert space adjoint of $\overline{\partial}_{p,q}$ is denoted by $\overline{\partial}^*_{p,q}$ or simply by $\overline{\partial}^*$. It is easy to see that $\mathcal{R}\,(\overline{\partial}_{p,q}) \subset \text{dom}\,(\overline{\partial}_{p,q+1})$ because of $\overline{\partial}_{p,q+1} \circ \overline{\partial}_{p,q} = 0$. This also implies $\mathcal{R}\,(\overline{\partial}^*_{p,q+1}) \subset \text{dom}\,(\overline{\partial}^*_{p,q})$.

In the sequel we shall verify that the system $\overline{\partial} \oplus \vartheta$ satisfies the assumptions of Theorem 4.6. To make the notations slightly simpler we restrict attention to $(0,q)$-forms. Let $f \in C^1_{0,q}(\overline{\Omega})$ satisfy (gCc) for $\overline{\partial}$. Then there exists a constant $C > 0$ such that

$$|(f, \vartheta g)_{L^2(\Omega)}| \le C\|g\|_{L^2(\Omega)} \quad \forall g \in C^\infty_{0,q+1}(\overline{\Omega}).$$

This means that in Green's formula (2.27) the boundary integral must vanish, that is

$$f \sim \overline{\partial} r$$

on $b\Omega$.

Let $z_0 \in b\Omega$ and $\omega_1, \ldots, \omega_n$ be a local orthonormal frame of $(1,0)$-forms in a neighbourhood of z_0, with $\omega_n = \partial r$. Then we have the uniquely defined decomposition

$$f = f_t + f_n \wedge \overline{\omega}_n, \tag{5.2}$$

where f_t, f_n do not contain $\overline{\omega}_n$. f_t is called the *tangential component* and f_n the *normal component* of f. The above condition means that

$f \in C^1_{0,q}(\overline{\Omega})$ *satisfies* (gCc) *for* $\overline{\partial}$ *if and only if* $f_t|_{b\Omega} = 0$ *(locally in any $z_0 \in b\Omega$).* (5.3)

Now let $f \in C_{0,q}^1(\overline{\Omega})$ satisfy (gCc) for ϑ. Then (see also Remark 3.7)

$$f \in \mathrm{dom}\,(\bar{\partial}^*).$$

That means that

$$|(f, \bar{\partial}g)_{L^2(\Omega)}| \leq C\|g\|_{L^2(\Omega)} \quad \text{for all } g \in C_{0,q-1}^\infty(\overline{\Omega}).$$

Green's formula (2.27) now tells us that

$$\iota_{b\Omega}^*(*\bar{f}) = 0.$$

And consequently we deduce

$f \in C_{0,q}^1(\overline{\Omega})$ *satisfies (gCc) for ϑ if and only if $f_n|_{b\Omega} = 0$ (locally in any $z_0 \in b\Omega$).* (5.4)

In particular if f satisfies (gCc) for $\bar{\partial} \oplus \vartheta$ then

$$f|_{b\Omega} = 0 \qquad \text{(Dirichlet condition)} \tag{5.5}$$

According to (3.9) the matrix A is given by the coefficients of the system

$$(\bar{\partial} + \vartheta)(rf) = 0 \qquad \text{on } b\Omega,$$

and A_0 is given by

$$\vartheta(rf) = 0 \qquad \text{on } b\Omega.$$

Since $\vartheta = - * \partial *$ the first system is equivalent to

$$\bar{\partial}r \wedge f - *(\partial r \wedge *f) = 0 \qquad \text{on } b\Omega$$

or

$$\bar{\partial}r \wedge f_t - *(\partial r \wedge *(f_n \wedge \bar{\partial}r)) = 0 \qquad \text{on } b\Omega$$

because of $\partial r \wedge *f_t = 0$.

The first term is normal and the second tangential by the properties of the $*$-operator (see Chapter IV). Thus the system is equivalent to

$$\bar{\partial}r \wedge f_t = 0$$
$$*(\partial r \wedge *(f_n \wedge \bar{\partial}r)) = 0 \qquad \text{on } b\Omega.$$

and to

$$f_t = 0$$
$$*(\partial r \wedge *(f_n \wedge \bar{\partial}r)) = 0 \qquad \text{on } b\Omega.$$

But $*$ is a linear isomorphism so we obtain equivalently

$$f_t = 0$$
$$\partial r \wedge *(f_n \wedge \bar{\partial}r) = 0 \qquad \text{on } b\Omega$$

or

$$f_t = 0, \quad f_n = 0 \qquad \text{on } b\Omega$$

because $\overline{*(f_n \wedge \overline{\partial} r)}$ is tangential.

If we consider

$$\vartheta f = 0 \qquad \text{on } b\Omega$$

alone we obtain by the same rearrangements

$$f_n = 0 \qquad \text{on } b\Omega.$$

So A and A_0 have constant maximal ranks on $b\Omega$. Therefore the conditions of the Frie-drichs-Hörmander extension lemma are satisfied.

Because we have for $f \in C^1_{0,q}(\overline{\Omega})$ by Remark (3.7): f satisfies (gCc) for $\vartheta \Leftrightarrow f \in \text{dom } \overline{\partial}^*$, it follows from the Friedrichs extension lemma and a partition of unity argument

Theorem 5.6 (Hörmander's density theorem) *Let $f \in L^2_{0,q}(\Omega) \cap \text{dom } \overline{\partial} \cap \text{dom } \overline{\partial}^*$. Then there exists a sequence $f_j \in C^\infty_{0,q}(\overline{\Omega}) \cap \text{dom } \overline{\partial}^*$ so that*

$$\lim_{j \to \infty} (\|f_j - f\|^2_{L^2} + \|\overline{\partial} f_j - \overline{\partial} f\|^2_{L^2} + \|\overline{\partial}^* f_j - \overline{\partial}^* f\|^2_{L^2}) = 0.$$

with respect to L^2-norms on Ω.

Remark A corresponding result is true for L^1-convergence if we define $\text{dom } \overline{\partial}^* \subset L^1$ as

$$\text{dom } \overline{\partial}^* = \{f \in L^1_{0q}(\Omega) : \vartheta \widehat{f} \in L^1_{0q-1}(X)\}$$

(where $\widehat{f}$ is the trivial extension of f).

In the preceding chapters we already introduced the complex Laplacian

$$\Box = \overline{\partial} \vartheta + \vartheta \overline{\partial}$$

as a second order differential operator defined on C^2-forms or even on forms with distribu-tion coefficients (on Ω resp. X). We now replace the formal adjoint ϑ of $\overline{\partial}$ by the Hilbert space adjoint $\overline{\partial}^*$ and set

$$\Box = \overline{\partial} \overline{\partial}^* + \overline{\partial}^* \overline{\partial}$$

with domain

$$
\begin{aligned}
\text{dom } \Box \;=\; & \text{dom } \Box_q \\
=\; & \{f \in L^2_{0q}(\Omega) : f \in \text{dom } \overline{\partial} \cap \text{dom } \overline{\partial}^*, \overline{\partial} f \in \text{dom } \overline{\partial}^*, \overline{\partial}^* f \in \text{dom } \overline{\partial}\}.
\end{aligned}
$$

It is this operator which will from now on be called the complex Laplacian; it is naturally the restriction of the previously defined Laplacian from forms with distribution coefficients to the more special forms in $\text{dom } \Box$; it could also be considered as the extension of the former operator defined on smooth forms satisfying the correct boundary conditions to the space dom $\square$. It is a densely defined operator because of $C_{0,q,c}^{\infty}(\Omega) \subset \text{dom}\,(\square)$.

The two conditions

$$f \in \text{dom}\,\bar{\partial}^*, \quad \text{i.e.} \quad \iota_{b\Omega}^*(*f) = 0$$

$$\bar{\partial}f \in \text{dom}\,\bar{\partial}^*, \quad \text{i.e.} \quad \iota_{b\Omega}^*(*\bar{\partial}f) = 0$$

are called the *first* and *second Neumann condition*.

Proposition 5.7 $\square$ *is a densely defined closed self-adjoint operator with* $(\square f, f) \geq 0$ *for all* $f \in \text{dom}\,(\square)$.

Proof   Let $\text{dom}\,(\square) \ni f_j$ be a sequence, $f, g \in L_{0,q}^2(\Omega)$ such that $\lim_{j \to \infty} f_j = f$, $\lim_{j \to \infty} \square f_j = g$. For all $h \in \text{dom}\,(\square)$ one has

$$(\square h, h)_{L^2(\Omega)} = \|\bar{\partial}h\|_{L^2(\Omega)}^2 + \|\bar{\partial}^* h\|_{L^2(\Omega)}^2. \tag{5.8}$$

So the last assertion is already shown.

Because $\{\square f_j\}_j$ and $\{f_j\}_j$ are Cauchy sequences we obtain from (5.8) and the closedness of $\bar{\partial}$ and $\bar{\partial}^*$ that

$$\lim_{j \to \infty} \bar{\partial}f_j = \bar{\partial}f, \quad \lim_{j \to \infty} \bar{\partial}^* f_j = \bar{\partial}^* f,$$

and $f \in \text{dom}\,\bar{\partial} \cap \text{dom}\,\bar{\partial}^*$.

We have for $h \in \text{dom}\,(\square)$

$$\|\square h\|_{L^2(\Omega)}^2 = \|\bar{\partial}\bar{\partial}^* h\|_{L^2(\Omega)}^2 + \|\bar{\partial}^* \bar{\partial} h\|_{L^2(\Omega)}^2$$

because of

$$(\bar{\partial}\bar{\partial}^* h, \bar{\partial}^* \bar{\partial} h)_{L^2(\Omega)} = 0.$$

So $\{\bar{\partial}\bar{\partial}^* f_j\}_j$, $\{\bar{\partial}^* \bar{\partial} f_j\}_j$ are Cauchy sequences. Because $\bar{\partial}f_j \to \bar{\partial}f, \bar{\partial}^* f_j \to \bar{\partial}^* f$ we obtain

$$\bar{\partial}^* f \in \text{dom}\,\bar{\partial}, \quad \bar{\partial}f \in \text{dom}\,\bar{\partial}^*$$

and

$$\bar{\partial}\bar{\partial}^* f_j \to \bar{\partial}\bar{\partial}^* f, \quad \bar{\partial}^* \bar{\partial} f_j \to \bar{\partial}^* \bar{\partial} f.$$

This implies that $g = \square f$, $f \in \text{dom}\,(\square)$.

Let now $f, g \in \text{dom}\,(\square)$. Then clearly

$$(\square f, g)_{L^2(\Omega)} = (f, \square g)_{L^2(\Omega)}. \tag{5.9}$$

We will now complete our proof of the self-adjointness of $\square$ by checking that $\text{dom}\,(\square) = \text{dom}\,(\square^*)$. Set

$$F = \text{id} + \square,$$

where id denotes the identical mapping. We need the following lemma of J. von Neumann.

Lemma 5.10 *Let* $A : V \to H$ *be a closed densely defined operator on a Hilbert space* H. *Set* $S = \mathrm{id} + A^*A$, $T = \mathrm{id} + AA^*$,

$$\mathrm{dom}\,(S) = \{x \in \mathrm{dom}\,A \mid Ax \in \mathrm{dom}\,A^*\},$$
$$\mathrm{dom}\,(T) = \{x \in \mathrm{dom}\,A^* \mid Ax \in \mathrm{dom}\,A\}.$$

Then $S : \mathrm{dom}\,(S) \to H$, $T : \mathrm{dom}\,(T) \to H$ *are linear bijective maps and* $S^{-1}, T^{-1} :$ $H \to H$ *are continuous self-adjoint operators.*

Proof   It suffices to show the assertion for S. Let $U : H \times H \to H \times H$ be the unitary transformation $(x, y) \mapsto (-y, x)$ for $x, y \in H$. Then it is easy to see that

$$\mathrm{graph}\,(A^*) = (U\,\mathrm{graph}\,(A))^\perp$$

in $H \times H$. Let $h \in H$. Then there exist uniquely defined elements $f \in \mathrm{dom}\,(A)$, $g \in \mathrm{dom}\,(A^*)$, with

$$\{h, 0\} = \{f, Af\} + \{A^*g, -g\}.$$

Here $\{x, y\}$ denotes the elements of $H \times H$.

Let B, resp. C, be the linear map which associates to h the element f, resp. g. From this it follows

$$\mathrm{id} = B + A^*C, \quad 0 = AB - C,$$

with $B : H \to \mathrm{dom}\,(A)$, $C : H \to \mathrm{dom}\,(A^*)$. Therefore we have

$$C = AB, \quad \mathrm{id} = SB. \tag{5.11}$$

$\{f, Af\}$ and $\{A^*g, -g\}$ are orthogonal in $H \times H$, hence the continuity of B and C follows from

$$\|h\|^2 = \|f\|^2 + \|g\|^2 + \|Af\|^2 + \|A^*g\|^2 \geq \|Bh\|^2 + \|Ch\|^2.$$

Let $h \in \mathrm{dom}\,(S)$. Then

$$\|Sh\|\|h\| \geq (Sh, h) = \|h\|^2 + \|Ah\|^2 \geq \|h\|^2.$$

Consequently $\ker(S) = 0$. (5.11) implies the surjectivity of S. So

$$B = S^{-1}.$$

Let now $u, v \in H$. We have

$$(Bu, v) = (Bu, SBv) = (Bu, Bv) + (Bu, A^*ABv)$$
$$= (Bu, Bv) + (ABu, ABv).$$

Moreover (5.11) implies that $Bu \in \mathrm{dom}\,(A^*A)$, hence

$$(Bu, v) = (Bu, Bv) + (A^*ABu, Bv) = (SBu, Bv) = (u, Bv).$$

$\square$

The self-adjointness of $\Box$ is now shown by ideas of Gaffney. For the convenience of the reader we present the proof as it is described in [FoK 72].

$F = \text{id} + \Box$ is a densely defined closed operator with

$$\|Ff\|_{L^2} \cdot \|f\|_{L^2} \geq (Ff, f)_{L^2} = \|f\|_{L^2}^2 + \|\bar{\partial}f\|_{L^2}^2 + \|\bar{\partial}^* f\|_{L^2}^2$$

on $\text{dom}\,(F)$. So $\ker(F) = 0$ and $\mathcal{R}\,(F)$ is closed. By the lemma we have that

$$(\text{id} + \bar{\partial}\bar{\partial}^*)^{-1} \text{ and } (\text{id} + \bar{\partial}^* \bar{\partial})^{-1}$$

are bounded self-adjoint operators. Thus

$$S := (\text{id} + \bar{\partial}\bar{\partial}^*)^{-1} + (\text{id} + \bar{\partial}^* \bar{\partial})^{-1} - \text{id}$$

is bounded and self-adjoint.

We show that F is surjective with $S = F^{-1}$. We have

$$(\text{id} + \bar{\partial}\bar{\partial}^*)^{-1} - \text{id} = (\text{id} - (\text{id} + \bar{\partial}\bar{\partial}^*))(\text{id} + \bar{\partial}\bar{\partial}^*)^{-1} = -\bar{\partial}\bar{\partial}^*(\text{id} + \bar{\partial}\bar{\partial}^*)^{-1} \quad (5.12)$$

$$(\text{id} + \bar{\partial}^* \bar{\partial})^{-1} - \text{id} = (\text{id} - (\text{id} + \bar{\partial}^* \bar{\partial}))(\text{id} + \bar{\partial}^* \bar{\partial})^{-1} = -\bar{\partial}^* \bar{\partial}(\text{id} + \bar{\partial}^* \bar{\partial})^{-1} \quad (5.13)$$

Therefore $\mathcal{R}\,((\text{id} + \bar{\partial}\bar{\partial}^*)^{-1}) \subset \text{dom}\,(\bar{\partial}\bar{\partial}^*)$ and $\mathcal{R}\,((\text{id} + \bar{\partial}^* \bar{\partial})^{-1}) \subset \text{dom}\,(\bar{\partial}^* \bar{\partial})$. Moreover from (5.12) it follows that

$$S = (\text{id} + \bar{\partial}^* \bar{\partial})^{-1} - \bar{\partial}\bar{\partial}^*(\text{id} + \bar{\partial}\bar{\partial}^*)^{-1}.$$

Because of $\bar{\partial}^2 = 0$ one has

$$\mathcal{R}\,(S) \subset \text{dom}\,(\bar{\partial}^* \bar{\partial})$$

and

$$\bar{\partial}^* \bar{\partial}S = \bar{\partial}^* \bar{\partial}(\text{id} + \bar{\partial}^* \bar{\partial})^{-1}.$$

Symmetrically it follows from (5.13)

$$\mathcal{R}\,(S) \subset \text{dom}\,(\bar{\partial}\bar{\partial}^*)$$

and

$$\bar{\partial}\bar{\partial}^* S = \bar{\partial}\bar{\partial}^*(\text{id} + \bar{\partial}\bar{\partial}^*)^{-1}.$$

Thus $\mathcal{R}\,(S) \subset \text{dom}\,(F)$ and

$$\begin{aligned}
FS &= S + \bar{\partial}^* \bar{\partial}S + \bar{\partial}\bar{\partial}^* S \\
&= (\text{id} + \bar{\partial}\bar{\partial}^*)^{-1} + (\text{id} + \bar{\partial}^* \bar{\partial})^{-1} - \text{id} + \bar{\partial}^* \bar{\partial}(\text{id} + \bar{\partial}^* \bar{\partial})^{-1} + \bar{\partial}\bar{\partial}^*(\text{id} + \bar{\partial}\bar{\partial}^*)^{-1} \\
&= \text{id}
\end{aligned}$$

on H. This shows that $\mathcal{R}\,(F) = H$ and $S = F^{-1}$. F and $\Box$ are therefore self-adjoint. $\quad\Box$

We can sum up these results as

Theorem 5.14 *Let $\Omega \subset X$ be a bounded domain with smooth boundary in a hermitian manifold X. Let $\overline{\partial}$ be the maximal closed extension of the $\overline{\partial}$ operator and $\overline{\partial}^*$ its adjoint operator in the $L^2(\Omega)$-space. Let*

$$\Box_q = \Box = \overline{\partial}\overline{\partial}^* + \overline{\partial}^*\overline{\partial}$$

be the complex Laplacian with

$$\text{dom}\,(\Box_q) = \{f \in L^2_{0,q}(\Omega) \cap \text{dom}\,(\overline{\partial}) \cap \text{dom}\,(\overline{\partial}^*) \mid \overline{\partial}f \in \text{dom}\,(\overline{\partial}^*),\, \overline{\partial}^*f \in \text{dom}\,(\overline{\partial})\}.$$

Then $\Box_q$ is self-adjoint and it holds the weak orthogonal decomposition

$$L^2_{0,q}(\Omega) = \ker(\Box) \oplus \overline{\mathcal{R}(\Box)}$$
$$= (\ker(\overline{\partial}) \cap \ker(\overline{\partial}^*)) \oplus \overline{\mathcal{R}(\overline{\partial})} \oplus \overline{\mathcal{R}(\overline{\partial}^*)}.$$

Proof   The first equality follows from the self-adjointness of $\Box$. $\ker(\Box) = \ker(\overline{\partial}) \cap \ker(\overline{\partial}^*)$ follows from (5.8).

$$\ker(\Box)^\perp = \overline{\mathcal{R}(\Box)} \subset \overline{\mathcal{R}(\overline{\partial})} + \overline{\mathcal{R}(\overline{\partial}^*)}.$$

Thus

$$L^2_{0,q}(\Omega) = (\ker\overline{\partial} \cap \ker\overline{\partial}^*) + \overline{\mathcal{R}(\overline{\partial})} + \overline{\mathcal{R}(\overline{\partial}^*)}.$$

$\mathcal{R}(\overline{\partial}) \perp \mathcal{R}(\overline{\partial}^*)$, hence $\overline{\mathcal{R}(\overline{\partial})} \perp \overline{\mathcal{R}(\overline{\partial}^*)}$. Let $f \perp \overline{\mathcal{R}(\overline{\partial})} \oplus \overline{\mathcal{R}(\overline{\partial}^*)}$. This implies $(f, \overline{\partial}g) = 0$, $(f, \overline{\partial}^*h) = 0$ for all $g \in \text{dom}\,\overline{\partial}$, $h \in \text{dom}\,\overline{\partial}^*$. Therefore $f \in \text{dom}\,\overline{\partial}^* \cap \text{dom}\,\overline{\partial}$ and

$$(\overline{\partial}^*f, g) = (\overline{\partial}f, h) = 0$$

for all $g, h \in L^2_{0,q}(\Omega)$. So $f \in \ker\overline{\partial} \cap \ker\overline{\partial}^*$. $\Box$

§6   The $\overline{\partial}$-Neumann Problem

We can now formulate the Neumann problem for the $\overline{\partial}$-equation. In order to make our presentation conceptually independent of the previous paragraphs we repeat some definitions and notations.

Ω is a relatively compact subdomain of a hermitian manifold X with smooth boundary $b\Omega$. The metric defines the scalar product

$$(f, g) = \int\limits_{\Omega} f \wedge *\overline{g}$$

on $(0, q)$-forms and the corresponding Hilbert spaces L_{0q}^2.

$$\bar{\partial} : L_{0q}^2 \to L_{0q+1}^2$$

is a densely defined closed linear operator (derivatives in the distributional sense) and therefore has a Hilbert space adjoint

$$\bar{\partial}^* : L_{0q+1}^2 \to L_{0q}^2$$

given by

$$(\bar{\partial} f, g) = (f, \bar{\partial}^* g),$$

whenever both sides of the equation are defined. $\bar{\partial}^*$ is again linear closed and densely defined; the smooth forms in its domain are singled out by the *first Neumann condition*

$$\text{dom } \bar{\partial}^* \cap C_{0q+1}^1(\overline{\Omega}) = \{f : *f|_{b\Omega} \equiv 0\};$$

the restriction is the pull-back $\iota^* f$ to the boundary. Where both operators are defined we have

$$\bar{\partial}^* = \vartheta = - * \partial * .$$

The complex Laplacian is

$$\square = \bar{\partial}\bar{\partial}^* + \bar{\partial}^*\bar{\partial},$$

with the smooth forms in the domain of definition given by the first and second Neumann condition:

$$\text{dom } \square \cap C_{0q}^2(\overline{\Omega}) = \{f : *f|_{b\Omega} \equiv 0, *\bar{\partial} f|_{b\Omega} \equiv 0\}.$$

The Neumann problem in its weakest form is to find a criterion for the closedness of $\mathcal{R}(\square)$. If $\mathcal{R}(\square)$ is closed, $\square$ is invertible in the following sense. By Theorem 5.14

$$L_{0,q}^2(\Omega) = \ker(\square) \oplus \mathcal{R}(\square).$$

So we can define a linear operator $N : L_{0,q}^2 \to L_{0,q}^2$ as follows:

$Nu = 0$ *for* $u \in \ker(\square)$ *and* Nu *is the uniquely defined preimage of* u *orthogonal to* $\ker(\square)$, *if* $u \in \mathcal{R}(\square)$.

Definition 6.1 *The above defined operator is called the $\bar{\partial}$-Neumann operator.*

Stronger versions of the Neumann problem ask for regularity of N with respect to certain norms. We shall discuss this for a strictly pseudoconvex domain Ω. Our main tool is the basic integral formula.

Theorem 6.2 *Let* $\Omega \subset X$ *be a strictly pseudoconvex domain in a hermitian manifold. Then* $\mathcal{R}(\bar{\partial})$, $\mathcal{R}(\bar{\partial}^*)$ *and* $\mathcal{R}(\square)$ *are closed in* $L_{0,q}^2(\Omega)$, *for all* $q \geq 0$. *For* $q \geq 0$ *the strong orthogonal decompositions holds:*

$$L_{0,q}^2(\Omega) = \ker(\square) \oplus \mathcal{R}(\square),$$
$$L_{0,q}^2(\Omega) = (\ker(\bar{\partial}) \cap \ker(\bar{\partial}^*)) \oplus \mathcal{R}(\bar{\partial}) \oplus \mathcal{R}(\bar{\partial}^*).$$

There exists a positive constant c such that for all $q \geq 0$

$$c\|f\|^2_{L^2(\Omega)} \leq \|\bar{\partial}f\|^2_{L^2(\Omega)} + \|\bar{\partial}^* f\|^2_{L^2(\Omega)}, \quad \text{for } f \in \text{dom}(\bar{\partial}) \cap \text{dom}(\bar{\partial}^*) \cap \ker(\Box)^\perp,$$

$$c\|f\|_{L^2(\Omega)} \leq \|\Box f\|_{L^2(\Omega)}, \quad \text{for } f \in \text{dom}(\Box) \cap \ker(\Box)^\perp.$$

In particular the Neumann operator exists and is bounded. Moreover $\bar{\partial}N$, $\bar{\partial}^* N$ and N are compact operators, and $\mathbb{H}^{0q} = \ker \Box_{0q}$ is finite-dimensional for $q \geq 1$.

Proof   We shall write throughout the proof $\|.\|$ instead of $\|.\|_{L^2(\Omega)}$.

i) We have proved in Chapter IV that $\mathcal{R}(\bar{\partial})$ is closed: this is a consequence of the basic integral formula

$$f = P_q f + T_q \bar{\partial} f + \bar{\partial} S_{q-1} f \tag{6.3}$$

and the compactness of P_q for $q \geq 1$. But functional analysis – see [Hör 65], Theorem 1.1.1 – now implies that $\mathcal{R}(\bar{\partial}^*)$ is closed.

ii) We now have the second strong decomposition for all $q \geq 0$. In particular we have

$$\overline{\mathcal{R}(\Box)} = \mathcal{R}(\bar{\partial}) \oplus \mathcal{R}(\bar{\partial}^*).$$

Let $f \in \mathcal{R}(\bar{\partial})$. Then $f = \bar{\partial}g$, with $g \in \ker(\bar{\partial})^\perp = \mathcal{R}(\bar{\partial}^*)$. Thus $g = \bar{\partial}^* h$, with $h \in \ker(\bar{\partial}^*)^\perp = \mathcal{R}(\bar{\partial})$. Hence $h \in \text{dom}(\Box)$ and

$$f = \bar{\partial}\bar{\partial}^* h = \Box h.$$

Analogously one shows for $f \in \mathcal{R}(\bar{\partial}^*)$ with a $h \in \text{dom}(\Box)$ that

$$f = \bar{\partial}^* \bar{\partial} h = \Box h.$$

Therefore $\mathcal{R}(\Box)$ is closed and the first strong decomposition is proven.

Next we show the existence of a bounded solution operator for $\bar{\partial}^*$. The corresponding proof for $\bar{\partial}$ is analogous. Let

$$X = \{f \in \text{dom } \bar{\partial}^* : f \perp \ker \bar{\partial}^*\}$$

be the Banach space equipped with the norm

$$\|f\|_X = \sqrt{\|f\|^2 + \|\bar{\partial}^* f\|^2}.$$

Then the map $A : X \to \mathcal{R}(\bar{\partial}^*)$, $f \mapsto \bar{\partial}^* f$, is a bounded linear isomorphism. Therefore

$$A^{-1} : \mathcal{R}(\bar{\partial}^*) \to \{f \in \text{dom } \bar{\partial}^* : f \perp \ker \bar{\partial}^*\} \tag{6.4}$$

is an L^2-bounded solution operator for $\bar{\partial}^*$. We denote the corresponding solution operator for $\bar{\partial}$ by B^{-1},

$$B^{-1} : \mathcal{R}(\bar{\partial}) \to \{f \in \text{dom } \bar{\partial} : f \perp \ker \bar{\partial}\}. \tag{6.5}$$

Now let $f \in \mathrm{dom}\ \bar{\partial} \cap \mathrm{dom}\ \bar{\partial}^*$, $f \perp \ker \square$. Then

$$f = f' + f'' \in \mathcal{R}(\bar{\partial}) \oplus \mathcal{R}(\bar{\partial}^*).$$

Obviously

$$f' \in \mathrm{dom}\ \bar{\partial}^*,\ f'' \in \mathrm{dom}\ \bar{\partial},\ \bar{\partial}^* f' = \bar{\partial}^* f,\ \bar{\partial} f'' = \bar{\partial} f. \tag{6.6}$$

Now $f' = A^{-1}(\bar{\partial}^* f)$, $f'' = B^{-1}(\bar{\partial} f)$, and the continuity of A^{-1} and B^{-1} imply

$$\|f\|^2 = \|f'\|^2 + \|f''\|^2 \le C(\|\bar{\partial}^* f\|^2 + \|\bar{\partial} f\|^2)$$

with a $C > 0$ independent of f.

Now the second inequality of 6.2 follows for $f \in \mathrm{dom}\ \square \cap \ker \square^\perp$ from the first one:

$$\begin{aligned} c\|f\|^2 &\le \|\bar{\partial} f\|^2 + \|\bar{\partial}^* f\|^2 \\ &= (\square f, f) \\ &\le \|\square f\|\ \|f\| \end{aligned}$$

In order to show compactness we first show that

$$\begin{aligned} B_{\bar{\partial}} &:= \{f \in \mathrm{dom}\,(\bar{\partial}) \cap \ker(\bar{\partial})^\perp \mid \|\bar{\partial} f\| \le 1\}, \\ B_{\bar{\partial}^*} &:= \{f \in \mathrm{dom}\,(\bar{\partial}^*) \cap \ker(\bar{\partial}^*)^\perp \mid \|\bar{\partial}^* f\| \le 1\}, \\ B_{\square} &:= \{f \in \mathrm{dom}\,(\square) \cap \ker(\square)^\perp \mid \|\square f\| \le 1\} \end{aligned}$$

are compact in $L^2_{0,q}(\Omega)$.

Let $\{f_n\}_n$ be a sequence in $B_{\bar{\partial}}$. $f_n \in \mathcal{R}(\bar{\partial}^*)$. Thus $\bar{\partial}^* f_n = 0$. By the first inequality of Theorem (6.2) we obtain that $\{f_n\}_n$ is bounded in $L^2_{0,q}(\Omega)$. Let $q \ge 1$. Applying Π' to (6.3) and passing to a subsequence we have because of $\Pi' f_n = f_n$ that $\lim_{n\to\infty} f_n = f$ exists. This in turn implies $f \in \mathrm{dom}\,((\bar{\partial}^*)^*) = \mathrm{dom}\,(\bar{\partial})$ because for all $g \in \mathrm{dom}\,(\bar{\partial}^*)$ we have

$$|(f, \bar{\partial}^* g)| = \lim_{n\to\infty} |(f_n, \bar{\partial}^* g)| \le \|g\|.$$

Thus $\|\bar{\partial} f\| \le 1$ and $f \in B_{\bar{\partial}}$. – For $q = 0$ this follows as above from (6.6). Analogously one treats $B_{\bar{\partial}^*}$.

Let $g \in B_\square$. Then $g \in \mathcal{R}(\bar{\partial}^*) \oplus \mathcal{R}(\bar{\partial})$ and $\|g\| \le c^{-1}$ because $c\|g\| \le \|\square g\|$. Hence

$$\|\bar{\partial} g\|^2 + \|\bar{\partial}^* g\|^2 = (g, \square g) \le c^{-1}.$$

Therefore $g \in c^{-1/2}(B_{\bar{\partial}} + B_{\bar{\partial}^*})$. The last set is compact. Let $\{f_n\}_n$ be a sequence in $B_\square$. Then there exists a subsequence, which is also denoted by $\{f_n\}_n$, with $f = \lim_{n\to\infty} f_n \in c^{-1/2}(B_{\bar{\partial}} + B_{\bar{\partial}^*})$.

Let $g \in \mathrm{dom}\,(\square)$. Then

$$|(f, \square g)| = \lim_{n\to\infty} |(f_n, \square g)| = \lim_{n\to\infty} |(\square f_n, g)| \le \|g\|.$$

Hence $f \in \text{dom}\,(\square^*) = \text{dom}\,(\square)$ and $\|\square f\| \leq 1$. Consequently $f \in B_\square$.

The compactness of N, $\overline{\partial} N$ and $\overline{\partial}^* N$ follows easily from the compactness of $B_\square$, $B_{\overline{\partial}^*}$ and $B_{\overline{\partial}}$ because of

$$c\|Nf\| \leq \|f\|,$$
$$c(\|\overline{\partial} N f\|^2 + \|\overline{\partial}^* N f\|^2) = c(\square N f, N f) \leq c\|f\| \cdot \|N f\| \leq \|f\|^2$$
$$\|\square N f\| \leq \|f\|,$$

and

$$\mathcal{R}\,(N) \subset \text{dom}\,(\square) \cap \text{ker}(\square)^\perp,$$
$$\mathcal{R}\,(\overline{\partial}^* N) \subset \text{dom}\,(\overline{\partial}) \cap \text{ker}(\overline{\partial})^\perp,$$
$$\mathcal{R}\,(\overline{\partial} N) \subset \text{dom}\,(\overline{\partial}^*) \cap \text{ker}(\overline{\partial}^*)^\perp.$$

From (6.2) and Ch. IV.6.5 it follows that, for $q \geq 1$,

$$\dim(\text{ker}(\square)) = \dim(\text{ker}\,\overline{\partial} \cap \text{ker}\,\overline{\partial}^*) = \dim(\text{ker}\,\overline{\partial}/\,\mathcal{R}\,(\overline{\partial})) < \infty.$$

$\square$

An important theorem of Kohn [FoK 72] asserts that compactness alone of N implies regularity in the following sense:

Theorem 6.7 *Let W_{pq}^s be the Sobolev space of order $s \geq 0$. If N_{pq} is compact from L_{pq}^2 into itself, then it sends W_{pq}^s continuously into itself.*

We shall prove different regularity theorems in the last chapter. Here we end with a definition (in case the Neumann operator exists):

Definition 6.8 *The operator $\overline{\partial}^* N$ is called the canonical solution operator to the Cauchy-Riemann equations.*

In fact, it follows from the above: if f is $\overline{\partial}$-closed and orthogonal to the harmonic space, then

$$f = \overline{\partial}\,\overline{\partial}^* N f,$$

and $\overline{\partial}^* N f$ has minimal L^2-norm among all solutions of $\overline{\partial} u = f$. But, naturally, since it depends on the metric, there are infinitely many canonical operators.

§7 Notes

Theorem 6.2 solves the Neumann problem on strictly pseudoconvex manifolds and immediately shows its compactness. The original solution of this problem is due to J. J. Kohn

[Koh 63, Koh 64]; it does not use integral formulae. Kohn proves, simultaneously with its existence, the regularity of the Neumann operator in Sobolev spaces. We will later-on prove regularity theorems in terms of C^k-Hölder spaces.

The first 5 paragraphs lay the ground for our further work: Theorem 2.6 [Fri 44] is all that is needed to justify our use of the homotopy formula in L^p spaces. The other main results, namely Hörmander's density theorem 5.6 and the self-adjointness of $\Box$, will be used in the following chapters when $\bar{\partial}^*$ appears in the integral formulae. We have followed, in §§2ff, Hörmander's paper [Hör 65] and Folland-Kohn [FoK 72].

Chapter VI

Integral Representations for the $\bar{\partial}$-Neumann Problem

We solved, in the last chapter, the Neumann problem on strictly pseudoconvex manifolds, more precisely, we proved existence and compactness of the Neumann operator. We will reach more in this chapter:

$1°$ We introduce the algebra of operators on L^2 which have asymptotic developments in terms of the previously introduced isotropic resp. admissible operators. This new algebra 3^a, the *algebra of asymptotic Z-operators*, contains (and is generated by) the isotropic and the admissible operators which were introduced before. It is again filtered by a type function, and we can define the notion of principal part in terms of this type function.

$2°$ The Bergman projector P (for any metric) is contained in 3^a. This follows fairly easily from our integral representations if Ω is Stein, but the proof in the general non-Stein case requires more powerful methods: existence of the Remmert reduction, invariance of cohomology dimension under perturbations, interior elliptic estimates — see 3.14 to 3.16. So the result is fairly deep.

$3°$ If one chooses the metric carefully (as a normalised Levi metric), then this algebra also contains the operators N, $\bar{\partial} N$ and $\bar{\partial}^* N$.

$4°$ For all these operators we succeed in computing and analyzing their principal parts.

The main tool is again the integral formula of chapter IV. But, as this formula does not invoke the adjoint operator $\bar{\partial}^*$ of $\bar{\partial}$ we must transform it into a shape more suitable for our purpose.

§1 The Basic Integral Representation

1. We consider again a strictly pseudoconvex subdomain $\Omega \subset\subset X$, where X is an n-dimensional hermitian manifold whose metric will be suitably chosen later on. The *Hodge $*$-operator* and the *geodesic distance function* ρ have been introduced before (Chapter IV,

§§1, 2 and 5). We still need a strictly plurisubharmonic C^∞ defining function r for Ω. Associated with these data are the following objects (see IV.5.46):

$$P(x,y) = \rho^2(x,y) + 2r(x)r(y), \tag{1.1}$$

the *extended norm function*, and

$$\Phi(x,y), \tag{1.2}$$

the *extended Levi polynomial*, whose essential properties are stated in Chapter IV, §5, 5.28ff. These functions are used to define *isotropic* and *admissible kernels of type* $\geq \lambda$ – see Chapter IV, §2 and 5.49. As before we will use the abbreviation

$$\mathcal{K} = \mathcal{E}_j \quad \text{or} \quad \mathcal{K} = \mathcal{A}_\lambda \tag{1.3}$$

for the statement: $\mathcal{K}$ *is an isotropic (resp. admissible) kernel of order* $\geq j$ *(and type* $\geq j + 2n$*) (resp. of type* $\geq \lambda$*)* and use $\mathcal{A}_\lambda$, $\mathcal{E}_j$ as a generic notation.

The following double differential forms are our starting point:

$$\beta(x,y) = \frac{\partial_x \rho^2(x,y)}{\rho^2(x,y)} \tag{1.4}$$

$$\alpha(x,y) = \xi(x) \frac{\partial r(x)}{\Phi(x,y)}, \tag{1.5}$$

where ξ is a smooth patching function which is $\equiv 1$ for $|r(x)| < \delta$ and 0 for $|r(x)| \geq \frac{3}{2}\delta$, $\delta > 0$ sufficiently small. Both forms were introduced in Chapter IV, 5.33 and in the lines preceding 5.19.

These forms are used to build up the *boundary kernels*

$$K_q = \Omega_q(\alpha) \tag{1.6}$$

and the *generalised Bochner-Martinelli-Koppelman kernels*

$$B_q = \Omega_q(\beta) \tag{1.7}$$

and the corresponding *transition kernels*

$$A_q = A_q(\alpha, \beta) = \sum a_{q\mu\nu} A_{q\mu\nu}(\alpha, \beta) \tag{1.8}$$

with typical term

$$A_{q\mu\nu}(\alpha, \beta) = \left(\frac{1}{2\pi i}\right)^n \alpha \wedge \beta \wedge (\bar{\partial}_x \alpha)^\mu \wedge (\bar{\partial}_x \beta)^{n-q-\mu-2} \wedge (\bar{\partial}_y \alpha)^\nu \wedge (\bar{\partial}_y \beta)^{q-\nu}. \tag{1.9}$$

In (1.8) and (1.9) we replace – as in Chapter IV, 5.46ff – the denominators ρ^2 by the extended norm function P and denote the resulting kernels again by A_q etc., and recall the basic formula Ch. IV.5.48, which we write in slightly different notation: *if $f \in C^1_{0,q}(\overline{\Omega})$, then*

$$\int_{b\Omega} f \wedge B_q = \int_\Omega \overline{\partial} f \wedge K_q + (-1)^q \int_\Omega f \wedge \overline{\partial}_x K_q + (-1)^{q+1} \int_\Omega \overline{\partial} f \wedge \overline{\partial}_x A_q$$

$$+ \int_\Omega \overline{\partial} f \wedge \overline{\partial}_y A_{q-1} + (-1)^q \int_\Omega f \wedge \overline{\partial}_y \overline{\partial}_x A_{q-1} + (\overline{\partial} f, \mathcal{E}_{2-2n})$$

$$+ (\overline{\partial} f, \mathcal{A}_2) + (f, \mathcal{E}_{1-2n}) + (f, \mathcal{A}_1), \quad (1.10)$$

with isotropic resp. admissible kernels of the indicated order resp. type. In order to write all the integrals as scalar products and, at the same time, to exhibit the role of the adjoint ϑ, we set

$$\mathcal{L}_q(x,y) = (-1)^{q+1} *_x \overline{A_q(x,y)} \qquad (1.11)$$

$$\vartheta = - * \partial * \qquad (1.12)$$

and obtain 1.10 in the following form:

Proposition 1.13 *For $f \in C^1_{0,q}(\overline{\Omega})$ and $y \in \Omega$ one has*

$$\int_{b\Omega} f(x) \wedge B_q(x,y) = (-1)^{q+1}(\overline{\partial} f(x), *\overline{K_q(x,y)}) + (f(x), *\partial_x \overline{K_q(x,y)})$$

$$+ (\overline{\partial} f(x), \vartheta_x \mathcal{L}_q(x,y) - \partial_y \mathcal{L}_{q-1}(x,y)) + (f(x), \vartheta_x \partial_y \mathcal{L}_{q-1}(x,y)) + (\overline{\partial} f(x), \mathcal{A}_2(x,y))$$

$$+ (f(x), \mathcal{A}_1(x,y)) + (\overline{\partial} f(x), \mathcal{E}_{2-2n}(x,y)) + (f(x), \mathcal{E}_{1-2n}(x,y)),$$

where $\mathcal{A}_2$ and $\mathcal{A}_1$ are admissible kernels of the indicated type, $\mathcal{E}_{2-2n}$ and $\mathcal{E}_{1-2n}$ isotropic of the corresponding order.

We now state the BMK formula (see IV.5.14 and 5.21).

Proposition 1.14 *For f and y as above,*

$$f(y) = \int_{b\Omega} f(x) \wedge B_q(x,y) + (\overline{\partial} f, \overline{\partial}_x \Gamma_{0,q}) + (\vartheta f, \vartheta_x \Gamma_{0,q})$$

$$+ \int_{b\Omega} \overline{\vartheta_x \Gamma_{0,q}(x,y)} \wedge *f(x) + (f, \mathcal{E}_{1-2n}) + (\overline{\partial} f, \mathcal{E}_{2-2n})$$

with isotropic kernels of the indicated order.

Recall that $\Gamma_{0q}(x,y)$ is the parametrix of Chapter IV for the complex Laplacian; it satisfies (see IV.5.15)

$$\Gamma_{0,q}(x,y) = \Gamma_{0,q}(x,y)^* \qquad (1.15)$$

and

$$\vartheta_x \Gamma_{0,q} = \partial_y \Gamma_{0,q-1} + \mathcal{E}_{2-2n} \quad (\text{for } q \geq 0). \qquad (1.16)$$

2.   Let us now suppose that $f \in C^1_{0,q}(\overline{\Omega})$ *satisfies the first Neumann condition*

$$*f|_{b\Omega} \equiv 0. \tag{1.17}$$

Then the second boundary integral in (1.14) vanishes, and the first can be replaced by (1.13). We introduce $\bar{\partial}^* f$ in (1.13) in the following way:

$$(f, \vartheta_x \partial_y \mathcal{L}_{q-1}) = (f, \vartheta_x \partial_y \mathcal{L}_{q-1} - (\vartheta_x \partial_y \mathcal{L}_{q-1})^*) + (f, (\vartheta_x \partial_y \mathcal{L}_{q-1})^*). \tag{1.18}$$

Recall that the *adjoint* of a kernel $\mathcal{K}(x,y)$ is

$$\mathcal{K}^*(x,y) = \overline{\mathcal{K}(y,x)}.$$

In particular,

$$(\vartheta_x \partial_y \mathcal{L}_{q-1})^* = \overline{\vartheta}_y \overline{\partial}_x \mathcal{L}^*_{q-1},$$

and consequently, by partial integration,

$$(f, (\vartheta_x \partial_y \mathcal{L}_{q-1})^*) = (\bar{\partial}^* f, \overline{\vartheta}_y \mathcal{L}^*_{q-1}) \tag{1.19}$$

because of (1.17). We insert (1.13), (1.18) and (1.19) into (1.14) and rearrange:

$$f(y) = (\overline{\partial} f, \vartheta_x \mathcal{L}_q - \partial_y \mathcal{L}_{q-1} + (-1)^{q+1} * \overline{K_q}) + (\overline{\partial} f, \overline{\partial}_x \Gamma_{0q})$$
$$+ (\bar{\partial}^* f, \overline{\vartheta}_y \mathcal{L}^*_{q-1}) + (\bar{\partial}^* f, \vartheta_x \Gamma_{0q}) + (f, *\partial_x \overline{K_q}) + (f, \vartheta_x \partial_y \mathcal{L}_{q-1} - (\vartheta_x \partial_y \mathcal{L}_{q-1})^*)$$
$$+ (\overline{\partial} f, \mathcal{E}_{2-2n} + \mathcal{A}_2) + (f, \mathcal{A}_1 + \mathcal{E}_{1-2n}) \tag{1.20}$$

We abbreviate the above kernels:

$$\begin{aligned}
T^a_q(x,y) &= \vartheta_x \mathcal{L}_q(x,y) - \partial_y \mathcal{L}_{q-1}(x,y), \qquad q \geq 1 \\
T^a_0(x,y) &= \vartheta_x \mathcal{L}_0(x,y) - *\overline{K}_0(x,y) \\
T^i_q(x,y) &= \overline{\partial}_x \Gamma_{0q}(x,y), \qquad q \geq 0 \\
T_q(x,y) &= T^a_q(x,y) + T^i_q(x,y) \\
P_q(x,y) &= Q_q(x,y) - Q^*_q(x,y) \\
&= \vartheta_x \partial_y \mathcal{L}_{q-1}(x,y) - (\vartheta_x \partial_y \mathcal{L}_{q-1}(x,y))^* \\
Q_q(x,y) &= \vartheta_x \partial_y \mathcal{L}_{q-1}(x,y), \qquad q \geq 1, \\
P_0(x,y) &= *\partial_x \overline{K_0(x,y)}.
\end{aligned} \tag{1.21}$$

Moreover, for $q \geq 1$ the kernel

$$K_q(x,y) = \mathcal{E}_0(x,y), \tag{1.22}$$

that is, it is infinitely differentiable on all of $\overline{\Omega} \times \overline{\Omega}$, because it contains $\overline{\partial}_y \alpha$ as a factor, and

$$\overline{\partial}_y \alpha(x,y) \equiv 0$$

near the boundary diagonal. For $q = 0$, we have, correspondingly,

$$\overline{\partial}_y K_0(x, y) = \mathcal{E}_0(x, y). \tag{1.23}$$

Next, we compute, for $q \geq 2$,

$$\begin{aligned}
(\overline{\partial}^* f, \mathcal{T}_{q-1}^*) &= (\overline{\partial}^* f, [\vartheta_x \mathcal{L}_{q-1} - \partial_y \mathcal{L}_{q-2} + \overline{\partial}_x \Gamma_{0,q-1}]^*) \\
&= (\overline{\partial}^* f, \overline{\partial}_y \mathcal{L}_{q-1}^*) - (\overline{\partial}^* f, \overline{\partial}_x \mathcal{L}_{q-2}^*) + (\overline{\partial}^* f, \partial_y \Gamma_{0,q-1}^*) \\
&= (\overline{\partial}^* f, \overline{\partial}_y \mathcal{L}_{q-1}^*) + (\overline{\partial}^* f, \vartheta_x \Gamma_{0,q}) + (\overline{\partial}^* f, \mathcal{E}_{2-2n}),
\end{aligned} \tag{1.24}$$

where we have used (1.17) and (1.16). Similarly, for $q = 1$,

$$\begin{aligned}
(\overline{\partial}^* f, \mathcal{T}_0^*) &= (\overline{\partial}^* f, \overline{\partial}_y \mathcal{L}_0^*) - (\overline{\partial}^* f, (*\overline{K_0})^*) + (\overline{\partial}^* f, \partial_y \Gamma_{00}^*) \\
&= (\overline{\partial}^* f, \overline{\partial}_y \mathcal{L}_0^*) - (f, \mathcal{E}_0^\infty) + (\overline{\partial}^* f, \vartheta_x \Gamma_{01}) + (\overline{\partial}^* f, \mathcal{E}_{2-2n}),
\end{aligned} \tag{1.25}$$

because of (1.16), (1.17) and (1.23).

Let us insert the above information into (1.20) and use

$$\mathcal{E}_j + \mathcal{E}_0 = \mathcal{E}_j \quad \text{for } j \leq 0.$$

(1.20) now becomes

$$\begin{aligned}
f(y) = (\overline{\partial} f, \mathcal{T}_q) + (\overline{\partial}^* f, \mathcal{T}_{q-1}^*) + (f, \mathcal{P}_q) \\
+ (\overline{\partial} f, \mathcal{A}_2 + \mathcal{E}_{2-2n}) + (\overline{\partial}^* f, \mathcal{E}_{2-2n}) + (f, \mathcal{A}_1 + \mathcal{E}_{1-2n}). \tag{1.20'}
\end{aligned}$$

We denote the operators with kernels $\mathcal{T}_q$, $\mathcal{P}_q$, $\mathcal{T}_q^a$, $\mathcal{T}_q^i$ by $\boldsymbol{T}_q$, $\boldsymbol{P}_q$, $\boldsymbol{T}_q^a$, $\boldsymbol{T}_q^i$ resp. and the operators with kernels $\mathcal{A}_2$ or $\mathcal{E}_{2-2n}$ by A_2, E_{2-2n} etc. and obtain

Theorem 1.26 (Basic integral representation) *The above integral operators* $\boldsymbol{T}_q = \boldsymbol{T}_q^a + \boldsymbol{T}_q^i$ *and* $\boldsymbol{P}_q$ *have the following properties:*

i. $\boldsymbol{T}_q^a$ *and* $\boldsymbol{P}_q$ *are admissible,* $\boldsymbol{T}_q^i$ *isotropic operators.*

ii. *There are admissible resp. isotropic operators of the indicated type resp. order, such that for* $f \in C_{0,q}^1(\overline{\Omega})$ *satisfying the first Neumann condition*

$$f = \boldsymbol{T}_q \overline{\partial} f + \boldsymbol{T}_{q-1}^* \overline{\partial}^* f + \boldsymbol{P}_q f + (A_2 + E_{2-2n})\overline{\partial} f + E_{2-2n}\overline{\partial}^* f + (A_1 + E_{1-2n})f$$

(The notation of operators is generic; so the two E_{2-2n} operators in 1.26 are different).

3. The above theorem has to be completed by a determination of the type of the operators $\boldsymbol{T}_q$ etc.

Theorem 1.27 *The operators* $\boldsymbol{T}_q^a$ *are admissible of type* ≥ 1; *the operators* $\boldsymbol{T}_q^i$ *are isotropic of type* ≥ 1 *(that is of order* $\geq 1 - 2n$*)*.

This implies the same property for the adjoint operators T_q^{a*} and T_q^{i*}, of course.

Theorem 1.28   *i. The operator P_0 is special admissible of type ≥ 0.*

   ii. The operator $P_0 - P_0^*$ is admissible of type ≥ 1.

   iii. The operator $\bar{\partial} P_0$ has a smooth kernel which vanishes near the boundary diagonal on $\bar{\Omega} \times \bar{\Omega}$.

Theorem 1.29   *i. The operators P_q for $q \geq 1$ are of the form*

$$P_q = Q_q - Q_q^*,$$

   where Q_q is admissible of type ≥ 0.

   ii. If the metric is a normalised Levi metric, then the operators P_q (for $q \geq 1$) are of type ≥ 1.

(See Definitions 2.16, 2.34 and Proposition 2.25.)

To prove the above statements we have to look at (1.21), the formula for the kernels which occur. This yields immediately the assertion for the T_q^i. Furthermore, near the boundary diagonal, there is a smooth function N such that

$$K_0(x, y) = \frac{N(x)}{\Phi(x, y)^n},$$

which is admissible of type ≥ 2; so all x-derivatives are special admissible of type ≥ 0, whereas all derivatives with respect to $\bar{y}$ are 0 near the boundary diagonal – which implies (1.28) (i) and (iii).

The kernel $\mathcal{L}_q$ is a sum of terms

$$\frac{\mathcal{E}_1}{\bar{\Phi}^{\mu+1} P^{n-\mu-1}}, \quad \mu \geq 0. \tag{1.30}$$

If $\mu > 0$, then (1.30) is of type ≥ 3, hence any first derivative is of type ≥ 1. For $\mu = 0$ and X an arbitrary vector field in x or y we have

$$X \frac{\mathcal{E}_1}{\bar{\Phi} P^{n-1}} = \frac{\mathcal{E}_0}{\bar{\Phi} P^{n-1}} + \frac{\mathcal{E}_1}{\bar{\Phi}^2 P^{n-1}} + \frac{\mathcal{E}_1 \cdot XP}{\bar{\Phi} P^n}. \tag{1.31}$$

Since

$$XP = X(\rho^2 + 2rr^*) = \mathcal{E}_1 + \mathcal{E}_0 r + \mathcal{E}_0 r^*, \tag{1.32}$$

all three terms in (1.31) are easily seen to be of type ≥ 1. This proves the remaining assertion in (1.27).

In order to determine the type of Q_q, we have to compute

$$\partial_y \partial_x \frac{\mathcal{E}_1}{\overline{\Phi}^{\mu+1} \mathrm{P}^{n-\mu-1}}, \qquad \mu \geq 0. \tag{1.33}$$

But since $\partial_y \overline{\Phi} = 0$ near the boundary diagonal, (1.33) becomes

$$\partial_x \frac{1}{\overline{\Phi}^{\mu+1}} \partial_y \frac{\mathcal{E}_1}{\mathrm{P}^{n-\mu-1}}, \tag{1.34}$$

and a straightforward computation as (1.31), using again (1.32), shows that (1.34) is of type ≥ 0.

By now we have proved all assertions (1.27)–(1.29) except (1.28)(ii) and (1.29)(ii). These two statements are much more delicate, in particular the last one; the whole of §2 is dedicated to their proof.

§2 Cancellation of Singularities

We now complete the proof of Theorems (1.28) and (1.29).

1. The kernel $\mathcal{P}_0$ can be computed using (1.21) and (1.6); the result is

$$\mathcal{P}_0(x, y) = *\partial_x \overline{K_0(x, y)} \tag{2.1}$$

$$= (-1)^n \left(\frac{1}{2\pi i}\right)^n \frac{n\xi^n(x)}{\overline{\Phi}^{n+1}(x, y)} * \partial r(x) \wedge \overline{\partial} r(x) \wedge (\partial \overline{\partial} r(x))^{n-1} + \mathcal{A}_1$$

$$= \frac{l(x)}{\overline{\Phi}^{n+1}(x, y)} + \mathcal{A}_1(x, y),$$

where we have used the definition of K_0 (see Chapter IV, 5.34) and the relation

$$\partial_x \overline{\Phi}(x, y) = -\partial r(x) + \mathcal{E}_1(x, y). \tag{2.2}$$

The function $l(x)$ is smooth and real-valued. Consequently,

$$\mathcal{P}_0(x, y)^* = \frac{l(y)}{\Phi(y, x)^{n+1}} + \mathcal{A}_1(x, y), \tag{2.3}$$

and we obtain

$$\mathcal{P}_0(x, y) - \mathcal{P}_0(x, y)^* = \frac{l(x) - l(y)}{\overline{\Phi}(x, y)^{n+1}} + l(y) \left[\frac{1}{\overline{\Phi}^{n+1}(x, y)} - \frac{1}{\Phi(y, x)^{n+1}} \right] + \mathcal{A}_1(x, y) \tag{2.4}$$

$$= \frac{\mathcal{E}_1(x, y)}{\overline{\Phi}(x, y)^{n+1}} + l(y) \left[\Phi(y, x) - \overline{\Phi}(x, y) \right] \cdot$$

$$\cdot \left[\frac{1}{\overline{\Phi}(x, y)^{n+1} \Phi(y, x)} + \cdots + \frac{1}{\overline{\Phi}(x, y) \Phi(y, x)^{n+1}} \right] + \mathcal{A}_1(x, y).$$

But

$$\Phi(y,x) - \overline{\Phi}(x,y) = \mathcal{E}_3(x,y),\tag{2.5}$$

and this implies that all terms on the right-hand side of (2.4) are $\mathcal{A}_1$. – So Theorem 1.28 is proved.

2.   Before we can turn to the proof of Theorem 1.29 we need some explicit information on the derivatives of ρ^2 and r. To this end, we denote points in $X \times X$ with (x,y) and local coordinates with ζ for x, z for y. Let us fix a point $x_0 \in b\Omega$ and choose an orthonormal frame $\omega^1, \ldots, \omega^n$ of smooth $(1,0)$-forms on a neighbourhood U of x_0 such that

$$\partial r(x) = h(x)\omega^n(x)\tag{2.6}$$

with a smooth nonvanishing function h on U. Let $L_1, \ldots, L_n$ be the dual frame of $(1,0)$ vector fields on U. On $U \times U$ these operators refer to the first variable x; when they are to refer to the second variable y they will be denoted by Θ^j resp. Λ_j.

Proposition 2.7   *i.* $L_n r(x) \equiv h(x)$

   ii. $L_j r(x) \equiv 0$ *for* $j < n$

   iii. $L_j \overline{L}_k \rho^2(x,y) = 2\delta_{jk} + \mathcal{E}_1(x,y)$

   iv. $\partial_x \overline{\partial}_y \rho^2(x,y) = 2\sum_j \omega^j \wedge \overline{\omega}^j + \mathcal{E}_1(x,y)$

   v. $\partial_y \overline{\partial}_x \rho^2(x,y) = -2\sum_j \Theta^j \wedge \overline{\omega}^j + \mathcal{E}_1(x,y)$

   vi. $\Lambda_j \rho^2(x,y) = -L_j \rho^2(x,y) + \mathcal{E}_2(x,y)$

   vii. $L_j L_k \rho^2(x,y) = \mathcal{E}_1(x,y)$

   viii. $\partial_x \rho^2(x,y) = \mathcal{E}_1(x,y)$

   ix. $L_n \overline{\Phi} = -1 + \mathcal{E}_1(x,y)$

Proposition 2.8

$$\vartheta\Big(\sum f_J \overline{\omega}^J\Big) = -\sum \varepsilon^J_{kK}(L_k f_J)\overline{\omega}^K + \ldots,$$

where ... *stands for a differential operator of order 0.*

Proof of 2.7. *i*) and *ii*) are immediate; let us look at *iii*). We fix y and choose local coordinates ζ near y such that

$$d\zeta^j(y) = \omega^j(y).\tag{2.9}$$

Then $ds^2 = \sum g_{jk}(\zeta)\,d\zeta^j\,d\overline{\zeta}^k$, and, setting

$$R^2(\zeta,z) = \sum g_{jk}(\zeta)(\zeta^j - z^j)(\overline{\zeta}^k - \overline{z}^k),\tag{2.10}$$

we have

$$\rho^2(x, y) = R^2(\zeta(x), z(y)) + \mathcal{E}_3(x, y). \tag{2.11}$$

Consequently,

$$L_j \overline{L}_k \rho^2 = L_j \overline{L}_k R^2 + \mathcal{E}_1(x, y). \tag{2.12}$$

But in view of (2.9) we have

$$L_j = \frac{\partial}{\partial \zeta^j} + \mathcal{E}_1 L$$

with a first order operator L, and so (2.12) becomes

$$L_j \overline{L}_k \rho^2 = \frac{\partial^2}{\partial \zeta^j \partial \overline{\zeta}^k} R^2 + \mathcal{E}_1 = g_{jk}(\zeta) + \mathcal{E}_1. \tag{2.13}$$

Now, since the $d\zeta^j$ are orthonormal at z we have

$$g_{jk}(\zeta) = 2\delta_{jk} + \mathcal{E}_1. \tag{2.14}$$

This yields iii).   □

The other claims in 2.7 and 2.8 are proved in the same way, always using the basic information (2.12) which we state once more as

Proposition 2.15 *If R^2 is defined in local coordinates as*

$$R^2(\zeta, z) = \sum g_{jk}(\zeta)(\zeta^j - z^j)(\overline{\zeta}^k - \overline{z}^k),$$

then

$$\rho^2 = R^2 + \mathcal{E}_3.$$

3. It is convenient to slightly improve 2.7i:

Definition 2.16 *The metric ds^2 is normalised (with respect to the boundary function r) if the length*

$$|\partial r| \equiv 1$$

at all points of the boundary.

Any metric is of course equivalent to a normalised metric. If ds^2 is normalised, then we can arrange, instead of 2.7i, for the more precise relation

$$L_n r \equiv 1 \qquad \text{on } b\Omega. \tag{2.17}$$

From now on we will assume the normalisation condition. We then have the following fairly delicate relations between the metric and the extended Levi polynomial Φ:

Proposition 2.18 $i.$ $\Lambda_n P = -2\overline{\Phi} + \mathcal{E}_2 + r(x)\mathcal{E}_1.$

  ii. $L_n P = -2\Phi^* + \mathcal{E}_2 + r(x)\mathcal{E}_1.$

Proposition 2.19

$$2P - \sum_{j<n} |L_j\rho^2|^2 = 4|\Phi|^2 + r(x)\mathcal{E}_2 + \mathcal{E}_3.$$

Proof of 2.18. *ii*) follows from *i*), so we prove *i*). By (2.17) we have

$$\Lambda_n P^2 = \Lambda_n \rho^2 + 2r(x) + r(x)\mathcal{E}_1. \tag{2.20}$$

Let us fix the point x and choose local coordinates z such that

$$dz^j(x) = \Theta^j(x). \tag{2.21}$$

Then

$$ds^2 = \sum g_{jk}(z)\, dz^j d\overline{z}^k$$
$$R^2(\zeta, z) = \sum g_{jk}(\zeta^j - z^j)(\overline{\zeta}^k - \overline{z}^k), \tag{2.22}$$

where ζ is the coordinate of x. (2.15) yields

$$\Lambda_n \rho^2 = \Lambda_n R^2 + \mathcal{E}_2, \tag{2.23}$$

and from (2.21) we have

$$\Lambda_n = \frac{\partial}{\partial z^n} + \mathcal{E}_1 \Lambda \tag{2.24}$$

with a first order differential operator Λ. So (2.23) reads as

$$\Lambda_n \rho^2 = \frac{\partial}{\partial z^n} R^2 + \mathcal{E}_2. \tag{2.25}$$

But (2.21) and the orthogonality of the Θ^j imply

$$g_{jk}(\zeta) = 2\delta_{jk}. \tag{2.26}$$

Putting all this together we obtain

$$\Lambda_n P = -2(\overline{\zeta}^n - \overline{z}^n) + 2r(\zeta) + r(\zeta)\mathcal{E}_1 + \mathcal{E}_2. \tag{2.27}$$

On the other hand, if $F(\zeta, z)$ is the Levi polynomial of r in the above coordinates, we have

$$\Phi(\zeta, z) = F(\zeta, z) - r(\zeta) + \mathcal{E}_2(\zeta, z) \tag{2.28}$$
$$= (\zeta^n - z^n) - r(\zeta) + \mathcal{E}_2(\zeta, z).$$

Taking conjugates and inserting (2.28) into (2.27) yields the claim. $\square$

Proof of 2.19. We fix x and choose local coordinates as in the previous proof. Then – cf. (2.24) –

$$\Lambda_j = \frac{\partial}{\partial z^j} + \mathcal{E}_1 \Lambda. \tag{2.29}$$

By 2.7vi, (2.23) and (2.26), we have

$$|L_j \rho^2|^2 = |\Lambda_j \rho^2|^2 + \mathcal{E}_3 = \left| \frac{\partial}{\partial z^j} \rho^2 \right|^2 + \mathcal{E}_3$$
$$= 4|\zeta^j - z^j|^2 + \mathcal{E}_3,$$

and so

$$2P - \sum_{j<n} |L_j \rho^2|^2 = 4|\zeta^n - z^n|^2 + 4r(\zeta)r(z) + \mathcal{E}_3. \tag{2.30}$$

On the other hand, (2.28) yields

$$\Phi\overline{\Phi} = [(\zeta^n - z^n) - r(\zeta) + \mathcal{E}_2]\overline{\Phi} \tag{2.31}$$
$$= (\zeta^n - z^n)[\overline{\zeta}^n - \overline{z}^n - r(\zeta) + \mathcal{E}_2] - r(\zeta)\overline{\Phi} + \mathcal{E}_2\overline{\Phi}$$
$$= |\zeta^n - z^n|^2 - r(\zeta)[\zeta^n - z^n + \overline{\Phi}] + r(\zeta)\mathcal{E}_2 + \mathcal{E}_3.$$

Now

$$\zeta^n - z^n + \overline{\Phi}(\zeta, z) = \zeta^n - z^n + \overline{\Phi}^*(\zeta, z) + \mathcal{E}_3 \tag{2.32}$$
$$= \zeta^n - z^n + \Phi(z, \zeta) + \mathcal{E}_3$$
$$= \zeta^n - z^n + z^n - \zeta^n - r(z) + \mathcal{E}_2$$
$$= -r(z) + \mathcal{E}_2.$$

Combining (2.31) and (2.32) gives

$$\Phi\overline{\Phi} = |\zeta^n - z^n|^2 + r(\zeta)r(z) + r(\zeta)\mathcal{E}_2 + \mathcal{E}_3; \tag{2.33}$$

(2.30) and (2.33) now yield the claim.   □

The normalisation condition 2.16 is not yet sufficient to establish Theorem 1.29.ii (except if $q = n - 1$); we need a closer correspondence between the hermitian geometry of the ambient space and the complex geometry of the boundary. This is provided for by a special choice of the metric.

Definition 2.34 *A hermitian metric ds^2 is called a Levi metric for Ω, if it is given in local coordinates near the boundary of Ω as*

$$ds^2 = f(\zeta) \sum_{j,k} \frac{\partial^2 r(\zeta)}{\partial \zeta^j \partial \overline{\zeta}^k} d\zeta^j d\overline{\zeta}^k$$

with a positive smooth function f. Here r is a strictly plurisubharmonic defining function for Ω.

Levi metrics always exist; any two such metrics are conformally equivalent on the holomorphic tangent bundle to the boundary. Moreover, we can normalise a given Levi metric by choosing the function f appropriately. In fact, we proved more in Chapter III, 1.14:

Proposition 2.35 *There is a strictly plurisubharmonic defining function r for Ω such that*

$$ds^2 = \sum \frac{\partial^2 r(\zeta)}{\partial \zeta^j \partial \bar{\zeta}^k} \, d\zeta^j \, d\bar{\zeta}^k$$

is a normalised Levi metric.

So the function f can be dispensed with.

If ds^2 is a Levi metric, then we have, in the expression 2.15 for ρ^2,

$$R^2(\zeta, z) = \sum f(\zeta) \frac{\partial^2 r(\zeta)}{\partial \zeta^j \partial \bar{\zeta}^k}(\zeta)(\zeta^j - z^j)(\bar{\zeta}^k - \bar{z}^k) \,,$$

$$\bar{\partial}_\zeta \partial_\zeta R^2(\zeta, z) = \sum f(\zeta) \frac{\partial^2 r(\zeta)}{\partial \zeta^j \partial \bar{\zeta}^k} \, d\zeta^j \, d\bar{\zeta}^k + \mathcal{E}_1 \,.$$

We note the result as

Proposition 2.36 *For any Levi metric, we have the relation*

$$\bar{\partial}_x \partial_x \rho^2(x, y) = f(x) \bar{\partial} \partial r(x) + \mathcal{E}_1(x, y)$$

near the boundary of Ω.

4. We now choose a fixed normalised Levi metric and compute the principal parts (i. e.terms of type 0) of the kernels 1.21

$$\mathcal{Q}_q = \vartheta_x \partial_y \mathcal{L}_{q-1}$$

which arise in Theorem 1.29, using, as our essential tools, the identities 2.18, 2.19 and 2.36. The relevant information lies in the formulae (1.11), (1.8) and (1.9), as well in (1.4) and (1.5):

$$\mathcal{L}_q = (-1)^{q+1} * \bar{A}_q \tag{2.37}$$

$$A_q = \sum_{\substack{\mu=0 \\ \nu=0,\ldots q}}^{n-q-2} a_{q\mu\nu} A_{q\mu\nu}(\alpha, \beta)$$

$$A_{q\mu\nu}(\alpha, \beta) = \left(\frac{1}{2\pi i}\right)^n \alpha \wedge \beta \wedge (\bar{\partial}_x \alpha)^\mu \wedge (\bar{\partial}_x \beta)^{n-q-\mu-2} \wedge (\bar{\partial}_y \alpha)^\nu \wedge (\bar{\partial}_y \beta)^{q-\nu} \tag{2.38}$$

Since $\bar{\partial}_y \alpha = 0$ near the boundary diagonal, we have, first of all

$$A_q = \sum_{\mu=0}^{n-q-2} a_{q\mu 0} A_{q\mu 0}(\alpha, \beta) + \mathcal{E}_0 \,. \tag{2.39}$$

Inserting (1.4) and (1.5) into (2.38) we get

$$A_{q\mu 0} = \left(\frac{1}{2\pi i}\right)^n \frac{1}{\overline{\Phi}^{\mu+1}\mathrm{P}^{n-\mu-1}} \partial r \wedge \partial \rho^2 \wedge (\overline{\partial}_x \partial_x r)^\mu \wedge (\overline{\partial}_x \partial_x \rho^2)^{n-q-\mu-2} \wedge (\overline{\partial}_y \partial_x \rho^2)^q$$

$$+ \mathcal{E}_0 . \quad (2.40)$$

We now use (2.7) and (2.36) to express (2.40) near the boundary diagonal in terms of the local base fields chosen at the beginning of this paragraph, and obtain

$$A_{q\mu 0} = \left(\frac{1}{2\pi i}\right)^n \frac{1}{\overline{\Phi}^{\mu+1}\mathrm{P}^{n-\mu-1}} \left[h_{q\mu}(x)\omega^n \wedge \sum_j (L_j \rho^2) \, \omega^j \right.$$

$$\left. \wedge \left(\sum_k \overline{\omega}^k \wedge \omega^k\right)^{n-q-2} \wedge \left(\sum_l \overline{\Theta}^l \wedge \omega^l\right)^q + \mathcal{E}_2 \right], \quad (2.41)$$

where the function $h_{q\mu}(x)$ incorporates powers of (-1) and factors occurring in 2.36 and is therefore real-valued. Introducing multiindices and carrying out the powers in (2.41) now yields

Lemma 2.42 *There are real-valued functions $g_{q\mu}(x)$ such that*

$$\mathcal{L}_q = \sum_{\mu=0}^{n-q-2} \left[g_{q\mu} C_{q\mu} + \frac{\mathcal{E}_2}{\overline{\Phi}^{\mu+1}\mathrm{P}^{n-\mu-1}} \right] + \mathcal{E}_0 ,$$

$$\text{with} \quad C_{q\mu} = \sum_{\substack{|Q|=q \\ j<n}} \frac{\overline{L}_j \rho^2}{\overline{\Phi}^{\mu+1}\mathrm{P}^{n-\mu-1}} \overline{\omega}^{njQ} \wedge \Theta^Q .$$

Proof   Pass to the complex conjugate in (2.41), do the powers, and note that the $*$-operator involves i^n which cancels against $(2\pi i)^{-n}$ to make all the factors real-valued. $\square$

From 2.42 we easily deduce

Lemma 2.43

$$\mathcal{Q}_{q+1} = \sum_{\mu=0}^{n-q-2} g_{q\mu}(x)[\vartheta_x \partial_y C_{q\mu} - (\vartheta_x \partial_y C_{q\mu})^*] + \mathcal{A}_1 .$$

In fact, $g_{q\mu}(y)$ can be replaced by $g_{q\mu}(x) + \mathcal{E}_1$, and the arising $\mathcal{E}_1$-term is incorporated in $\mathcal{A}_1$. – So Theorem 1.29 is proved once we establish

Proposition 2.44 *Let $C_{q\mu}$ be given in 2.42. Then*

$$\vartheta_x \partial_y C_{q\mu} - (\vartheta_x \vartheta_y C_{q\mu})^* = \mathcal{A}_1.$$

This will be proved in the remainder of this paragraph.

5.   The form $\vartheta_x \partial_y C_{q\mu}$ is of double type $(0, q+1; q+1, 0)$ and so has a representation

$$\vartheta_x \partial_y C_{q\mu} = \sum_{\substack{|K|=q+1 \\ |L|=q+1}} A^\mu_{KL} \overline{\omega}^K \wedge \Theta^L \tag{2.45}$$

in the local base fields ω and Θ; the summation extends over strictly increasing $q+1$-tuples K and L. Therefore

$$(\vartheta_x \partial_y C_{q\mu})^* = \sum_{\substack{|K|=q+1 \\ |L|=q+1}} A^{\mu *}_{LK} \overline{\omega}^K \wedge \Theta^L,$$

and we have to show

$$A^\mu_{KL} - A^{\mu *}_{LK} = \mathcal{A}_1. \tag{2.46}$$

Before going into the bulky computations we introduce some simplifications. We shall namely neglect in all formulae terms which eventually lead to $\mathcal{A}_1$-kernels. So all equations of the form

$$\mathcal{R} = \mathcal{S}$$

are to be understood as

$$\mathcal{R} = \mathcal{S} + \mathcal{T},$$

where $\mathcal{T}$ leads in the further calculations to an $\mathcal{A}_1$-operator. For instance, when we use (2.8) we shall drop the order zero operator, or, in 2.7.iii we skip the $\mathcal{E}_1$-term.

To compute the A^μ_{KL} we work out (2.45) using (2.42):

$$\vartheta_x \partial_y C_{q\mu} = \vartheta_x \partial_y \sum_{\substack{|Q|=q \\ j<n}} \frac{1}{\overline{\Phi}^{\mu+1}} \frac{\overline{L}_j \rho^2}{\mathrm{P}^{n-\mu-1}} \overline{\omega}^{njQ} \wedge \Theta^Q \tag{2.47}$$

$$= \vartheta_x \sum_{\substack{|Q|=q \\ j<n \\ k}} \frac{1}{\overline{\Phi}^{\mu+1}} \Lambda_k \left(\frac{\overline{L}_j \rho^2}{\mathrm{P}^{n-\mu-1}} \right) \overline{\omega}^{njQ} \wedge \Theta^{kQ} .$$

Let us now introduce

$$\mathcal{M}^\mu_{kj} = \frac{1}{\overline{\Phi}^{\mu+1}} \Lambda_k \left(\frac{\overline{L}_j \rho^2}{\mathrm{P}^{n-\mu-1}} \right) . \tag{2.48}$$

Then (2.47) continues as

$$\vartheta_x \sum_{\substack{|Q|=q \\ j<n \\ k}} \mathcal{M}_{kj}^\mu \overline{\omega}^{njQ} \wedge \Theta^{kQ} \tag{2.49}$$

$$= - \sum_{\substack{|Q|=q \\ j<n \\ k,m \\ |K|=q+1}} \varepsilon_{mK}^{njQ} L_m(\mathcal{M}_{kj}^\mu) \overline{\omega}^K \wedge \Theta^{kQ}$$

$$= - \sum_{\substack{|K|=q+1 \\ |L|=q+1 \\ |Q|=q \\ j<n \\ k,m}} \varepsilon_{kQ}^L \varepsilon_{mK}^{njQ} L_m(\mathcal{M}_{kj}^\mu) \overline{\omega}^K \wedge \Theta^L.$$

Consequently,

$$A_{KL}^\mu = - \sum_{\substack{|Q|=q \\ j<n \\ k,m}} \varepsilon_{kQ}^L \varepsilon_{mK}^{njQ} L_m \mathcal{M}_{kj}^\mu. \tag{2.50}$$

The signs ε_S^R, ε_{aS}^{aT} etc. have been explained in Chapter I, as well as the multiindex-notation.

A straightforward computation yields

$$\mathcal{M}_{kj}^\mu = \frac{1}{\overline{\Phi}^{\mu+1}} \left[\frac{-2\delta_{kj}}{P^{n-\mu-1}} + \frac{n-\mu-1}{P^{n-\mu}} (L_k\rho^2)(\overline{L}_j\rho^2) \right], \quad k < n \tag{2.51}$$

$$\mathcal{M}_{nj}^\mu = \frac{2(n-\mu-1)\overline{L}_j\rho^2}{\overline{\Phi}^\mu P^{n-\mu}}, \quad j < n \tag{2.52}$$

(We have used 2.7 and 2.18, and the above made convention.)

We start with the case $\mu = 0$ and compute $L_m\mathcal{M}_{kj}^0$:

Lemma 2.53 *Let $m, k, j < n$. Then*

    i.

$$L_m\mathcal{M}_{kj}^0 = \frac{2(n-1)}{\overline{\Phi}P^n} \left[\delta_{kj}L_m\rho^2 + \delta_{mj}L_k\rho^2 \right] - \frac{n(n-1)}{\overline{\Phi}P^{n+1}}(L_m\rho^2)(L_k\rho^2)\overline{L}_j\rho^2,$$

    ii.

$$L_m\mathcal{M}_{nj}^0 = 2(n-1) \left[\frac{2\delta_{mj}}{P^n} - \frac{n}{P^{n+1}}(L_m\rho^2)(\overline{L}_j\rho^2) \right],$$

    iii.

$$L_n\mathcal{M}_{kj}^0 = -\frac{2\delta_{kj}}{\overline{\Phi}^2 P^{n-1}} + \frac{(n-1)(L_k\rho^2)(\overline{L}_j\rho^2)}{\overline{\Phi}^2 P^n}$$

$$+ \left[\frac{2n(n-1)(L_k\rho^2)(\overline{L}_j\rho^2)}{P^{n+1}} - \frac{4(n-1)\delta_{kj}}{P^n} \right] \frac{\Phi^*}{\overline{\Phi}},$$

iv.

$$L_n \mathcal{M}_{nj}^0 = 4n(n-1) \frac{\Phi^*}{\mathrm{P}^{n+1}} \overline{L}_j \rho^2 \,.$$

Proof   *i)*

$$\begin{aligned}
L_m \mathcal{M}_{kj}^0 &= \frac{-L_m \overline{\Phi}}{\overline{\Phi}^2} \left[\frac{-2\delta_{kj}}{\mathrm{P}^{n-1}} + \frac{n-1}{\mathrm{P}^n} (L_k \rho^2)(\overline{L}_j \rho^2) \right] \\
&\quad + \frac{1}{\overline{\Phi}} \left[2(n-1)\delta_{kj} \frac{L_m \mathrm{P}}{\mathrm{P}^n} - \frac{n(n-1)}{\mathrm{P}^{n+1}} (L_m \mathrm{P})(L_k \rho^2)(\overline{L}_j \rho^2) \right. \\
&\quad \left. + \frac{n-1}{\mathrm{P}^n} L_m L_k \rho^2 \overline{L}_j \rho^2 + \frac{n-1}{\mathrm{P}^n} (L_k \rho^2) L_m \overline{L}_j \rho^2 \right] \\
&= \frac{1}{\overline{\Phi}} \left[\frac{2(n-1)\delta_{kj} L_m \rho^2}{\mathrm{P}^n} - \frac{n(n-1)}{\mathrm{P}^{n+1}} (L_m \rho^2)(L_k \rho^2)(\overline{L}_j \rho^2) + \cdots \right. \\
&\quad \left. \cdots + \frac{n-1}{\mathrm{P}^n} L_k \rho^2 2\delta_{mj} \right],
\end{aligned}$$

which gives *i)*. *ii)* follows as *i)*.

iii)

$$\begin{aligned}
L_n \mathcal{M}_{kj}^0 &= \frac{-L_n \overline{\Phi}}{\overline{\Phi}^2} \left[-\frac{2\delta_{kj}}{\mathrm{P}^{n-1}} + \frac{n-1}{\mathrm{P}^n} (L_k \rho^2)(\overline{L}_j \rho^2) \right] \\
&\quad + \frac{1}{\overline{\Phi}} \left[2(n-1)\delta_{kj} \frac{L_n \mathrm{P}}{\mathrm{P}^n} - n(n-1) \frac{L_n \mathrm{P}}{\mathrm{P}^{n+1}} (L_k \rho^2)(\overline{L}_j \rho^2) \right. \\
&\quad \left. + \frac{n-1}{\mathrm{P}^n} (L_n L_k \rho^2)(\overline{L}_j \rho^2) + \frac{n-1}{\mathrm{P}^n} (L_k \rho^2)(L_n \overline{L}_j \rho^2) \right] \\
&= \frac{1}{\overline{\Phi}^2} \left[-\frac{2\delta_{kj}}{\mathrm{P}^{n-1}} + \frac{n-1}{\mathrm{P}^n} (L_k \rho^2)(\overline{L}_j \rho^2) \right] \\
&\quad + \frac{1}{\overline{\Phi}} \left[2(n-1)\delta_{kj} \frac{-2\Phi^*}{\mathrm{P}^n} - n(n-1) \frac{-2\Phi^*}{\mathrm{P}^{n+1}} (L_k \rho^2)(\overline{L}_j \rho^2) \right] \\
&= \frac{-2\delta_{kj}}{\overline{\Phi}^2 \mathrm{P}^{n-1}} + \frac{n-1}{\overline{\Phi}^2 \mathrm{P}^n} (L_k \rho^2)(\overline{L}_j \rho^2) \\
&\quad + \frac{2(n-1)\Phi^*}{\overline{\Phi}} \left[\frac{-2\delta_{kj}}{\mathrm{P}^n} + \frac{n}{\mathrm{P}^{n+1}} (L_k \rho^2)(\overline{L}_j \rho^2) \right] \,.
\end{aligned}$$

v) is proved accordingly.   □

To compute A_{KL}^0 we distinguish four cases according to the occurrence of n in K and L:

Case 1. $n \in K$ and $n \in L$

Then from (2.50) we deduce: $n \notin Q$, so $k = n$ and $m < n$. There is exactly one Q such that $L = kQ$ as sets. Therefore, in this case

$$
\begin{aligned}
A_{KL}^0 &= - \sum_{j,m<n} \varepsilon_{nQ}^L \varepsilon_{mK}^{njQ} L_m \mathcal{M}_{nj}^0 \\
&= \sum_{j,m<n} \varepsilon_{nQ}^L \varepsilon_{mK}^{jnQ} L_m \mathcal{M}_{nj}^0 \\
&= \sum_{j,m<n} \varepsilon_{mK}^{jL} L_m \mathcal{M}_{nj}^0 \,.
\end{aligned}
$$

On the other hand,

$$
\begin{aligned}
A_{LK}^{0*} &= \sum_{j,m<n} \varepsilon_{mL}^{jK} (L_m \mathcal{M}_{nj}^0)^* \\
&= \sum_{j,m<n} \varepsilon_{mK}^{jL} (L_j \mathcal{M}_{nm}^0)^* \,.
\end{aligned}
$$

So it suffices to establish

$$
L_m \mathcal{M}_{nj}^0 - (L_j \mathcal{M}_{nm}^0)^* = \mathcal{A}_1,
$$

which immediately follows from 2.53.ii.

Case 2. $n \notin K$ and $n \notin L$

This implies $m = n$, so

$$
A_{KL}^0 = - \sum_{\substack{|Q|=q \\ j<n \\ k}} \varepsilon_{kQ}^L \varepsilon_K^{jQ} L_n \mathcal{M}_{kj}^0 \,.
$$

Furthermore, $k \in L$ and Q is determined by k and L, which leads to

$$
A_{KL}^0 = - \sum_{\substack{j \in K \\ k \in L}} \varepsilon_{kQ}^L \varepsilon_K^{jQ} L_n \mathcal{M}_{kj}^0 \,.
$$

$$
\begin{aligned}
A_{LK}^{0*} &= - \sum_{\substack{j \in L \\ k \in K}} \varepsilon_{kQ}^K \varepsilon_L^{jQ} (L_n \mathcal{M}_{kj}^0)^* \\
&= - \sum_{\substack{j \in K \\ k \in L}} \varepsilon_{jQ}^K \varepsilon_L^{kQ} (L_n \mathcal{M}_{jk}^0)^* \,.
\end{aligned}
$$

So we have to check

$$
L_n \mathcal{M}_{kj}^0 - (L_n \mathcal{M}_{jk}^0)^* = \mathcal{A}_1,
$$

which follows from the symmetry of 2.53.iii in j and k and the fact that

$$
\Phi = \Phi^* + \mathcal{E}_3.
$$

The asymmetric cases

Case 3. $n \notin K$ and $n \in L$
Case 4. $n \in K$ and $n \notin L$

have naturally to be treated simultaneously. We start with computing A^0_{KL} in these cases.

Case 3. $n \notin K$ and $n \in L$

Necessarily $m = n$, and, since $n \notin Q$, we must have $k = n$. So now

$$A^0_{KL} = -\varepsilon^L_{nQ} \sum_{j<n} \varepsilon^{njQ}_{nK} (L_n \mathcal{M}^0_{nj}) \tag{2.54}$$

$$= -\varepsilon^L_{nQ} \sum_{j<n} \varepsilon^{jQ}_K \overline{L}_j \rho^2 \frac{4n(n-1)\Phi^*}{P^{n+1}}$$

in view of (2.53).iv.

Case 4. $n \in K$ and $n \notin L$

This case is fairly delicate. We set $K = n \cup J$ as sets and obtain from (2.50)

$$A^0_{KL} = -\varepsilon^K_{nJ} \sum_{\substack{j,k,m<n \\ |Q|=q}} \varepsilon^L_{kQ} \varepsilon^{njQ}_{mnJ} (L_m \mathcal{M}^0_{kj}) \tag{2.55}$$

$$= \varepsilon^K_{nJ} \sum_{\substack{j,k,m<n \\ |Q|=q}} \varepsilon^L_{kQ} \varepsilon^{jQ}_{mJ} L_m \mathcal{M}^0_{kj}.$$

In (2.55) we distinguish the terms with $m = j$ from those with $m \neq j, j \notin J$.

Let $m = j$. This implies $Q = J \subset L$, and k is determined by Q and L.

If $m \neq j$ then $Q \neq J$, whereas as sets $jQ = mJ$, which means $Q = m \cup M, J = j \cup M$. We have

$$\varepsilon^L_{kQ} \varepsilon^K_{nJ} \varepsilon^{jQ}_{mJ} = \varepsilon^L_{kmM} \varepsilon^K_{njM} \varepsilon^{jmM}_{mjM} \tag{2.56}$$

$$= -\varepsilon^L_{kmM} \varepsilon^K_{njM}.$$

We can now rewrite (2.55), using (2.56), as

$$A^0_{KL} = \varepsilon^K_{nJ} \sum_{k\in L} \varepsilon^L_{kJ} \sum_{j\notin J} L_j \mathcal{M}^0_{kj} - \sum_{\substack{m,k\in L \\ j\in J \\ m\neq j}} \varepsilon^L_{kmM} \varepsilon^K_{njM} L_m \mathcal{M}^0_{kj}. \tag{2.57}$$

Let us now also form the terms of the second sum with $m = j$:

$$\sum_{\substack{j,k\in L \\ j\in J}} \varepsilon^L_{kjM} \varepsilon^K_{njM} L_j (\mathcal{M}^0_{kj}) \tag{2.58}$$

and note

$$\varepsilon^L_{kjM}\varepsilon^K_{njM} = \varepsilon^L_{kJ}\varepsilon^K_{nJ}. \tag{2.59}$$

So (2.58) becomes

$$\sum_{\substack{j\in J \\ k\in L}} \varepsilon^L_{kJ}\varepsilon^K_{nJ}L_j\mathcal{M}^0_{kj}. \tag{2.60}$$

Now we add (2.60) to the first sum in (2.57) and subtract it from the second sum to obtain

$$A^0_{KL} = \varepsilon^K_{nJ}\sum_{k\in L}\varepsilon^L_{kJ}\sum_{j<n}L_j\mathcal{M}^0_{kj} - \sum_{\substack{m,k\in L \\ j\in J}}\varepsilon^L_{kmM}\varepsilon^K_{njM}L_m\mathcal{M}^0_{kj}. \tag{2.61}$$

But from 2.53.i we get

$$L_k\mathcal{M}^0_{mj} = L_m\mathcal{M}^0_{kj}, \tag{2.62}$$

and this shows that the second sum in (2.61) is zero because interchanging m and k changes the sign of the corresponding term.

To deal with the first sum we set $m = j$ in 2.53.i:

$$L_j\mathcal{M}^0_{kj} = \frac{2(n-1)}{\overline{\Phi}\mathrm{P}^n}[\delta_{kj}L_j\rho^2 + L_k\rho^2] - \frac{n(n-1)}{\overline{\Phi}\mathrm{P}^{n+1}}|L_j\rho^2|^2 L_k\rho^2$$

$$\sum_{j<n}L_j\mathcal{M}^0_{kj} = \frac{2n(n-1)}{\overline{\Phi}\mathrm{P}^n}L_k\rho^2 - \frac{n(n-1)}{\overline{\Phi}\mathrm{P}^{n+1}}L_k\rho^2\sum_{j<n}|L_j\rho^2|^2. \tag{2.63}$$

By (2.19) this can be rewritten as

$$\frac{(n-1)n}{\overline{\Phi}\mathrm{P}^{n+1}}L_k\rho^2[2\mathrm{P} - \sum_{j<n}|L_j\rho^2|^2] = \frac{n(n-1)}{\overline{\Phi}\mathrm{P}^{n+1}}4|\Phi|^2.$$

Putting everything together leads to

$$A^0_{KL} = 4n(n-1)\frac{\Phi}{\mathrm{P}^{n+1}}\varepsilon^K_{nJ}\sum_{k<n}\varepsilon^L_{kJ}L_k\rho^2. \tag{2.64}$$

We can now compare A^0_{KL} and A^{0*}_{LK} in the cases 3 and 4. Recall

Case 3. $n\notin K, n\in L.$

By (2.54) and (2.64)

$$A^0_{KL} = -\varepsilon^L_{nQ}\sum_{j<n}\varepsilon^{jQ}_K\overline{L}_j\rho^2\frac{4n(n-1)\Phi^*}{\mathrm{P}^{n+1}},$$

$$A^0_{LK} = \varepsilon^L_{nJ}\sum_{k\in K}\varepsilon^K_{kJ}L_k\rho^2\frac{4n(n-1)\Phi}{\mathrm{P}^{n+1}}.$$

Comparing both formulae we note that $Q = J$. Q is determined by L unless $A^0_{KL} = A^0_{LK} = 0$. So only the case $Q \subset K$ remains. But then $k = j$. We obtain

$$A^0_{KL} = -\varepsilon^L_{nQ}\varepsilon^{kQ}_K 4n(n-1)\overline{L}_k\rho^2 \frac{\Phi^*}{\mathbf{P}^{n+1}}$$

$$A^0_{LK} = \varepsilon^L_{nJ}\varepsilon^K_{kJ} 4n(n-1)L_k\rho^2 \frac{\Phi}{\mathbf{P}^{n+1}},$$

and $A^0_{KL} - (A^0_{LK})^* = 0$. (Note that $(L_k\rho^2)^* = -L_k\rho^2 + \mathcal{E}_1$.)

To finally treat

Case 4. $n \in K, n \notin L,$

we reduce it to the previous case by writing

$$A^0_{KL} - A^{0*}_{LK} = -(A^0_{LK} - A^{0*}_{KL})^*.$$

We now turn to the easier case $\mu > 0$.

Lemma 2.65   *If $\mu > 0$ then*

$$L_m\mathcal{M}^\mu_{kj} = \mathcal{A}_1$$

except in the case $m = n$, and $k, j < n$.

Here we have

$$L_n\mathcal{M}^\mu_{kj} = \frac{\mu+1}{\overline{\Phi}^{\mu+2}}\left[\frac{-2\delta_{jk}}{\mathbf{P}^{n-\mu-1}} + \frac{n-1}{\mathbf{P}^{n-\mu}}L_k\rho^2\overline{L}_j\rho^2\right].$$

The proof is immediate from a computation of the derivatives as in 2.53 and determination of the types. The higher power of $\overline{\Phi}$ accounts for the higher type.

Looking at (2.50) in case $\mu > 0$ we now see by (2.65)

$$A^\mu_{KL} = 0 \quad \text{(i.e.}\,\mathcal{A}_1\text{) if } n \in K \text{ or } n \in L.$$

So only the case $n \notin K, n \notin L$ remains.

Here we have

$$A^\mu_{KL} = -\sum_{\substack{j,k<n \\ |Q|=q}} \varepsilon^L_{kQ}\varepsilon^{jQ}_K L_n\mathcal{M}^\mu_{kj},$$

$$A^\mu_{LK} = -\sum_{\substack{j,k<n \\ |Q|=q}} \varepsilon^K_{jQ}\varepsilon^{kQ}_L L_n\mathcal{M}^\mu_{jk}.$$

The symmetry of 2.65 in j and k now immediately gives

$$A^\mu_{KL} - A^{\mu*}_{LK} = 0.$$

(Recall again our convention: $\mathcal{A}_1$-kernels are considered as 0).

This concludes the proof of 1.29, and concludes this paragraph.

§3 The Bergman Projection

1. We choose an arbitrary hermitian metric and use the case $q = 0$ of 1.26 in order to analyse the orthogonal projection

$$P : L^2(\Omega) \longrightarrow L^2(\Omega) \cap \mathcal{O}(\Omega),$$

the *Bergman projection*. (1.26) now reads as

$$f = T_0 \bar{\partial} f + P_0 f + (A_2 + E_{2-2n}) \bar{\partial} f + (A_1 + E_{1-2n}) f \tag{3.1}$$

for $f \in \mathrm{dom}\ \bar{\partial}$. Let us apply this to $f - Pf$:

$$f - Pf = T_0 \bar{\partial} f + P_0(f - Pf) + (A_2 + E_{2-2n}) \bar{\partial} f + (A_1 + E_{1-2n})(f - Pf). \tag{3.2}$$

Now by 1.28i we have

$$P_0 - P_0^* = \tilde{A}_1, \tag{3.3}$$

an admissible operator of type ≥ 1. This gives

$$\begin{aligned} f - Pf = {} & T_0 \bar{\partial} f + P_0^*(f - Pf) + \tilde{A}_1(f - Pf) \\ & + (A_2 + E_{2-2n}) \bar{\partial} f + (A_1 + E_{1-2n})(f - Pf). \end{aligned} \tag{3.4}$$

To analyse the second term on the right-hand side we use the Neumann decomposition

$$f - Pf = \bar{\partial}^* N \bar{\partial} f, \tag{3.5}$$

with N the Neumann operator on $(0, 1)$-forms. A few paragraphs later in this section we shall show that there exists an infinitely smooth kernel $\mathcal{E}_0$ with

$$\bar{\partial}_x \mathcal{E}_0 = \bar{\partial}_x P_0^*. \tag{3.6}$$

Assuming this we have

$$\begin{aligned} P_0^*(f - Pf) &= (f - Pf, \mathcal{P}_0^*) \tag{3.7} \\ &= (\bar{\partial}^* N \bar{\partial} f, \mathcal{P}_0^*) \\ &= (N \bar{\partial} f, \bar{\partial} \mathcal{P}_0^*) \\ &= (N \bar{\partial} f, \bar{\partial} \mathcal{E}_0) \\ &= (\bar{\partial}^* N \bar{\partial} f, \mathcal{E}_0) \\ &= E_0(f - Pf), \end{aligned}$$

where E_0 is the operator with kernel $\mathcal{E}_0$.

From (3.4) and (3.7) we obtain

Theorem 3.8 *For the Bergman projector P there are operators of the indicated type such that, if $f \in \mathrm{dom}\ \overline{\partial}$,*

$$f - Pf = (T_0 + A_2 + E_{2-2n})\overline{\partial}f + (A_1 + E_{1-2n})(f - Pf).$$

(We have used the generic notation

$$A_1 + \widetilde{A}_1 = A_1, \quad E_{1-2n} + E_0 = E_{1-2n}\quad)$$

We note, moreover, that the same letters A_1 etc. in 3.1 and 3.4 denote the same operators; so we can subtract 3.4 from 3.1 to get

Theorem 3.9

$$Pf = P_0 f + A_1 f + (A_1 + E_{1-2n})Pf.$$

2.   We still have to show (3.6). Let $P_0(x, y)$ be the kernel of P_0 and set

$$\widetilde{\mathcal{P}}(x, y) = *_x \overline{P_0(x, y)}. \tag{3.10}$$

Then $\widetilde{\mathcal{P}}(x, y)$ is smooth on $\overline{\Omega} \times \overline{\Omega} - \Lambda$ and

$$\mathcal{K}(x, y) = \overline{\partial}_y \widetilde{\mathcal{P}}(x, y) \tag{3.11}$$

is smooth and $\overline{\partial}_y$-closed on $\overline{\Omega} \times \overline{\Omega}_1$, where Ω_1 is a strictly pseudoconvex neighbourhood of $\overline{\Omega}$. Consider the $\overline{\partial}$-Neumann decomposition of $\mathcal{K}(x, y)$ on Ω_1:

$$\mathcal{K}(x, y) = H_y \mathcal{K}(x, y) + \overline{\partial}_y \overline{\partial}_y^* N_y \mathcal{K}(x, y). \tag{3.12}$$

The operators in (3.12) refer to the y-variable; $x \in \overline{\Omega}$ is a parameter. By interior regularity of the elliptic system $\overline{\partial} + \overline{\partial}^*$ both terms on the right-hand side of (3.12) are smooth on $\overline{\Omega} \times \overline{\Omega}$. Moreover, since $\mathcal{K}(x, y)$ is $\overline{\partial}$-exact on $\overline{\Omega} \times \Omega_0$ for any $\Omega_0 \subset\subset \Omega$, it follows that $\mathcal{K}(x, y)$ is $\overline{\partial}$-exact on $\overline{\Omega} \times \overline{\Omega}_1$ (see the last paragraph of section 3). So $H_y \mathcal{K}(x, y) = 0$, and we have

$$\mathcal{K}(x, y) = \overline{\partial}_y \widetilde{\mathcal{S}}(x, y) \tag{3.13}$$

with

$$\widetilde{\mathcal{S}}(x, y) = \overline{\partial}_y^* N_y \mathcal{K}(x, y)\big|_{\overline{\Omega} \times \overline{\Omega}},$$

a smooth solution on $\overline{\Omega} \times \overline{\Omega}$. This implies (3.6).

3.   We have to justify the arguments leading to (3.13). The proof that $H_y \mathcal{K}(x, y) = 0$ is unnecessary if X is a Stein manifold. The general case is covered by the following isomorphism theorem.

Theorem 3.14 *Let $\Omega = \{x \in X : r(x) < 0\}$ be strictly pseudoconvex, r strictly plurisubharmonic in a neighbourhood U of $b\Omega$, with $dr \neq 0$. Let $\Omega_\varepsilon = \{x : r(x) < \varepsilon\}$, $-\delta < \varepsilon < \delta$. Then, if δ is small enough, the restriction maps*

$$H_2^{0q}(\Omega_{\varepsilon_1}) \longrightarrow H_2^{0q}(\Omega_{\varepsilon_2}),$$

for $\varepsilon_1 \geq \varepsilon_2$ and $q \geq 1$, are bijective.

Proof   We only show injectivity – which is all we have used.

If δ is sufficiently small, $\Omega_{2\delta}$ is still strictly pseudoconvex, we call it X. Then X is holomorphically convex, and its *Remmert reduction* is a Stein space $\tilde{X}$ with an isolated singularity $\tilde{x}_0$:

$$X, A \xrightarrow{\pi} \tilde{X}, \tilde{x}_0; \tag{3.15}$$

where π is proper holomorphic, biholomorphic from $X - A \to \tilde{X} - \tilde{x}_0$, and $\pi(A) = \tilde{x}_0$. A is a compact analytic subset of X which is necessarily contained in $\Omega_{-\delta}$.

Now suppose that f is a $\overline{\partial}$-closed form on Ω_{ε_1} which is $\overline{\partial}$-exact on Ω_{ε_2}. By the Dolbeault isomorphism f gives rise to a cocycle F in $Z^q(\Omega_{\varepsilon_1}, \mathcal{O}_{\Omega_{\varepsilon_1}})$; since f is $\overline{\partial}$-exact on Ω_{ε_2} and $A \subset \Omega_{\varepsilon_2}$, F can be defined with respect to an open cover $\mathfrak{U} = \{\Omega_{\varepsilon_2}\} \cup \mathfrak{U}'$ of Ω_{ε_1}, where the elements of $\mathfrak{U}'$ are disjoint from A; consequently F yields a cocycle $\tilde{F}$ in $Z^q(\tilde{\Omega}_{\varepsilon_1}, \mathcal{O}_{\tilde{\Omega}_{\varepsilon_1}})$, where $\tilde{\Omega}_\varepsilon = \pi(\Omega_\varepsilon)$. Cartan's Theorem B tells us that $\tilde{F}$ is a coboundary:

$$\tilde{F} = \delta\tilde{U};$$

$U = \tilde{U} \circ \pi$ is then a cochain in $C^{q-1}(\Omega_{\varepsilon_1}, \mathcal{O}_{\Omega_{\varepsilon_1}})$ with

$$\delta U = F.$$

Applying once more the Dolbeault isomorphism gives a solution u with $\overline{\partial}u = f$. $\qquad\square$

Remark   The existence of the *Remmert reduction* depends on Grauert's direct image theorem which is the deepest result of sheaf theory. But we need less: *we only need* some *Stein space $\tilde{X}, \tilde{x}_0$ in 3.15 such that π, A, $\tilde{x}_0$ have the above properties.* Such a space was constructed independently of Grauert's direct image theorem by Lieb in [Lie 66]. An approach which instead of possibly singular spaces uses a Runge type approximation theorem can be found in [HeL 88]. Our approach should also work for vector-valued forms.

It remains to show the smoothness of the decomposition 3.12. This amounts to the following

Proposition 3.16   *Let $g(x,y)$ be a smooth form on $\overline{B} \times \overline{\Omega}$, where B is a ball in $\mathbb{C}^n$ and Ω a strictly pseudoconvex domain in a hermitian manifold. Suppose $g(x,y)$ is, for each x in $\overline{B}$, $\overline{\partial}_y$-exact and set $f(x,y) = \overline{\partial}_y^* N_y g(x,y)$, with N_y the Neumann operator of Ω and $\overline{\partial}_y^*$ the adjoint of $\overline{\partial}_y$. Then $f(x,y)$ is smooth on $\overline{B} \times \Omega$.*

Proof   a)   We need the following

Interior elliptic estimates   *Let $f \in L^2_{0q}(\Omega)$ with compact support. Then*

$$\|f\|_{L^2_{k+1}} \lesssim \|\overline{\partial}f\|_{L^2_k} + \|\vartheta f\|_{L^2_k} + \|f\|_{L^2}, \tag{3.17}$$

provided the right-hand side makes sense. The constant turning up in (3.17) depends on k and the support of f.

These estimates are easily proved with the methods of Chapter VII (see, in particular, proofs of 4.3, 4.5 and 4.6 in that chapter).

b)   Let now $\chi(y)$ be a smooth function with compact support in Ω. We will show the estimate

$$\|\chi(y)f(x,y)\|_{L^2_{k+1}(\Omega)} \lesssim \|g(x,y)\|_{L^2_k(\Omega)} \tag{3.18}$$

($x \in \overline{B}$ remains fixed, and the constant does not depend on x). We have

$$\overline{\partial}_y(\chi(y)f(x,y)) = \overline{\partial}_y\chi(y) \wedge f(x,y) + \chi(y)g(x,y) \tag{3.19}$$
$$\vartheta_y(\chi(y)f(x,y)) = a(f(x,y)),$$

where a is a zero order operator with smooth coefficients in y independent of x. We have used

$$\overline{\partial}_y f(x,y) = g(x,y)$$
$$\vartheta_y f(x,y) = 0.$$

From (3.19) we get

$$\|\overline{\partial}_y(\chi(y)f(x,y))\|_{L^2(\Omega)} \leq \mathrm{const}\|f(x,y)\|_{L^2(\Omega)} + \|g(x,y)\|_{L^2(\Omega)}$$
$$\lesssim \|g(x,y)\|_{L^2(\Omega)} \tag{3.20}$$
$$\|\vartheta_y(\chi(y)f(x,y))\|_{L^2(\Omega)} \lesssim \|g(x,y)\|_{L^2(\Omega)},$$

since $\overline{\partial}^* N_y$ is an L^2-continuous operator. But now (3.20) and (3.17) give (3.18) for $k = 0$.

Now suppose (3.18) proved for $k - 1$ (and all compactly supported functions $\chi(y)$), and choose $\chi_0(y) \equiv 1$ in a neighbourhood of the support of $\chi(y)$, compactly supported again. Then (3.19) can be expressed as follows:

$$\overline{\partial}_y(\chi(y)f(x,y)) = \overline{\partial}_y\chi(y) \wedge \chi_0(y)f(x,y) + \chi(y)g(x,y) \tag{3.19'}$$
$$\vartheta_y(\chi(y)f(x,y)) = a(\chi_0(y)f(x,y)),$$

and so, by the induction assumption, applied to $\chi_0 f$:

$$\|\overline{\partial}_y(\chi(y)f(x,y))\|_{L^2_k} \leq \mathrm{const}\|\chi_0(y)f(x,y)\|_{L^2_k} + \|g(x,y)\|_{L^2_k}$$
$$\lesssim \|g(x,y)\|_{L^2_{k-1}} + \|g(x,y)\|_{L^2_k} \tag{3.21}$$
$$\lesssim \|g(x,y)\|_{L^2_k}, \tag{3.22}$$

and similarly

$$\|\vartheta_y(\chi(y)f(x,y))\|_{L^2_k} \lesssim \|g(x,y)\|_{L^2_k}.$$

Combination of (3.21) with (3.17) gives

$$\|\chi(y)f(x,y)\|_{L^2_{k+1}} \lesssim \|g(x,y)\|_{L^2_k} + \|\chi(y)f(x,y)\|_{L^2}$$
$$\lesssim \|g(x,y)\|_{L^2_k},$$

as desired.

c) Now let $x_j \in \overline{B}$ be a sequence converging to x_0 in $\overline{B}$. We apply (3.18) to

$$\chi(y)f(x_j, y) - \chi(y)f(x_0, y)$$

and obtain, for each k,

$$\|\chi(y)f(x_j, y) - \chi(y)f(x_0, y)\|_{L^2_{k+1}} \lesssim \|g(x_j, y) - g(x_0, y)\|_{L^2_k} \longrightarrow 0.$$

Consequently, $f(x_j, y)$ converges to $\chi(y)f(x_0, y)$ in each $L^2_k(\Omega)$-norm, and therefore uniformly with all derivatives with respect to y. In particular, each y-derivative of $f(x, y)$ is continuous on $\overline{B} \times \Omega$.

d) Now let D be a differentiation of order r with respect to x. The continuity of $\overline{\partial}^*_y N_y$ in $L^2(\Omega)$ implies

$$Df(x, y) \equiv \overline{\partial}^*_y N_y Dg(x, y)$$

(in the sense of distributions). Because of c), all y-derivatives of the right-hand side are continuous in x and y simultaneously. But then $f(x, y) \in C^\infty(\overline{B} \times \Omega)$, as desired. □

§4 *Z*-operators

1. In order to express results like (3.8) more comfortably we introduce an algebra of continuous operators on

$$L^2(\Omega) = \bigoplus_{0 \le q \le n} L^2_{0q}(\Omega) \tag{4.1}$$

which contains all the operators which we have so far considered, by the following rules:

4.2 *An isotropic kernel $\mathcal{E}$ of order $\ge 1 - 2n$ is a Z-kernel.*

4.3 *An admissible kernel $\mathcal{A}$ of type ≥ 1 is a Z-kernel.*

4.4 *A special admissible kernel $\mathcal{A}$ of type ≥ 0 is a Z-kernel.*

4.5 *Let $\mathcal{A}$ be an admissible kernel of type ≥ 0 and $\varepsilon > 0$. Then the kernels*

$$|r(x)|^\varepsilon \mathcal{A}(x, y) \quad or \quad |r(y)|^\varepsilon \mathcal{A}(x, y)$$

are Z-kernels of type ≥ 0.

4.6 *An operator Z with a Z-kernel $\mathcal{Z}$ is a Z-operator.*

4.7 *The algebra $\mathfrak{Z}$ of all Z-operators is the algebra generated by the operators (4.6) (in the algebra of all endomorphisms of the Hilbert space $L^2(\Omega)$).*

We now introduce the *type filtration* on $\mathfrak{Z}$:

4.8 *If a Z-operator Z is of the form A_j or E_{j-2n}, then it is of type $\ge j$.*

4.9 *If Z^1 and Z^2 are Z-operators of type $\geq j$ resp. $\geq k$, then*

$$Z^1 + Z^2 \text{ is of type } \geq \min(j, k)$$
$$Z^1 \circ Z^2 \text{ is of type } \geq j + k.$$

The type is always a non-negative integer; if Z is of type $\geq j$ but not of type $\geq j+1$ then j is the type of Z. Special admissible kernels of type 0 (which are not of type ≥ 1) define Z-operators of type 0, similarly operators defined by the kernel 4.5. We write, generically,

$$Z = Z_j \quad \text{or} \quad K = Z_j \tag{4.10}$$

to express that Z or K are Z-operators of the indicated type, and use the generic notation

$$Z_j \circ Z_k = Z_{j+k} \tag{4.11}$$

(in particular: $Z_1 \circ Z_0 = Z_1$) or, for instance,

$$Z_1 + Z_2 = Z_1 \tag{4.12}$$

(meaning, of course, that the sum on the left hand side is a Z-operator of type ≥ 1). Note, in particular, that operators with smooth kernels are Z-operators of type $\geq 2n$.

2. The algebra $\mathfrak{Z}$ is not yet sufficient for our purposes. The following definition will allow us to connect Z-operators with abstractly defined operators.

Definition 4.13 *An operator S on $L^2(\Omega)$ is an asymptotic Z-operator of type $\geq j$ if for each $l \in \mathbb{N}$ there exist a Z-operator Z_j^l of type $\geq j$, a finite number k of Z-operators of type $\geq l$, $Z_l^1, \ldots, Z_l^k$, (with k depending on l) and k L^2-bounded operators $K_1^l, \ldots K_k^l$ such that*

$$S = Z_j^l + \sum_{\varkappa=1}^{k} Z_l^\varkappa \circ K_\varkappa^l.$$

If S is of type $\geq j$ for all $j \in \mathbb{N}$ we call S of type ∞.

It is clear that asymptotic Z-operators again form a subalgebra $\bar{\mathfrak{Z}}$ of End $L^2(\Omega)$; we have $\mathfrak{Z} \subset \bar{\mathfrak{Z}}$. This algebra is again filtered by the type function $j : \bar{\mathfrak{Z}} \to \mathbb{N}_0 \cup \{\infty\}$.

3. We finally introduce *principal parts of asymptotic Z-operators* .

Let $\bar{\mathfrak{Z}}_j$ be the ideal of asymptotic Z-operators of type $\geq j$, $j = 0, 1, 2, \ldots$ Consequently

$$\bar{\mathfrak{Z}} = \bar{\mathfrak{Z}}_0 \supset \bar{\mathfrak{Z}}_1 \supset \bar{\mathfrak{Z}}_2 \supset \ldots$$

Definition 4.14 *Denote, for an asymptotic Z-operator S, the image of S in $\bar{\mathfrak{Z}}/\bar{\mathfrak{Z}}_{j+1}$ by $S(j)$. The number*

$$j_s = \min\{j : S \notin \bar{\mathfrak{Z}}_{j+1}\} \quad (\leq \infty)$$

is the exact type of S, and $S(j_s)$ is the principal part of S if $j_s < \infty$. If $j_s = \infty$, then 0 is a principal part of S.

The definition implies the following facts:

    i. *S itself represents its principal part.*

    ii. Suppose *T has the same principal part as S.* This means that, if j_s is the exact type of S, then $T - S \in \overline{3}_{j_s+1}$. If T is a Z-operator of type j_s such that $S - T \in \overline{3}_{j_s+1}$, then we will also call T itself *"the" principal part of S.* But T is not uniquely determined; moreover, there need not exist a Z-operator of type j_s whose principal part coincides with the principal part of S.

4.   We can now look again at the results of §3. Theorem 3.8 reads, firstly, as

$$f - Pf = (\boldsymbol{T}_0 + Z_2)\overline{\partial}f + Z_1(f - Pf). \tag{4.15}$$

This gives more simply

$$f - Pf = Z_1\overline{\partial}f + Z_1(f - Pf). \tag{4.16}$$

(3.9) can be rewritten as

$$Pf = \boldsymbol{P}_0 f + Z_1 f + Z_1 Pf. \tag{4.17}$$

Here $\boldsymbol{P}_0$ is the explicit admissible operator of type 0 with kernel $\mathcal{P}_0$ from §3, $\boldsymbol{P}_0 + Z_1$ is again a Z-operator of type 0, and by plugging (4.17) into the last term of (4.17) and iteration, we have

Theorem 4.18 *The Bergman projector P has the asymptotic development*

$$P = \boldsymbol{P}_0 + Z_1 + Z_l P, \quad l = 1, 2, 3, \ldots$$

It is an asymptotic Z-operator of type $\geq$ 0, whose principal part is represented by the explicit A_0-operator $\boldsymbol{P}_0$. (The Z_1-operator Z_1 depends on l).

The theorem can be more concisely stated as

$$P = \boldsymbol{P}_0 + Z_1^a.$$

Here and in the sequel asymptotic Z-operators will be generically denoted by Z^a.

Remark According to (3.13) there is an operator E_0 with everywhere smooth kernel such that

$$\overline{\partial}\boldsymbol{P}_0 + \overline{\partial}E_0 = 0.$$

Replacing $\boldsymbol{P}_0$ by $\widehat{\boldsymbol{P}}_0 = E_0 + \boldsymbol{P}_0$ leads (with new operators Z_1^a) to the representation

$$P = \widehat{\boldsymbol{P}}_0 + Z_1^a$$
$$\overline{\partial}\widehat{\boldsymbol{P}}_0 = 0$$

So one can choose, as a principal part of P, an (almost explicit) Z_0-operator $\widehat{\boldsymbol{P}}_0$ which maps L^2 into the holomorphic functions $L^2 \cap \mathcal{O}$.

5.   To finish this paragraph let us introduce the *notion of type and principal part* also *for kernels*: If a kernel $\mathcal{Z}$ yields a Z-operator Z of type $\geq j$,   then $\mathcal{Z}$ is called of type $\geq j$; if this operator has Z^p as representation of its principal part, and Z^p has $\mathcal{Z}^p$ as a kernel, then $\mathcal{Z}^p$ is "the" principal part of $\mathcal{Z}$.

§5   The Structure of the Kernels $\mathcal{T}_q$

1.   The kernels $\mathcal{T}_q$ are $\mathcal{Z}_1$-kernels whose structure appears to be very complicated. We will now show that their principal parts can be more easily understood as

$$\mathcal{T}_q = \bar{\partial}_x \mathcal{N}_q + \mathcal{Z}_2, \tag{5.1}$$

where the $\mathcal{N}_q$ are $\mathcal{Z}_2$-kernels with a fairly simple "diagonal" structure.

The proof of 5.1 is a long calculation which seems to succeed by pure luck. To show that more than good luck is involved, we start with a heuristic argument. The metric will always be normalised Levi.

2.   Let $f \in \mathcal{R}(\bar{\partial})$ be a $\bar{\partial}$-exact $(0, q)$-form with $q \geq 2$. Then

$$f = \bar{\partial}\bar{\partial}^* N_q f. \tag{5.2}$$

We apply Theorem 1.26 to the $(0, q-1)$-form $\bar{\partial}^* N_q f$:

$$\bar{\partial}^* N_q f = \boldsymbol{T}_{q-1}\bar{\partial}\bar{\partial}^* N_q f + Z_2 \bar{\partial}\bar{\partial}^* N_q f + Z_1 \bar{\partial}^* N_q f,$$

which is, in view of 5.2,

$$\bar{\partial}^* N_q f = \boldsymbol{T}_{q-1} f + Z_2 f + Z_1 \bar{\partial}^* N_q f. \tag{5.3}$$

If (5.3) were valid on the whole space L^2_{0q}, it would simply tell us that $\bar{\partial}^* N_q$ is an asymptotic Z_1-operator with principal part (represented by) the explicit operator $\boldsymbol{T}_{q-1}$. We shall show later that we have, in fact, the

Theorem 5.4 *The operators $\boldsymbol{T}_{q-1}$ and $\boldsymbol{T}_q^*$ are "the" principal parts of $\bar{\partial}^* N_q$ resp. $\bar{\partial}N_q$, that is*

$$\bar{\partial}^* N_q = \boldsymbol{T}_{q-1} + Z_2^a$$
$$\bar{\partial}N_q = \boldsymbol{T}_q + Z_2^a$$

with asymptotic type 2 operators Z_2^a.

(The proof will be given in §6 where it will also be noted that the $\mathcal{Z}_2$-kernels arising in the above representation satisfy $\bar{\partial}_x \mathcal{Z}_2 = \mathcal{Z}_1$).

Now suppose that we have an exact kernel $\mathcal{N}_q$ of the Neumann operator, in the sense that

$$N_q f = (f, \mathcal{N}_q) \tag{5.5}$$

for $f \in L^2_{0q}(\Omega)$ and $*_x \mathcal{N}_q|_{b\Omega} \equiv 0$. (So we assume $\mathcal{N}_q$ smooth off the diagonal.) This implies

$$\overline{\partial}^* N_{q+1} f = N_q \overline{\partial}^* f = (\overline{\partial}^* f, \mathcal{N}_q) = (f, \overline{\partial}_x \mathcal{N}_q)$$
$$\overline{\partial} N_{q-1} f = N_q \overline{\partial} f = (\overline{\partial} f, \mathcal{N}_q) = (f, \overline{\partial}^*_x \mathcal{N}_q),$$

and, together with 5.4,

$$\overline{\partial} \mathcal{N}_q = \mathcal{T}_q + \mathcal{Z}_2 \tag{5.6}$$
$$\overline{\partial}^* \mathcal{N}_q = \mathcal{T}^*_{q-1} + \mathcal{Z}_2.$$

To make the heuristic argument easier we ignore the fact that the error terms in 5.4 are only asymptotic Z-operators. So we can reasonably hope that there is some $\mathcal{Z}_2$-kernel $\mathcal{N}_q$ which yields $\mathcal{T}_q$ and $\mathcal{T}^*_{q-1}$ via (5.6), that is up to error terms of type 2. Note that this would imply, in particular, the relation

$$\overline{\partial}_x \mathcal{T}_q = \mathcal{Z}_1. \tag{5.7}$$

It can be verified, by a long and difficult computation which is again based on some cancellation of singularities, that 5.7 does hold; this gives additional weight to our hope. This ends our heuristic discussion.

3. We will attack (5.6) directly. In other words, we will study the system

$$\overline{\partial}_x \mathcal{N}_q = \mathcal{T}_q + \mathcal{Z}_2 \tag{5.8}$$
$$\overline{\partial}^*_x \mathcal{N}_q = \mathcal{T}^*_{q-1} + \mathcal{Z}_2$$

subject to the conditions

$$\mathcal{N}_q = \mathcal{N}^*_q + \mathcal{Z}_3 \tag{5.9}$$
$$*_x \mathcal{N}_q|_{b\Omega} \equiv 0.$$

Theorem 5.10 *There are explicit $\mathcal{Z}_2$-kernels $\mathcal{N}_q$, $0 \le q \le n-1$, which satisfy (5.9) and*

$$\mathcal{N}_q = \mathcal{N}^*_q + \mathcal{Z}_3$$
$$\overline{\partial}_x \mathcal{N}_q = \mathcal{T}_q + \mathcal{Z}_2.$$

Moreover, for $1 \le q \le n-2$, we also have

$$\overline{\partial}^*_x \mathcal{N}_q = \mathcal{T}^*_{q-1} + \mathcal{Z}_2.$$

The $\mathcal{Z}_2$-kernels in the above relations satisfy

$$\overline{\partial}_x \mathcal{Z}_2 = \mathcal{Z}_1.$$

Note that, for $q \neq n, n-1$, the forms

$$\widehat{\mathcal{T}}^*_{q-1} = \overline{\partial}^*_x \mathcal{N}_q$$

satisfy

$$\mathcal{T}^*_{q-1} = \widehat{\mathcal{T}}^*_{q-1} + \mathcal{Z}_2, \quad \overline{\partial}_x \mathcal{Z}_2 = \mathcal{Z}_1$$

$$*\widehat{\mathcal{T}}^*_{q-1}|_{b\Omega} \equiv 0 \tag{5.11}$$

$$-\overline{\partial}^*_x \widehat{\mathcal{T}}^*_{q-1} = \mathcal{Z}_1 \quad \text{(in fact, it is 0.)}$$

Theorem 5.12 *For $q = n$ or $q = n - 1$, there are explicit $\mathcal{Z}_1$-kernels $\widehat{\mathcal{T}}_q$ which satisfy* (5.11).

But these latter kernels will not be constructed as $\overline{\partial}^*_x$-derivatives of some other kernels because in this case there is no elementary solution to (5.11). For more details see [LiR 93].

4.   The remainder of this paragraph is dedicated to the proof of Theorems 5.10 and 5.12. Theorem 5.4 will be deduced from these results in the next paragraph.

The Levi form of r defines our metric near the boundary of Ω; from the results of Chapter III we know that we can arrange matters such that

$$|\partial r| \equiv 1 \tag{5.13}$$

at the boundary, where the length is measured in terms of the Levi metric itself (see (2.34)). We shall always work under these strong normalisation conditions.

Let us choose, near a boundary point, the local vector fields and forms L_j, Λ_j, ω^j, Θ^j defined in (2.6). In particular, $\omega^n = \partial r(x)$ at the boundary. Then, from the construction of Φ and P and from (2.7), (2.18), (2.19), we note the following facts.

Lemma 5.14

$$r(x) = -\Phi(x, y) + \mathcal{E}_1(x, y),$$
$$\Phi(x, y) = \Phi^*(x, y) + \mathcal{E}_3(x, y).$$

Lemma 5.15   *i.*

$$\Lambda_n \Phi = -1 + \mathcal{E}_1, \quad \overline{L_n}\Phi = -1 + \mathcal{E}_1$$

ii.

$$\left.\begin{array}{r} \overline{\Lambda_j \Phi} \\ L_j \Phi \end{array}\right\} = \mathcal{E}_2 \quad \text{for all } j$$

iii.

$$\left.\begin{array}{r} \Lambda_j \Phi \\ L_j \overline{\Phi} \end{array}\right\} = \mathcal{E}_1 \quad \text{for } j < n.$$

Lemma 5.16   *i.*

$$\Lambda_n P = -2\overline{\Phi} + \mathcal{E}_2 + r\mathcal{E}_1$$
$$L_n P = -2\Phi + \mathcal{E}_2 + r\mathcal{E}_1$$

ii.

$$L_j \overline{\Phi} = L_j \rho^2 + \mathcal{E}_2 = L_j P + \mathcal{E}_2 \quad for \; j < n,$$

iii.

$$2P - \sum_{j<n} |L_j \rho^2|^2 = 4|\Phi|^2 + \mathcal{E}_2 r(x) + \mathcal{E}_3.$$

Only ii. needs a closer look. It is enough to check it for the euclidean metric and the special choice of coordinates such that $L_j = \frac{\partial}{\partial z^j}$; in that case the statement is trivial.

We still introduce the forms $-\frac{1}{2}\overline{\partial}_x \partial_y \rho^2$ and its normal and tangential component (with respect to the level surfaces of r and the Levi metric) as follows: there are double forms $\mathcal{E}_1$ of the appropriate type such that

$$(5.17) \quad -\tfrac{1}{2}\overline{\partial}_x \partial_y \rho^2 = \overline{\omega}^n \wedge \Theta^n + \overline{\omega}^n \wedge \mathcal{E}_1 + \mathcal{E}_1 \wedge \Theta^n + \tau,$$

where the form $\tau(x, y)$ is pure tangential, i.e. does not contain $\overline{\omega}^n$ or Θ^n terms.

(5.18) $\tau(x, y)$ *is the tangential component,* $v(x, y) = \overline{\omega}^n \wedge \Theta^n + \overline{\omega}^n \wedge \mathcal{E}_1 + \mathcal{E}_1 \wedge \Theta^n$ *the normal component of* $-\frac{1}{2}\overline{\partial}_x \partial_y \rho^2$.

So we get

Lemma 5.19   *i.* $v(x, y) = \overline{\omega}^n \wedge \Theta^n + \mathcal{E}_1$

   ii. $\tau(x, y) = \sum_{j<n} \overline{\omega}^j \wedge \Theta^j + \mathcal{E}_1,$

where $\mathcal{E}_1$ is strictly tangential.

5.   Now, by (1.21),

$$\mathcal{T}_q = \vartheta_x \mathcal{L}_q - \partial_y \mathcal{L}_{q-1} + \overline{\partial}_x \Gamma_{0q}, \quad q \geq 1 \tag{5.20}$$

$$\mathcal{T}_0 = \vartheta_x \mathcal{L}_0 - *\overline{K}_0 + \overline{\partial}_x \Gamma_{00}, \tag{5.21}$$

where $\mathcal{L}_q$ is defined in (1.11) and (1.8), K_0 in (1.6), and Γ_{0q} is the parametrix for the complex Laplacian.

We have to compute the principal part of these kernels.

Lemma 5.22   *For $0 \leq q \leq n - 2$, one has*

$$\mathcal{L}_q = c_{nq} \sum_{\substack{0 \leq r \leq n-q-2 \\ j < n \\ |L| = q}} \binom{n - 2 - r}{q} \frac{\overline{L}_j \rho^2}{\overline{\Phi}^{r+1} P^{n-r-1}} \overline{\omega}^{njL} \wedge \Theta^L + \mathcal{A}_3$$

with

$$c_{nq} = 2^{n-2}\left(\frac{1}{2\pi}\right)^n q!(n-q-2)!.$$

Moreover,

$$\vartheta_x \mathcal{A}_3 = \mathcal{A}_2, \ \partial_y \mathcal{A}_3 = \mathcal{A}_2, \ \overline{\partial}_x \vartheta_x \mathcal{A}_3 = \mathcal{A}_1, \ \overline{\partial}_x \partial_y \mathcal{A}_3 = \mathcal{A}_1.$$

We have used the multiindex notation of §2. – The proof is a straightforward computation, as is the proof of

Lemma 5.23

$$\vartheta_x \mathcal{L}_q = -c_{nq} \sum_{\substack{k \\ j\neq n \\ |K|=q+1 \\ |L|=q}} \binom{n-2-r}{q} L_k \left(\frac{\overline{L}_j \rho^2}{\overline{\Phi}^{r+1}\mathbf{P}^{n-r-1}}\right) \varepsilon_{njL}^{kK} \overline{\omega}^K \wedge \Theta^L + \mathcal{A}_2$$

$$\partial_y \mathcal{L}_{q-1} = c_{n,q-1} \sum_{\substack{j,k,r \\ |K|=q+1 \\ |L|=q}} \binom{n-2-r}{q-1} \Lambda_k \left(\frac{\overline{L}_j \rho^2}{\overline{\Phi}^{r+1}\mathbf{P}^{n-r-1}}\right) \varepsilon_K^{njQ} \varepsilon_L^{kQ} \overline{\omega}^K \wedge \Theta^L + \mathcal{A}_2$$

To simplify the above formulas we set

$$\mathcal{M}_{kj}^r = \Lambda_k \left(\frac{\overline{L}_j \rho^2}{\overline{\Phi}^{r+1}\mathbf{P}^{n-r-1}}\right) \tag{5.24}$$

$$\widetilde{\mathcal{M}}_{kj}^r = L_k \left(\frac{\overline{L}_j \rho^2}{\overline{\Phi}^{r+1}\mathbf{P}^{n-r-1}}\right) \tag{5.25}$$

(Note that 5.24 agrees with 2.48 up to $\mathcal{A}_3$-terms, so we can use 2.51 and 2.52 without ambiguity.) Let us now separate, in 5.23, the terms with $n \in K$ from those with $n \notin K$. This leads to

$$\vartheta_x \mathcal{L}_q - \partial_y \mathcal{L}_{q-1} = -\sum_{\substack{j<n \\ n\notin K \\ L}} c_{nq} \sum_r \binom{n-r-2}{q} \widetilde{\mathcal{M}}_{nj}^r \varepsilon_{jL}^K \overline{\omega}^K \wedge \Theta^L$$

$$+\sum_{\substack{j,k<n \\ n\notin J \\ n\notin L}} \left[c_{nq} \sum_r \binom{n-r-2}{q} \widetilde{\mathcal{M}}_{kj}^r \varepsilon_{jL}^{kJ} - c_{n,q-1}\sum_r \binom{n-r-2}{q-1} \mathcal{M}_{kj}^r \varepsilon_J^{jQ} \varepsilon_L^{kQ} \right] \overline{\omega}^{nJ} \wedge \Theta^L$$

$$-\sum_{\substack{j<n \\ n\notin J \\ n\notin Q}} c_{n,q-1} \sum_r \binom{n-r-2}{q-1} \mathcal{M}_{nj}^r \varepsilon_J^{jQ} \overline{\omega}^{nJ} \wedge \Theta^{nQ} + \mathcal{A}_2. \tag{5.26}$$

Now, $\mathcal{M}_{kj}^r$ and $\widetilde{\mathcal{M}}_{kj}^r$ can be easily computed: for $\mathcal{M}_{kj}^r$ we have (2.51) and (2.52), and

$$\widetilde{\mathcal{M}}_{kj}^r = \frac{2\delta_{jk}}{\overline{\Phi}^{r+1}\mathrm{P}^{n-r-1}} \tag{5.27}$$

$$- (n - r - 1)\frac{(L_k\rho^2)(\overline{L}_j\rho^2)}{\overline{\Phi}^{r+1}\mathrm{P}^{n-r}} + \mathcal{A}_2, \quad k < n,$$

$$\widetilde{\mathcal{M}}_{nj}^r = (r + 1)\frac{\overline{L}_j\rho^2}{\overline{\Phi}^{r+2}\mathrm{P}^{n-r-1}} + (n - r - 1)\frac{2\Phi(\overline{L}_j\rho^2)}{\overline{\Phi}^{r+1}\mathrm{P}^{n-r}} + \mathcal{A}_2. \tag{5.28}$$

So the types of the explicit terms are ≥ 1 in all cases, and, for $r > 0$, many of the terms are even of type ≥ 2. This will be tacitly used in the following arguments.

Let us now write

$$\vartheta_x \mathcal{L}_q - \partial_y \mathcal{L}_{q-1} = \sum_{|L|=q} \mathcal{H}_L \wedge \Theta^L = \mathcal{H}_q \tag{5.29}$$

and compute the $(0, q + 1)$-forms $\mathcal{H}_L$ up to $\mathcal{A}_2$-terms. For $n \in L$ we easily get

$$\mathcal{H}_{nQ} = -2^{n-1}\left(\frac{1}{2\pi}\right)^n (n - 1)!\frac{1}{\mathrm{P}^n}\sum_{\substack{j<n \\ j\notin Q}} \overline{L}_j\rho^2\overline{\omega}^{njQ} + \mathcal{A}_2. \tag{5.30}$$

If $n \notin L$ we distinguish three cases:

   a) $K = lL$, with $l < n$, up to order

   b) $K = nL$

   c) $K = nJ, J \neq L$

Recall that K refers to the exponent of $\overline{\omega}$ in (5.26). Case a) yields

$$\mathcal{H}_L = -\sum_{\substack{j<n \\ j\notin L}} c_{nq}\left[\sum_r \binom{n - r - 2}{q}\frac{(r + 1)\overline{L}_j\rho^2}{\overline{\Phi}^{r+2}\mathrm{P}^{n-r-1}} + (n - 1)\binom{n - 2}{q}\frac{2\Phi\overline{L}_j\rho^2}{\overline{\Phi}\mathrm{P}^n}\right]\overline{\omega}^{jL} + \mathcal{A}_2 \tag{5.31}$$

Case b) is treated using 5.16 iii:

$$\mathcal{H}_L = \frac{2^{n-2}}{(2\pi)^n}(n - 1)!\frac{4\Phi}{\mathrm{P}^n}\overline{\omega}^{nL} + \mathcal{A}_2. \tag{5.32}$$

In case c) we have

$$\mathcal{H}_L = \mathcal{A}_2. \tag{5.33}$$

To sum up:

Lemma 5.34 $\vartheta_x \mathcal{L}_q - \partial_y \mathcal{L}_{q-1} = \sum_{|L|=q} \mathcal{H}_L \wedge \Theta^L$ with

$$\mathcal{H}_{nQ} = -2^{n-1} \left(\frac{1}{2\pi} \right)^n (n-1)! \frac{1}{\mathbf{P}^n} \sum_{\substack{j<n \\ j \notin Q}} \overline{L}_j \rho^2 \overline{\omega}^{njQ} + \mathcal{A}_2,$$

and for $n \notin L$:

$$\mathcal{H}_L = -\sum_{\substack{j<n \\ j \neq L}} c_{nq} \left[\sum_r \binom{n-r-2}{q} \frac{(r+1)\overline{L}_j \rho^2}{\overline{\Phi}^{r+2} \mathbf{P}^{n-r-1}} + \binom{n-2}{q} \frac{2(n-1)\Phi \overline{L}_j \rho^2}{\overline{\Phi} \mathbf{P}^n} \right] \overline{\omega}^{jL}$$

$$+ \frac{2^{n-2}}{(2\pi)^n} (n-1)! \frac{4\Phi}{\mathbf{P}^n} \overline{\omega}^{nL} + \mathcal{A}_2.$$

6. We now turn to solving (5.8) subject to the condition (5.9).

So we want a $\mathcal{Z}_2$-kernel $\mathcal{N}_q$ such that

$$\begin{aligned} \mathcal{N}_q &= \mathcal{N}_q^* + \mathcal{Z}_3 \\ *_x \mathcal{N}_q &= 0 \qquad \text{on } b\Omega \\ \overline{\partial}_x \mathcal{N}_q &= \mathcal{T}_q + \mathcal{Z}_2 \end{aligned} \tag{5.35}$$

and

$$\overline{\partial}_x^* \mathcal{N}_q = \mathcal{T}_{q-1}^* + \mathcal{Z}_2. \tag{5.36}$$

Since the isotropic part of $\mathcal{T}_q$ is $\overline{\partial}\Gamma_{0q}$, we set

$$\mathcal{N}_q = \mathcal{G}_q + \Gamma_{0q} \tag{5.37}$$

and determine $\mathcal{G}_q$. Let us first work locally:

$$\mathcal{G}_q = \sum_{|L|=q} \mathcal{G}_L \wedge \Theta^L, \quad \mathcal{G}_q = \mathcal{G}_q^* + \mathcal{A}_3 \tag{5.38}$$

and solve

$$\overline{\partial}_x \mathcal{G}_L = \mathcal{H}_L + \mathcal{A}_2, \tag{5.39}$$

where $\mathcal{H}_L$ is given in 5.34. We have to distinguish the cases $q = n-1$ and $q < n-1$. In the latter case we verify immediately (5.39) if we choose

$$\mathcal{G}_{nQ} = -\frac{2^{n-1}(n-2)!}{(2\pi)^n} \frac{1}{\mathbf{P}^{n-1}} \overline{\omega}^{nQ}, \quad L = nQ, \tag{5.40}$$

$$\mathcal{G}_L = c_{nq} \left[\sum_r \binom{n-r-2}{q} \frac{r+1}{n-r-2} \frac{1}{\overline{\Phi}^{r+2} \mathbf{P}^{n-r-2}} + \binom{n-2}{q} \frac{2\Phi}{\overline{\Phi} \mathbf{P}^{n-1}} \right] \overline{\omega}^L, \ n \notin L.$$

For $q = n - 1$ and $n > 2$ we set

$$\mathcal{G}_{n-1} = \frac{2^{n-2}(n-2)!}{(2\pi)^n} \left[\left(\frac{2\Phi}{\overline{\Phi}P^{n-1}} + \frac{1}{n-2} \frac{1}{\overline{\Phi}^2 P^{n-2}} \right) \overline{\omega}^{N'} \wedge \Theta^{N'} - \cdots \right.$$

$$\left. \cdots - \frac{2}{P^{n-1}} \sum_{\substack{n \in L \\ |L|=n-1}} \overline{\omega}^L \wedge \Theta^L \right] \tag{5.41}$$

and again obtain (5.39) and (5.38). Here $N' = \{1, \ldots, n-1\}$.

Let us now globalise this construction using the forms τ and v – see 5.17 and 5.18. We have

$$(\overline{\partial}_x \partial_y \rho^2)^q = (-2)^q (\tau^q + q\tau^{q-1} \wedge v) + \mathcal{E}_1$$

$$\tau^q = q! \sum_{\substack{|L|=q \\ n \notin L}} \overline{\omega}^L \wedge \Theta^L + \mathcal{E}_1 \tag{5.42}$$

$$\tau^{q-1} \wedge v = (q-1)! \sum_{|Q|=q-1} \overline{\omega}^{nQ} \wedge \Theta^{nQ} + \mathcal{E}_1.$$

We insert (5.42) into (5.40) and (5.41) and get

Proposition 5.43 *Let the differential forms $\mathcal{N}_q$ be given by the following formulae:*

$$\mathcal{N}_0 = \frac{2^{(n-2)}(n-2)!}{(2\pi)^n} \left[\sum_{0 \le r \le n-2} \frac{r+1}{n-r-2} \frac{1}{\overline{\Phi}^{r+2} P^{n-r-2}} - (n-1)\frac{\log P}{\overline{\Phi}^n} + \frac{2\Phi}{\overline{\Phi}P^{n-1}} \right]$$

$$+ \frac{1}{2\pi^n}(n-2)!\frac{1}{\rho^{2n-2}}$$

$$\mathcal{N}_q = 2^{n-2} \left(\frac{1}{2\pi} \right)^n (n-q-2)! \left[\sum_{0 \le r \le n-q-2} \binom{n-r-2}{q} \frac{r+1}{n-r-2} \frac{1}{\overline{\Phi}^{r+2} P^{n-r-2}} \right.$$

$$\left. + \binom{n-2}{q} \frac{2\Phi}{\overline{\Phi}P^{n-1}} \right] \tau^q$$

$$+ \frac{(n-2)!}{q!2\pi^n} \frac{1}{\rho^{2n-2}} \tau^q + \frac{-2^{n-1}(n-2)!}{(q-1)!(2\pi)^n} \frac{1}{P^{n-1}} \tau^{q-1} \wedge v$$

$$+ \frac{(n-2)!}{(q-1)!2\pi^n} \frac{1}{\rho^{2n-2}} \tau^{q-1} \wedge v$$

for $1 \le q \le n-2$,

$$\mathcal{N}_{n-1} = \frac{2^{n-2}}{(n-1)(2\pi)^n} \left[\frac{2\Phi}{\overline{\Phi}P^{n-1}} + \frac{1}{n-2} \frac{1}{\overline{\Phi}^2 P^{n-2}} \right] \tau^{n-1}$$

$$+ \frac{1}{(n-1)2\pi^n} \frac{1}{\rho^{2n-2}} \tau^{n-1} + \frac{-2^{n-2}}{(2\pi)^n} \frac{2}{P^{n-1}} \tau^{n-2} \wedge v + \frac{1}{2\pi^n} \frac{1}{\rho^{2n-2}} \tau^{n-2} \wedge v$$

(here we assume $n \geq 3$).

Then the $\mathcal{N}_q$ are $\mathcal{Z}_2$-kernels which satisfy, for $q = 0, \ldots, n-1$,

   i. $\overline{\partial}_x \mathcal{N}_q = \mathcal{T}_q + \mathcal{Z}_2$ with $\overline{\partial}\mathcal{Z}_2 = \mathcal{Z}_1$

   ii. $*_x \mathcal{N}_q|_{b\Omega} \equiv 0$

   iii. $\mathcal{N}_q = \mathcal{N}_q^* + \mathcal{Z}_3$

Proof   This follows immediately from our previous calculations (for $q > 0$, the case $q = 0$ has to be verified accordingly). The remark about the error terms is true because they only involve $\overline{\Phi}$ in the denominator.   □

Remark.   The kernel $\frac{\log P}{\overline{\Phi}^n}$ is, strictly speaking, not an admissible kernel. But its properties are essentially those of an $\mathcal{A}_2$-kernel (almost). As we are not particularly interested in $\mathcal{N}_0$ we do not further extend the class of kernels which we study systematically, in order to also incorporate this isolated example.

7.   We now verify the identity

$$\overline{\partial}_x^* \mathcal{N}_q = \mathcal{T}_{q-1}^* + \mathcal{Z}_2. \tag{5.44}$$

It turns out that this is only true for $1 \leq q \leq n-2$. (The case $q = 0$ is trivial.)

It amounts to showing

$$\vartheta_x \mathcal{G}_q = \mathcal{H}_{q-1}^* + \mathcal{A}_2, \quad 1 \leq q \leq n-2, \tag{5.45}$$

with ϑ_x the formal adjoint of $\overline{\partial}_x$ (which coincides, of course, with $\overline{\partial}_x^*$, but the boundary condition is satisfied only for $\mathcal{N}_q$, not for $\mathcal{G}_q$ alone).

$\mathcal{G}_q$ and $\mathcal{H}_q$ have been defined in (5.40) and (5.29). Everything boils down again to a lengthy computation. For the right-hand side in (5.45) we get from Lemma 5.34 in the usual local frames

$$\mathcal{H}_{q-1}^* = \frac{2^{n-1}}{(2\pi)^n}(n-1)!\left[\sum_{\substack{n \in L \\ j < n \\ |L|=q-1}} \frac{\Lambda_j \rho^2}{P^n}\overline{\omega}^L \wedge \Theta^{jL} + \sum_{\substack{n \notin L \\ |L|=q-1}} \frac{2\Phi}{P^n}\overline{\omega}^L \wedge \Theta^{nL}\right.$$

$$\left. - \sum_{\substack{n \notin L \\ j < n \\ |L|=q-1}} \frac{\Phi \Lambda_j \rho^2}{\overline{\Phi} P^n}\overline{\omega}^L \wedge \Theta^{jL}\right]$$

$$- \sum_{\substack{n \notin L \\ j < n \\ |L|=q-1}} c_{n,q-1} \sum_r \binom{n-r-2}{q-1} \frac{(r+1)\Lambda_j \rho^2}{\overline{\Phi}^{r+2} P^{n-r-1}}\overline{\omega}^L \wedge \Theta^{jL} + \mathcal{A}_2 \tag{5.46}$$

(This is true for $q - 1 \geq 1$; the formula for $q - 1 = 0$ is similar, but simpler.) On the other hand,

$$\vartheta_x \mathcal{G}_q = \frac{2^{n-1}(n-2)!}{(2\pi)^n} \sum_{\substack{n \in J \\ j,L}} L_j \left(\frac{1}{\mathbf{P}^{n-1}} \right) \varepsilon_J^{jL} \overline{\omega}^L \wedge \Theta^J$$

$$- \sum_{\substack{n \notin J \\ j,L}} c_{nq} \left[\sum_r \binom{n-r-2}{q} \frac{r+1}{n-r-2} L_j \left(\frac{1}{\overline{\Phi}^{r+2} \mathbf{P}^{n-r-2}} \right) \right.$$

$$\left. + \binom{n-2}{q} L_j \left(\frac{2\Phi}{\overline{\Phi}\mathbf{P}^{n-1}} \right) \right] \varepsilon_J^{jL} \overline{\omega}^L \wedge \Theta^J + \mathcal{A}_2 \quad (5.47)$$

(see 5.43 and 5.37) which yields

$$\vartheta_x \mathcal{G}_q = \frac{2^{n-1}(n-1)!}{(2\pi)^n} \left[\sum_{\substack{n \in J \\ L}} \left(\sum_{j<n} \frac{\Lambda_j \rho^2}{\mathbf{P}^n} \overline{\omega}^L \wedge \varepsilon_J^{jL} \Theta^J \right) + \frac{2\Phi}{\mathbf{P}^n} \overline{\omega}^L \wedge \varepsilon_J^{nL} \Theta^J \right]$$

$$- \sum_{\substack{n \notin J \\ j,L}} c_{nq} \left[\sum_r \binom{n-r-2}{q} \left\{ \frac{(r+1)(r+2)}{n-r-2} \frac{\Lambda_j \rho^2}{\overline{\Phi}^{r+3} \mathbf{P}^{n-r-2}} + \frac{(r+1)\Lambda_j \rho^2}{\overline{\Phi}^{r+2} \mathbf{P}^{n-r-1}} \right\} \right.$$

$$\left. + \binom{n-2}{q}(n-1)\frac{2\Phi\Lambda_j \rho^2}{\overline{\Phi}\mathbf{P}^n} \right] \overline{\omega}^L \wedge \varepsilon_J^{jL} \Theta^J + \mathcal{A}_2.$$

Now,

$$c_{nq} \binom{n-2}{q} 2(n-1) = \frac{2^{n-1}}{(2\pi)^n}(n-1)!, \quad (5.48)$$

so comparison of (5.47) and (5.48) reduces to establishing the following identity

$$- \sum_{\substack{n \notin L \\ j<n \\ |L|=q-1}} c_{n,q-1} \sum_{0 \leq r \leq n-q-1} \binom{n-r-2}{q-1} \frac{(r+1)\Lambda_j \rho^2}{\overline{\Phi}^{r+2} \mathbf{P}^{n-r-1}} \overline{\omega}^L \wedge \Theta^{jL} =$$

$$- \sum_{\substack{n \notin L \\ j<n \\ |L|=q-1}} c_{nq} \sum_{0 \leq r \leq n-q-2} \binom{n-r-2}{q} \left[\frac{(r+1)(r+2)\Lambda_j \rho^2}{(n-r-2)\overline{\Phi}^{r+3} \mathbf{P}^{n-r-2}} \right.$$

$$\left. + \frac{(r+1)\Lambda_j \rho^2}{\overline{\Phi}^{r+2} \mathbf{P}^{n-r-1}} \right] \overline{\omega}^L \wedge \Theta^{jL} + \mathcal{A}_2. \quad (5.49)$$

This can again be explicitly checked.

So we have verified 5.44 and finished our proof of Theorem (5.10).

8.   We only sketch the proof of Theorem 5.12.

An explicit and long computation shows that for $q = n - 1$ or $n - 2$,

$$\vartheta_x \mathcal{H}_q^* = \mathcal{A}_1 \tag{5.50}$$

($\mathcal{H}_q$ as defined in 5.29). Since $\mathcal{T}_q^* = \mathcal{H}_q^* + \vartheta_x \Gamma_{0,q+1}$, this implies $\vartheta_x \mathcal{T}_q^* = \mathcal{A}_1$. We now express $\Gamma_{0,q+1}$ in local orthonormal frames and see that locally, $\mathcal{T}_q^*$ coincides, up to an $\mathcal{E}_{2-2n}$-kernel, with a kernel $\widehat{\mathcal{T}}_{q,\mathrm{loc}}^*$ satisfying the Neumann condition. The local kernels $\widehat{\mathcal{T}}_{q,\mathrm{loc}}^*$ are patched together with a partition of unity to eventually yield $\widehat{\mathcal{T}}_q^*$ with the desired properties.

This concludes our investigation of the kernels $\mathcal{T}_q$ and $\mathcal{T}_q^*$; we shall apply the above information 5.10 and 5.12 in the next paragraph.

§6   Asymptotic Development of the Neumann Operator

1.   We can now return to the programme we started in §3 and give asymptotic developments for the operators N, $\bar{\partial}^* N$ and $\bar{\partial} N$ arising in the Neumann decomposition. First we need an integral formula involving the complex Laplacian $\square$.

2.   Let $f \in \mathrm{dom}\, \square \subset L_{0q}^2(\Omega)$, $1 \le q \le n - 2$. Then from 1.26 we have

$$f = (\bar{\partial} f, \mathcal{T}_q) + (\bar{\partial}^* f, \mathcal{T}_{q-1}^*) + (\bar{\partial} f, \mathcal{Z}_2) + (\bar{\partial}^* f, \mathcal{Z}_2) + (f, \mathcal{Z}_1). \tag{6.1}$$

Here and in the sequel, all operators turning up in scalar products $(\cdot, \cdot)$ refer to the x-variable. In view of (5.10) the last expression becomes

$$f = (\bar{\partial} f, \bar{\partial} \mathcal{N}_q) + (\bar{\partial}^* f, \bar{\partial}^* \mathcal{N}_q) + (\bar{\partial} f, \mathcal{Z}_2) + (\bar{\partial}^* f, \mathcal{Z}_2) + (f, \mathcal{Z}_1). \tag{6.2}$$

Moreover, the unspecified $\mathcal{Z}_2$-kernels satisfy

$$\bar{\partial}_x \mathcal{Z}_2 = \mathcal{Z}_1. \tag{6.3}$$

Since $\mathcal{N}_q$ and $\bar{\partial} f$ are in the domain of $\bar{\partial}^*$ we can rewrite 6.2 as

$$f = (\square f, \mathcal{N}_q) + \mathcal{Z}_2 \bar{\partial} f + \mathcal{Z}_2 \bar{\partial}^* f + \mathcal{Z}_1 f. \tag{6.4}$$

We now have to deal with the error terms in this formula.

Let us first apply 1.26 to $\bar{\partial} f$:

$$\bar{\partial} f = (\bar{\partial}^* \bar{\partial} f, \mathcal{T}_q^*) + (\bar{\partial}^* \bar{\partial} f, \mathcal{Z}_2) + (\bar{\partial} f, \mathcal{Z}_1) \tag{6.5}$$

with $\bar{\partial}_x \mathcal{Z}_2 = \mathcal{Z}_1$. For $1 \le q \le n - 3$ we have

$$\bar{\partial}^* \mathcal{N}_{q+1} = \mathcal{T}_q^* + \mathcal{Z}_2$$

by (5.10), so for these q

$$\bar{\partial} f = (\bar{\partial}^* \bar{\partial} f, \bar{\partial}^* \mathcal{N}_{q+1}) + (\bar{\partial}^* \bar{\partial} f, \mathcal{Z}_2) + (\bar{\partial} f, \mathcal{Z}_1) \tag{6.6}$$
$$= (\square f, \bar{\partial}^* \mathcal{N}_{q+1}) - (\bar{\partial} \bar{\partial}^* f, \bar{\partial}^* \mathcal{N}_{q+1}) + (\bar{\partial} f, \bar{\partial} \mathcal{Z}_2) + (\bar{\partial} f, \mathcal{Z}_1).$$

The second term on the right-hand side is $(\bar{\partial}^* f, \bar{\partial}^* \bar{\partial}^* \mathcal{N}_{q+1}) = 0$, which leaves us with

$$\bar{\partial} f = (\square f, \bar{\partial}^* \mathcal{N}_{q+1}) + (\bar{\partial} f, \mathcal{Z}_1). \tag{6.7}$$

If $q = n - 2$ we use 5.12 to transform (6.5):

$$\mathcal{T}_q^* = \widehat{\mathcal{T}}_q^* + \mathcal{Z}_2, \quad \bar{\partial}_x \mathcal{Z}_2 = \mathcal{Z}_1, \tag{6.8}$$
$$\bar{\partial}^* \widehat{\mathcal{T}}_q^* = \mathcal{Z}_1,$$

where $\widehat{\mathcal{T}}_q^*$ satisfies the 1st Neumann condition, and obtain

$$\bar{\partial} f = (\bar{\partial}^* \bar{\partial} f, \widehat{\mathcal{T}}_q^*) + (\bar{\partial}^* \bar{\partial} f, \mathcal{Z}_2) + (\bar{\partial} f, \mathcal{Z}_1) \tag{6.9}$$
$$= (\square f, \widehat{\mathcal{T}}_q^*) - (\bar{\partial} \bar{\partial}^* f, \widehat{\mathcal{T}}_q^*) + (\bar{\partial} f, \mathcal{Z}_1)$$
$$= (\square f, \mathcal{T}_q^*) - (\bar{\partial}^* f, \bar{\partial}^* \mathcal{T}_q^*) + (\bar{\partial} f, \mathcal{Z}_1)$$
$$= (\square f, \mathcal{T}_q^*) + (\bar{\partial}^* f, \mathcal{Z}_1) + (\bar{\partial} f, \mathcal{Z}_1).$$

To sum up we have

Lemma 6.10 *If $f \in \mathrm{dom}\, \square$, then*

$$\bar{\partial} f = (\square f, \mathcal{Z}_1) + (\bar{\partial} f, \mathcal{Z}_1) + (\bar{\partial}^* f, \mathcal{Z}_1).$$

A similar representation holds for $\bar{\partial}^* f$ — we simply give the result as

Lemma 6.11 *For f as above,*

$$\bar{\partial}^* f = (\square f, \mathcal{Z}_1) + (\bar{\partial} f, \mathcal{Z}_1) + (\bar{\partial}^* f, \mathcal{Z}_1) + (f, \mathcal{E}_0).$$

Recall that $\mathcal{E}_0$ is simply an everywhere smooth kernel. It occurs for $q = 1$, where $\bar{\partial}^* f$ is a function, and is just the kernel $\bar{\partial}_x P_0$ – cf. 1.26.

We can now easily prove our first main result.

Theorem 6.12 *For $1 \le q \le n - 2$ and for $l, m, p \in \mathbb{N}$ arbitrarily chosen there are Z-operators of the indicated type such that for all $f \in \mathrm{dom}\, \square \subset L^2_{0q}(\Omega)$ one has*

$$f = (\square f, \mathcal{N}_q) + Z_3 \square f + Z_l \bar{\partial} f + Z_m \bar{\partial}^* f + Z_p f.$$

Proof Insert 6.4 into itself and 6.10 and 6.11 into 6.4 and repeat this process as often as necessary. Note that the Z_3-operator depends on l, m and p. □

3.   We now consider the orthogonal projector onto the harmonic space:

$$H_q : L^2_{0q} \longrightarrow \mathbb{H}^q, \quad q \geq 1.$$

Since, for $f \in \mathbb{H}^q$, $\bar{\partial}f$ and $\bar{\partial}^* f$ vanish, (6.1) reads

$$f = Z_1 f, \tag{6.13}$$

which, by iteration, yields

$$f = Z_m f \tag{6.14}$$

for arbitrary m. Consequently, the identity is a compact operator on $\mathbb{H}^q$; moreover,

$$H_q f = Z_m H_q f, \quad m \text{ arbitrary.} \tag{6.15}$$

So:

Theorem 6.16 *For $q \geq 1$ the harmonic spaces $\mathbb{H}^q$ are finite-dimensional, and for $1 \leq q \leq n - 2$, H_q is an asymptotic Z-operator of infinite type (consequently with principal part 0).*

The regularity results of the next chapter will allow to deduce the inclusion

$$\mathbb{H}^q \subset C^\infty_{0q}(\overline{\Omega}) \tag{6.17}$$

4.   **Theorem 6.18** *For $1 \leq q \leq n - 2$, the Neumann operator N_q is an asymptotic Z-operator of type 2 whose principal part is represented by the operator N^0_q with kernel $\mathcal{N}_q$:*

$$N_q = N^0_q + Z^a_3.$$

Proof   We apply Theorem 6.12 to $N_q f$:

$$N_q f = (\Box N_q f, \mathcal{N}_q) + Z_3 \Box N_q f + Z_l \bar{\partial} N_q f + Z_m \bar{\partial}^* N_q f + Z_p N_q f. \tag{6.19}$$

Now,

$$\Box N_q f = f - H_q f, \tag{6.20}$$

and $H_q = Z^a_\infty$ in view of 6.16. So (6.19) becomes

$$\begin{aligned}
N_q f &= (f, \mathcal{N}_q) - (H_q f, \mathcal{N}_q) + Z_3 f \\
&\quad - Z_3 H_q f + Z_l \bar{\partial} N_q f + Z_m \bar{\partial}^* N_q f + Z_p N_q f \\
&= (f, \mathcal{N}_q) + Z_3 f + Z^a_\infty f + Z_l \bar{\partial} N_q f + Z_m \bar{\partial}^* N_q f + Z_p N_q f \\
&= N^0_q f + Z^a_3 f,
\end{aligned}$$

as desired.   □

5. To conclude this line of arguments we can now identify the principal part of $\overline{\partial}^* N$ and $\overline{\partial} N$. Recall that the *conjecture*

$$\overline{\partial}^* N_q = T_{q-1} + Z_2^a \tag{6.21}$$

had been our starting point in §5 – it will now be proved.

Theorem 6.22 *The operators* $\overline{\partial}^* N_q$ *and* $\overline{\partial} N_q$ *are, for each* q *with* $1 \leq q \leq n$, *asymptotic Z-operators of type 1 with principal parts (represented by)* T_{q-1} *and* T_q^*, *respectively:*

$$\overline{\partial}^* N_q = T_{q-1} + Z_2^a$$
$$\overline{\partial} N_q = T_q^* + Z_2^a.$$

Proof We give the proof for $q \neq n-2, n-1$; the missing cases can be easily handled using 5.12 and are left to the reader. The case $q = n$ is trivial.

As usual, we apply 1.28 to $\overline{\partial} N_q$ and $\overline{\partial}^* N_q$:

$$\overline{\partial} N_q f = T_q^* \overline{\partial}^* \overline{\partial} N_q f + Z_2 \overline{\partial}^* \overline{\partial} N_q f + Z_1 \overline{\partial} N_q f, \quad 0 \leq q \leq n-1 \tag{6.23}$$

$$\overline{\partial}^* N_1 f = P_0 \overline{\partial}^* N_1 f + T_0 \overline{\partial} \overline{\partial}^* N_1 f + Z_2 \overline{\partial} \overline{\partial}^* N_1 f + Z_1 \overline{\partial}^* N_1 f \tag{6.24}$$

$$\overline{\partial}^* N_q f = T_{q-1} \overline{\partial} \overline{\partial}^* N_q f + Z_2 \overline{\partial} \overline{\partial}^* N_q f + Z_1 \overline{\partial}^* N_q f, \quad 2 \leq q \leq n \tag{6.25}$$

The last identity gives, in view of 5.10,

$$\overline{\partial}^* N_q f = T_{q-1} \Box N_q f - T_{q-1} \overline{\partial}^* \overline{\partial} N_q f \tag{6.26}$$
$$+ Z_2 \Box N_q f - Z_2 \overline{\partial}^* \overline{\partial} N_q f + Z_1 \overline{\partial}^* N_q f$$
$$= T_{q-1} f - T_{q-1} H_q f - (\overline{\partial}^* \overline{\partial} N_q f, \overline{\partial} \mathcal{N}_{q-1})$$
$$- (\overline{\partial}^* \overline{\partial} N_q f, \mathcal{Z}_2) + Z_2 f - Z_2 H_q f$$
$$- (\overline{\partial}^* \overline{\partial} N_q f, \mathcal{Z}_2) + Z_1 \overline{\partial}^* N_q f$$
$$= T_{q-1} f + Z_\infty^a f + Z_1 \overline{\partial} N_q f + Z_2 f$$
$$+ Z_2^a f + Z_1 \overline{\partial} N_q f + Z_1 \overline{\partial}^* N_q f$$
$$= T_{q-1} f + Z_2^a f + Z_1 \overline{\partial} N_q f + Z_1 \overline{\partial}^* N_q f$$

$(2 \leq q \leq n$; but 6.24 also yields the same formula for $q = 1$ since $\overline{\partial}_x^* P_0 = \mathcal{E}_0$ and $P_0 - P_0^* = \mathcal{Z}_1.)$

We now combine 6.23 with 5.10 in the same way:

$$\overline{\partial} N_q f = (\overline{\partial}^* \overline{\partial} N_q f, T_q^*) + (\overline{\partial}^* \overline{\partial} N_q f, \mathcal{Z}_2) + Z_1 \overline{\partial} N_q f \tag{6.27}$$
$$= (\overline{\partial}^* \overline{\partial} N_q f, \overline{\partial}^* \mathcal{N}_{q+1}) + (\overline{\partial}^* \overline{\partial} N_q f, \mathcal{Z}_2) + (\overline{\partial}^* \overline{\partial} N_q f, \mathcal{Z}_2) + Z_1 \overline{\partial} N_q f$$
$$= (\Box N_q f, \overline{\partial}^* \mathcal{N}_{q+1}) + Z_1 \overline{\partial} N_q f$$
$$= (f, \overline{\partial}^* \mathcal{N}_{q+1}) - (H_q f, \overline{\partial}^* \mathcal{N}_{q+1}) + Z_1 \overline{\partial} N_q f$$
$$= T_q^* f + Z_2 f + Z_\infty^a f + Z_1 \overline{\partial} N_q f$$
$$= T_q^* f + Z_2^a f + Z_1 \overline{\partial} N_q f$$

(where we have assumed $q \neq n - 2, n - 1$). The operators in the scalar products again refer to the x-variable.

Iteration of (6.27) gives

$$\overline{\partial} N_q f = \boldsymbol{T}_q^* f + Z_2^a f; \tag{6.28}$$

we plug this into (6.26) to obtain

$$\overline{\partial}^* N_q f = \boldsymbol{T}_{q-1} f + Z_2^a f + Z_1 (\boldsymbol{T}_q^* f + Z_2^a f) + Z_1 \overline{\partial}^* N_q f \tag{6.29}$$
$$= \boldsymbol{T}_{q-1} f + Z_2^a f + Z_1 \overline{\partial}^* N_q f,$$

and iterate to get

$$\overline{\partial}^* N_q f = \boldsymbol{T}_{q-1} f + Z_2^a f, \tag{6.30}$$

as desired. □

6. At this point we want to sum up the main results.

The operators occurring in the Neumann problem,

$$P, \overline{\partial} N, \overline{\partial}^* N, N, H$$

are all members of the algebra of asymptotic Z-operators, of type $0, 1, 1, 2, 2n$ respectively. Their principal parts can be given by explicit Z-operators. In particular, the principal part kernel of P is special admissible of type 0, the principal part of N is a diagonal operator which is even scalar in tangent directions (there is one dimension $q = n - 1$ not covered by the above), and the principal part kernels of $\overline{\partial} N, \overline{\partial}^* N$, though they look more complicated than the Neumann principal part $\mathcal{N}$, are in fact simply the $\overline{\partial}^*$ resp. $\overline{\partial}$-derivative of $\mathcal{N}$.

The regularity theorems of the next chapter will show that the mapping properties of the operators are those of their principal parts, and essentially governed by their type: the larger the type the more regular the operator.

§7 Notes

Most of the chapter is based on the work of Lieb and Range [LiR 86$_1$, LiR 86$_2$, LiR 87], the germ of the ideas presented here lies, however, in the work of Kerzman/Stein [KeS 78] and Ligocka [Lig 84] who have been first to exploit the crucial symmetry relation 2.5. The use of the normalised Levi metric, which gives rise to the additional symmetry expressed in theorem 1.29 and which is based on proposition 2.36, is due to Lieb/Range [LiR 86$_1$, LiR 86$_2$]. They also make use of an "approximate" solution of the system (5.8) (5.9) to find the principal part of the Neumann operator. It is clear that the diagonal structure of this operator is due to the presence of the Levi metric. Whereas we can fairly completely analyze $N, \overline{\partial} N$ and $\overline{\partial}^* N$ in this case, an analogous structure theorem for these operators in the case of general metrics is still unknown.

The case of codimension-one forms is somewhat exceptional for the Neumann operator N (and particularly easy for the canonical solution operator $\bar{\partial}^* N$); we have not included N_{n-1} in our presentation, but refer to [LiR 93]. The case $q = n$, finally, is covered by the classical theory of harmonic forms because the Neumann conditions, in this case, reduce to Dirichlet boundary conditions. To deal with the Bergman projection on general strictly pseudoconvex manifolds one needs the delicate isomorphism theorem 3.14; the use of the Remmert reduction in this context seems to be new. The reduction theory of complex manifolds and spaces was developed by Remmert and Stein between 1952 and 1960; see [Car 60]. A proof of the direct image theorem can be found in [GrR 84].

Asymptotic developments in terms of integral operators of the operators of the $\bar{\partial}$-Neumann theory, which are an essential aim of this book, are only one possiblity of getting precise information on the Neumann problem. Another way of dealing with this problem, which has, moreover, the advantage of being independent of the metric, was taken by Greiner/Stein [GrS 77] and Beals/Greiner/Stanton [BGS 87]. They exhibit the Neumann operator as a composition of simpler operators.

Chapter VII

Regularity Properties of Admissible Operators

We now turn to a systematic study of the regularity properties of Z-operators in various function spaces. Some of the results we obtain have already been proved in previous chapters, in particular in Chapters I and III; they will appear again in this chapter in a more general context.

The first paragraph is based on Krantz' work [Kra 76] and describes the function spaces we shall be working with; it centers around the "Hardy-Littlewood lemmas" 1.4 and 1.8. The essential estimates for admissible kernels are summed up in Lemma 2.1; many of the following statements simply interpret this lemma in the language of various norms. The estimates of §3 which refer to derivatives, depend essentially on the commutator relations 3.6 and 3.7 between kernels and vector fields and consequently require the operators to be of commutator type. The main content of the chapter is independent of the theory of singular integrals; the only (but important) exception is Theorem 4.4 on isotropic operators, where we make essential use of Stein's results [Ste 70, Ste 93].

§1   Spaces of Functions and Differential Forms

Let $D \subset\subset X$ be a bounded domain in a hermitian manifold with C^∞ boundary which is given by a defining function $r : U \to \mathbb{R}$, with $D \subset\subset U$.

At the beginning we want to define some Banach spaces of functions (or differential forms) on D. The corresponding norms will depend on arbitrary choices (as for example finite open coordinate coverings, the metric or the defining function r) but different choices will give rise to equivalent norms. We mention this fact here once and for all.

In particular this implies that all our norms will have a genuine local character which means that they all come from local constructions on finitely many domains in $\mathbb{C}^n$ and therefore can be obtained by using a geometric (euclidean) approach.

First of all we define *weighted L^p-spaces*.

Definition 1.1 *Let* $0 \leq \delta$ *and* $1 \leq p \leq \infty$. *We define for complex valued measurable functions* f *on* D

$$\|f\|_{L^{p,\delta}(D)} = \|f\|_{L^{p,\delta}} = \left(\int\limits_D |f(x)|^p |r(x)|^\delta \, dV(x) \right)^{1/p}$$

if $p < \infty$ ($dV(x)$ *denotes the volume element on* X),

$$\|f\|_{L^{\infty,\delta}} = \inf\{c \geq 0 \mid for \, a.\,e.\,x \in D : |f(x)||r(x)|^\delta \leq c\}$$

and set

$$L^{p,\delta}(D) = \{f \mid \|f\|_{L^{p,\delta}} < \infty\}.$$

Next we introduce *weighted Sobolev spaces*. We denote by $X, X_1, X_2, \ldots C^\infty$ vector fields on $\overline{D}$, i.e. smooth sections of the complexified tangent bundle. If applied to non regular functions they are taken in the distributional sense.

We cover $\overline{D}$ by finitely many open sets $V_1, \ldots, V_N \subset\subset X$ such that for any i there exists biholomorphic maps $\varphi_i : V_i \to \tilde{V}_i \subset\subset \mathbb{C}^n$ onto bounded convex domains in $\mathbb{C}^n$ which extend C^∞ smooth to $\overline{V}_i$.

The push forward of a vector field X via φ_i has a representation in canonical euclidean coordinates. Then it is clear how to define C^k norms of $X|_{V_i \cap D}$ and a posteriori of X. We denote this C^k norm by $\|X\|_{C^k}$ and set $\|X\| = \|X\|_{C^0}$.

Definition 1.2 (Sobolev spaces) *Let* $0 \leq \delta$, $k \in \mathbb{N}_0$, $1 \leq p \leq \infty$. *Set*

$$L_k^{p,\delta}(D) = L_k^{p,\delta} = \{f \in L^{p,\delta}(D) \mid \|f\|_{L_k^{p,\delta}} < \infty\}$$

with

$$\|f\|_{L_k^{p,\delta}} = \sup\{\|X_1 \ldots X_j f\|_{L^{p,\delta}} \mid 0 \leq j \leq k, X_1, \ldots, X_j \, vector \, fields \, with \, \|X_j\| \leq 1\}.$$

In the following notations we do not distinguish for a function f on D between $f|_{V_i \cap D}$ and $f \circ \varphi_i^{-1}$.

Definition 1.3 (Lipschitz spaces) *Let* $\varphi_i : V_i \to \tilde{V}_i$, $i = 1, \ldots, N$ *be the above covering. We set for* $0 < \alpha < 1$

$$\Lambda_\alpha = \Lambda_\alpha(D)$$
$$= \{f \in L^{\infty,0}(D) \mid \|f\|_{\Lambda_\alpha} := \|f\|_{L^{\infty,0}} + \sup_i \sup_{\substack{x,y \in \varphi_i(D \cap V_i) \\ x \neq y}} \frac{|f(x) - f(y)|}{|x - y|^\alpha} < \infty\}$$

and

$$\Lambda_1 = \Lambda_1(D) = \left\{ f \in L^{\infty,0}(D) \mid \|f\|_{\Lambda_1} := \right.$$

$$\left. \|f\|_{L^{\infty,0}} + \sup_i \sup_{\substack{x,y \in \varphi_i(D \cap V_i) \\ x \neq y}} \frac{2|f(x) + f(y) - 2f\left(\frac{x+y}{2}\right)|}{|x - y|} < \infty \right\}.$$

Now we are prepared to show the following embedding theorem (for the proof and the following compare also Krantz [Kra 76]).

Theorem 1.4 *Let* $0 < \delta < 1$. *Then every* $\tilde{f} \in L_1^{\infty,\delta}(D)$ *is represented by an uniquely defined locally Lipschitz continuous function* f *with the following properties:*

There exists a positive constant C *not depending on* $\tilde{f}$ *such that*

$$|f(x) - f(y)| \leq C \cdot \|\tilde{f}\|_{L_1^{\infty,\delta}} \rho(x,y)^{1-\delta}$$

for all $x, y \in D$ *and*

$$|\text{grad } f(x)| \leq C \cdot \|\tilde{f}\|_{L_1^{\infty,\delta}} |r(x)|^{-\delta}$$

for all $x \in D$ *where* $\text{grad } f(x)$ *exists.*

In particular $f \in \Lambda_{1-\delta}(D)$ *and*

$$L_1^{\infty,\delta}(D) \hookrightarrow \Lambda_{1-\delta}(D)$$

is a continuous embedding.

Proof   By using a partition of unity and the maps $\varphi_i : V_i \to \tilde{V}_i$ we can assume that $D \subset\subset \mathbb{C}^n$ and ρ is the euclidean distance. Extend $\tilde{f}$ to $\mathbb{C}^n \setminus D$ by setting $\tilde{f}(x) = 0$. Let $\tilde{\varepsilon} > 0$ and set

$$D_{\tilde{\varepsilon}} = \{x \in D \mid \delta(x) > \tilde{\varepsilon}\},$$

with $\delta(x) = \text{dist}\,(x, bD)$. $\tilde{f}$ is locally integrable in $\mathbb{C}^n$ such that we can regularise $\tilde{f}$ as in Chapter V. Let $\varphi \geq 0$ be a test function with support in the unit ball, $\varphi(x) = \varphi(-x)$ and $\int_{\mathbb{C}^n} \varphi(x)\,dx = 1$.

Set for $\varepsilon > 0$

$$\tilde{f}_\varepsilon = \tilde{f} * \varphi_\varepsilon.$$

Then for $\varepsilon < \tilde{\varepsilon}$ $\tilde{f}_\varepsilon \in C^\infty(\overline{D_{\tilde{\varepsilon}}})$. Let $y \in D_{\tilde{\varepsilon}}$ and $\varepsilon \leq \frac{\tilde{\varepsilon}}{2}$. Since we have with a constant $C \geq 0$, independent of ε and $\tilde{f}$,

$$|\tilde{f}_\varepsilon(y)| \leq \int_{\mathbb{C}^n} |\tilde{f}(x)| \varphi_\varepsilon(x - y)\,dV$$

$$\leq C \int_{\mathbb{C}^n} |\tilde{f}(x)| |r(x)|^\delta \delta(x)^{-\delta} \varphi_\varepsilon(x - y)\,dV$$

$$\leq C \|\tilde{f}\|_{L^{\infty,\delta}} \left(\frac{\tilde{\varepsilon}}{2}\right)^{-\delta}$$

and since convolution commutes with differentiation we obtain

$$\|\tilde{f}_\varepsilon\|_{L_1^{\infty,0}(D_{\tilde{\varepsilon}})} \leq C\|\tilde{f}\|_{L_1^{\infty,\delta}(D)}\tilde{\varepsilon}^{-\delta}$$

if $\varepsilon \leq \frac{\tilde{\varepsilon}}{2}$.

This inequality in turn and the smoothness of bD implies that there exists a positive constant $C \geq 0$, independent of ε and $\tilde{f}$, such that for $y_1, y_2 \in \overline{D}_{\tilde{\varepsilon}}$, $y_1 \neq y_2$, and $\varepsilon \leq \frac{\tilde{\varepsilon}}{2}$

$$\frac{|\tilde{f}_\varepsilon(y_1) - \tilde{f}_\varepsilon(y_2)|}{|y_1 - y_2|} \leq C\|\tilde{f}\|_{L_1^{\infty,\delta}(D)}\tilde{\varepsilon}^{-\delta}.$$

We fix $\tilde{\varepsilon}$. Let $(\varepsilon_n)_n$ be a sequence of positive numbers $\varepsilon_n \leq \frac{\tilde{\varepsilon}}{2}$ converging to 0. Then by the Ascoli-Arzelà theorem there exists a subsequence which converges uniformly on $\overline{D}_{\tilde{\varepsilon}}$ to a function $f_{\tilde{\varepsilon}}$. The last inequality above is preserved by passing to the limit.

It is clear that for $0 < \tilde{\varepsilon}_1 < \tilde{\varepsilon}_2$ we have $f_{\tilde{\varepsilon}_2}|_{D_{\tilde{\varepsilon}_1}} = f_{\tilde{\varepsilon}_1}$ because $f_{\tilde{\varepsilon}_1} = f_{\tilde{\varepsilon}_2} = \tilde{f}$ a. e.everywhere on $D_{\tilde{\varepsilon}_1}$.

Therefore this defines a global function and we obtain if we drop the subscript $\tilde{\varepsilon}$

$$\|f\|_{L^{\infty,0}(D_{\tilde{\varepsilon}})} \leq C\|\tilde{f}\|_{L_1^{\infty,\delta}(D)}\tilde{\varepsilon}^{-\delta},$$

$$\frac{|f(y_1) - f(y_2)|}{|y_1 - y_2|} \leq C\|\tilde{f}\|_{L_1^{\infty,\delta}(D)}\tilde{\varepsilon}^{-\delta}.$$

for all $0 < \tilde{\varepsilon}$ and all $y_1, y_2 \in D_{\tilde{\varepsilon}}$, $y_1 \neq y_2$.

So $\tilde{f}$ is represented on D by f, f is locally Lipschitz continuous and therefore differentiable almost everywhere.

Let f be differentiable in $x \in D$. Then

$$|\text{grad } f(x))| \leq C\|\tilde{f}\|_{L_1^{\infty,\delta}} \cdot \tilde{\varepsilon}^{-\delta}$$

for all $\tilde{\varepsilon}$ with $x \in D_{\tilde{\varepsilon}}$. Thus

$$|\text{grad } f(x)| \leq C\|\tilde{f}\|_{L_1^{\infty,\delta}}\delta(x)^{-\delta}.$$

In the last part of the proof we show that $f \in \Lambda_{1-\delta}$ and that $L_1^{\infty,\delta} \hookrightarrow \Lambda_{1-\delta}$ is continuous.

It suffices to show these properties near the boundary. Locally D is still a domain in $\mathbb{C}^n$. There exists a neighbourhood $\tilde{U} \subset\subset U$ of the boundary such that for every $x \in \tilde{U}$ there is an uniquely determined nearest point to the boundary denoted by $\pi(x)$. Moreover by shrinking $\tilde{U}$ we can achieve that $\pi : \tilde{U} \to bD$ is C^∞ and that for any $x \in \tilde{U}$ $\pi^{-1}(\pi(x))$ is an open line segment on the normal to bD in $\pi(x)$.

Denote euclidean balls with radius η and center x by $B(x, \eta)$. Let $U' \subset\subset \tilde{U}$ be a neighbourhood of bD and $\eta > 0$ a constant such that we have $B(x, \eta) \subset\subset \tilde{U}$ for all $x \in U'$.

Our goal is to prove

$$|f(y_1) - f(y_2)| \leq C\|f\|_{L_1^{\infty,0}}|y_1 - y_2|^{1-\delta} \tag{$*$}$$

for all $y_1, y_2 \in D \cap U'$.

Set $c = \frac{1}{3}\min(\text{dist}\,(\tilde{U}\setminus U', bD), \eta)$. It suffices to show ($*$) for $y_1, y_2 \in D\cap U'$, $|y_1 - y_2| < c$.

For suppose $|y_1 - y_2| \geq c$. Denote the outward normal to bD in $x \in bD$ by $n(x)$. Now for $y \in U' \cap D$

$$f(y) - f(y - \eta n(\pi(y))) = \int_0^1 \frac{d}{dt}f((1-t)y + t(y - \eta n(\pi(y))))\,dt$$

and therefore

$$|f(y) - f(y - \eta n(\pi(y)))| \leq \int_0^1 \left|\frac{d}{dt}f(y - t\eta n(\pi(y)))\right|\,dt$$

$$\leq \eta C\|f\|_{L_1^{\infty,\delta}} \int_0^1 (\delta(y) + t\eta)^{-\delta}\,dt$$

$$\leq \frac{C}{1-\delta}\|f\|_{L_1^{\infty,\delta}}\left((\delta(y) + \eta)^{1-\delta} - \delta(y)^{1-\delta}\right)$$

$$\leq C'\|f\|_{L_1^{\infty,\delta}}.$$

Since $y - \eta n(\pi(y))$ is varying in a relatively compact subset of D we obtain

$$|f(y)| \leq C'\|f\|_{L_1^{\infty,\delta}}$$

with a constant $C' \geq 0$ independent of $y \in U' \cap D$ and f. Thus

$$|f(y_1) - f(y_2)| \leq 2C'\|f\|_{L_1^{\infty,\delta}} \leq \frac{2C'}{C}\|f\|_{L_1^{\infty,\delta}}|y_1 - y_2|.$$

So let $|y_1 - y_2| < c$. Set

$$y_1' = y_1 - |y_1 - y_2|n(\pi(y_1)), \quad y_2' = y_2 - |y_1 - y_2|n(\pi(y_2)).$$

Then $y_1', y_2' \in \tilde{U}$.

$$|f(y_1) - f(y_2)| \leq |f(y_1) - f(y_1')| + |f(y_1') - f(y_2')| + |f(y_2') - f(y_2)|$$
$$= T_1 + T_2 + T_3.$$

Here the main idea, stemming from Hardy and Littlewood, is to push the points y_1, y_2 towards the interior.

$$T_1 \leq \int_0^1 \left| \frac{d}{dt} f(y_1 + t(y_1' - y_1)) \right| dt$$

$$\leq |y_1' - y_1| \int_0^1 |\text{grad } f(y_1 + t(y_1' - y_1))| dt$$

$$\leq |y_1' - y_1| C' \|f\|_{L_1^{\infty,\delta}} \int_0^1 (|y_1 - \pi(y_1)| + t|y_1' - y_1|)^{-\delta} dt$$

$$\leq \frac{C}{1-\delta} \|f\|_{L_1^{\infty,\delta}} \left((|y_1 - \pi(y_1)| + |y_1' - y_1|)^{1-\delta} - |y_1 - \pi(y_1)|^{1-\delta} \right)$$

$$\leq C' \|f\|_{L_1^{\infty,\delta}} |y_1' - y_1|^{1-\delta}$$

$$\leq \widetilde{C} \|f\|_{L_1^{\infty,\delta}} |y_1 - y_2|^{1-\delta}.$$

T_3 is estimated in the same way.

We have dist $(y_i', bD) \geq |y_1 - y_2|$. Therefore

$$|f(y_1') - f(y_2')| \leq C \|f\|_{L_1^{\infty,\delta}} |y_1 - y_2|^{-\delta} |y_1' - y_2'|$$

$$\leq C' \|f\|_{L_1^{\infty,\delta}} |y_1 - y_2|^{1-\delta},$$

because of

$$|y_1' - y_2'| = |y_1 - y_2 - |y_1 - y_2| (n(\pi(y_1)) - n(\pi(y_2)))|$$

$$\leq 3|y_1 - y_2|.$$

Thus

$$|f(y_1) - f(y_2)| \leq C \|f\|_{L_1^{\infty,\delta}} |y_1 - y_2|^{1-\delta}$$

with a constant $C \geq 0$ independent of f. This concludes the proof. $\square$

In the following part we want to refine the above proof by taking into account different behaviour in normal and tangential directions.

Definition 1.5 *Let W be a C^∞ vector field on $\overline{D}$ with values in $\mathbb{C}TX$. We call W allowable if for all $x \in bD$*

$$W_x \in T_x^{1,0}(bD) \oplus T_x^{0,1}(bD).$$

By using coordinate patches we can define C^k-norms $\|W\|_{C^k}$ for allowable vector fields.

Now we define for $1 \leq p \leq \infty, 0 \leq \delta \leq 1, k \in \mathbb{N}$ the so-called *Sobolev-Stein spaces*

$$S_k^{p,\delta} = \{f \in L^{p,\delta} \mid W_1 W_2 \ldots W_j f \in L^{p,\delta}, W_i \text{ allowable}, 1 \leq j \leq k\}$$

equipped with the norm

$$\|f\|_{S_k^{p,\delta}} = \sup\{\|W_1 W_2 \ldots W_j f\|_{L^{p,\delta}} \mid W_i \text{ allowable}, \|W_i\|_{C^1} \leq 1, 0 \leq j \leq k\}.$$

Remark 1.6 If W_1 and W_2 are allowable with $W_{1,x} \in T_x^{1,0}$, $W_{2,x} \in T_x^{0,1}$ then

$$W_{1,x} W_{2,x} - W_{2,x} W_{1,x}$$

is tangential to bD in x. If D is strictly pseudoconvex, allowable vector fields of this type contain a component transversal to $T_x^{1,0}(bD) \oplus T_x^{0,1}(bD)$. *Thus* $\|f\|_{S_k^{p,\delta}}$ *takes into account* $(k-1)$ *derivatives in all tangential directions and* k *derivatives in holomorphic or antiholomorphic tangential directions.*

Let X be an arbitrary C^∞ vector field on $\overline{D}$. Then rX is allowable. *Therefore normal derivatives are weighted with a factor which is comparable to the boundary distance.*

Let $0 \leq \delta \leq 1, 0 \leq \delta^\# \leq 1$. We equip $L_1^{\infty,\delta} \cap S_1^{\infty,\delta^\#}$ with the norm $\|f\|_{L_1^{\infty,\delta}} + \|f\|_{S_1^{\infty,\delta^\#}}$.
We know from Theorem 1.4 that $f \in L_1^{\infty,\delta} \cap S_1^{\infty,\delta^\#}$ can be given by a locally Lipschitz function $f \in \Lambda_{1-\delta}$ with

$$|\text{grad } f(x)| \leq C\|f\|_{L_1^{\infty,\delta}} \text{ dist } (x, bD)^{-\delta}.$$

It follows from the proof of the same theorem that we have for an allowable vector field W

$$|Wf(x)| \leq C\|W\|_{C^1}(\|f\|_{L_1^{\infty,\delta}} + \|f\|_{S_1^{\infty,\delta^\#}}) \text{ dist } (x, bD)^{-\delta^\#}$$

in all points where f is differentiable.

Now we want to investigate the regularity properties of such a function f near the boundary. In order to do this we introduce *non-isotropic Lipschitz spaces*. Let $\gamma = (\gamma_1, \ldots, \gamma_n) : I \to V_i \cap \overline{D}$ be a C^∞ curve contained in a coordinate patch V_i. We call γ *allowable* if

$$\sup_I \left| \frac{d\gamma(t)}{dt} \right| = |\gamma|_{C^1} \leq 1$$

and

$$\left| \sum_{i=1}^n \frac{\partial r(\gamma(t))}{\partial z_i} \frac{d\gamma_i(t)}{dt} \right| \leq |r(\gamma(t))| \quad \text{for all } t \in I.$$

That means that $\frac{d\gamma(t)}{dt}$ is an element of the holomorphic tangent space at bD if $\gamma(t) \in bD$.

Definition 1.7 *Let $0 \leq \alpha \leq 1$. We denote by Γ_α the space of all functions $f : \overline{D} \to \mathbb{C}$ such that $f \circ \gamma \in \Lambda_\alpha(I)$ for all allowable curves γ. We equip Γ_α with the norm*

$$\|f\|_{\Gamma_\alpha} = \sup\{\|f \circ \gamma\|_{\Lambda_\alpha(I)} \mid \gamma \, allowable\}.$$

If $k < \alpha \leq k+1$, $k \in \mathbb{N}$, we set inductively

$$\Gamma_\alpha = \{f \mid W_1 \ldots W_j f \in \Gamma_{\alpha-k}, W_i \, allowable, 0 \leq j \leq k\}$$

equipped with the norm

$$\|f\|_{\Gamma_\alpha} = \sup\{\|W_1 W_2 \ldots W_j f\|_{\Gamma_{\alpha-k}} \mid W_i \, allowable, \|W_i\|_{C^1} \leq 1, 0 \leq j \leq k\}.$$

The spaces $\Gamma_{\alpha,\beta} = \Lambda_\alpha \cap \Gamma_\beta$, $\alpha, \beta > 0$, will be equipped with the norm

$$\|f\|_{\Gamma_{\alpha,\beta}} = \|f\|_{\Lambda_\alpha} + \|f\|_{\Gamma_\beta}.$$

Theorem 1.8 *Let $0 \leq \delta < \frac{1}{2}$, $\delta < \delta^\# < 1$, $\alpha = \frac{1}{2} - \delta$. Then*

$$L_1^{\infty,\delta+\frac{1}{2}} \cap S_1^{\infty,\delta^\#} \subset \Gamma_{\alpha,\alpha/(\alpha+\delta^\#)}$$

and this inclusion is a continuous embedding if the elements of the left hand side are represented by locally Lipschitz functions.

Proof We can assume that $D \subset\subset \mathbb{C}^n$ is equipped with the euclidean metric. Let $f \in L_1^{\infty,\delta+1/2} \cap S_1^{\infty,\delta^\#}$ be locally Lipschitz with

$$|\mathrm{grad}\, f(x)| \leq C\|f\|_{L_1^{\infty,\delta+\frac{1}{2}}} \mathrm{dist}\,(x,bD)^{-(\delta+1/2)},$$

$$|Wf(x)| \leq C(\|f\|_{L_1^{\infty,\delta+\frac{1}{2}}} + \|f\|_{S_1^{\infty,\delta^\#}})\|W\|_{C^1} \mathrm{dist}\,(x,bD)^{-\delta^\#}$$

for allowable vector fields W in all points where grad $f(x)$ exists.

Let $\gamma : I \to \overline{D}$ be an allowable curve, that means with

$$\sup_I \left|\frac{d\gamma(t)}{dt}\right| \leq 1$$

and

$$\left|\sum_{i=1}^n \frac{\partial r}{\partial z_i}(\gamma(t))\frac{d\gamma_i(t)}{dt}\right| \leq \mathrm{dist}\,(\gamma(t), bD)$$

for all $t \in I$.

Here we have replaced $r(\gamma(t))$ by dist $(\gamma(t), bD)$. Obviously this does not change the involved function spaces and gives rise to an equivalent norm.

We have to show that $f \circ \gamma \in \Lambda_{\alpha/(\alpha+\delta^\#)}(I)$. Since $f \circ \gamma$ is locally Lipschitz, it suffices to analyse $f \circ \gamma$ in a neighbourhood of a boundary point.

If this neighbourhood is chosen sufficiently small, there are vector fields $W_1, \ldots, W_{n-1}$ and N of type $(1,0)$ such that $\{W_1, \ldots, W_{n-1}\}$ is an orthogonal base of the holomorphic tangent space of the level sets of dist $(\cdot, bD)$ and N is orthogonal to $W_1, \ldots, W_{n-1}$. Moreover we can assume that $\|W_i\|_{C^1} \le 1, \|N\|_{C^1} \le 1$. Since γ is allowable there are functions $a, a_i : I \to \mathbb{C}$ such that

$$\sum_{k=1}^{n} \frac{d\gamma_k(t)}{dt} \frac{\partial}{\partial z_k} = a(t)N(\gamma(t)) + \sum_{i=1}^{n-1} a_i(t)W_i(\gamma(t))$$

with $|a(t)|, |a_i(t)| \le C$ and $|a(t)| \le C$ dist $(\gamma(t), bD)$ with an universal constant C. Therefore

$$\frac{d(f \circ \gamma)(t)}{dt} = \sum_{k=1}^{n} \frac{\partial f}{\partial z_k}(\gamma(t)) \frac{d\gamma_k(t)}{dt} + \sum_{k=1}^{n} \frac{\partial f}{\partial \bar{z}_k}(\gamma(t)) \overline{\frac{d\gamma_k(t)}{dt}}$$

$$= \left(a(t)N(\gamma(t)) + \sum_{i=1}^{n-1} a_i(t)W_i(\gamma(t)) \right) f(\gamma(t))$$

$$+ \overline{\left(a(t)N(\gamma(t)) + \sum_{i=1}^{n-1} a_i(t)W_i(\gamma(t)) \right)} f(\gamma(t))$$

and

$$\left| \frac{d(f \circ \gamma)(t)}{dt} \right| \le C |\text{grad } f(\gamma(t))| \text{ dist } (\gamma(t), bD) + C \sum_{i=1}^{n-1} |W_i f(\gamma(t))|$$

$$\le C \left\{ \|f\|_{L_1^{\infty,\delta+\frac{1}{2}}} + \|f\|_{S_1^{\infty,\delta\#}} \right\} \left(\text{dist } (\gamma(t), bD)^\alpha + \text{dist } (\gamma(t), bD)^{-\delta^\#} \right)$$

$$\le C \left\{ \|f\|_{L_1^{\infty,\delta+\frac{1}{2}}} + \|f\|_{S_1^{\infty,\delta\#}} \right\} \text{dist } (\gamma(t), bD)^{-\delta^\#}.$$

Now let $x_0 = \gamma(t_0)$, $x_1 = \gamma(t_1)$, $t_0, t_1 \in I$, be two points of the curve. We set with $h = |t_1 - t_0|$, $\Theta = 1/(\alpha + \delta^\#)$

$$\eta(t) = \gamma(t) - h^\Theta n(\pi(\gamma(t))).$$

We have

$$\left| \sum_{i=1}^{n} \frac{\partial r}{\partial z_i}(\eta(t)) \frac{d\eta(t)}{dt} \right| \le \left| \sum_{i=1}^{n} \frac{\partial r}{\partial z_i}(\gamma(t)) \frac{d\gamma(t)}{dt} \right| + Ch^\Theta$$

$$\le C \left(\text{dist } (\gamma(t), bD) + h^\Theta \right)$$

$$\le C \text{ dist } (\eta(t), bD).$$

Therefore when we replace $\eta(t)$ by $\eta(\beta t)$, with a positive constant $\beta < 1$, independent of γ, we obtain an allowable curve. Since $\gamma(\beta t)$ is allowable, we can suppose that $\beta = 1$.

Set $x_0' = \eta(t_0)$, $x_1' = \eta(t_1)$. Then

$$|f(x_0) - f(x_1)| \leq |f(x_0) - f(x_0')| + |f(x_0') - f(x_1')| + |f(x_1') - f(x_1)|$$
$$= T_1 + T_2 + T_3.$$

$$T_2 \leq \int_{t_0}^{t_1} \left| \frac{d}{dt} f(\eta(t)) \right| \, dt$$

$$\leq C \left\{ \|f\|_{L_1^{\infty,\delta+\frac{1}{2}}} + \|f\|_{S_1^{\infty,\delta^\#}} \right\} \int_{t_0}^{t_1} \operatorname{dist}(\eta(t), bD)^{-\delta^\#} \, dt$$

$$\leq C \left\{ \|f\|_{L_1^{\infty,\delta+\frac{1}{2}}} + \|f\|_{S_1^{\infty,\delta^\#}} \right\} h^{-\Theta\delta^\#} h$$

$$\leq C \left\{ \|f\|_{L_1^{\infty,\delta+\frac{1}{2}}} + \|f\|_{S_1^{\infty,\delta^\#}} \right\} h^{\alpha/(\alpha+\delta^\#)}.$$

$$T_1 \leq \int_0^1 \left| \frac{d}{d\tau} f((1-\tau)x_0 + \tau x_0') \right| \, d\tau$$

$$\leq C\|f\|_{L_1^{\infty,\delta+\frac{1}{2}}} h^\Theta \int_0^1 \operatorname{dist}((1-\tau)x_0 + \tau x_0', bD)^{-(\delta+1/2)} \, d\tau$$

$$\leq C\|f\|_{L_1^{\infty,\delta+\frac{1}{2}}} h^{\Theta(1-\frac{1}{2}-\delta)}$$

$$\leq C\|f\|_{L_1^{\infty,\delta+\frac{1}{2}}} h^{\alpha/(\alpha+\delta^\#)}.$$

T_3 is estimated analogously. Thus we have obtained

$$\|f \circ \gamma\|_{\Lambda_{\alpha/(\alpha+\delta^\#)}(I)} \leq C \left\{ \|f\|_{L_1^{\infty,\delta+\frac{1}{2}}} + \|f\|_{S_1^{\infty,\delta^\#}} \right\}$$

for all allowable curves γ. This proves the theorem. $\qquad\qquad\qquad\qquad\qquad\square$

Differential forms of bidegree (r, s) with coefficients in one of the above function spaces V are defined by using a fixed finite coordinate covering of $\overline{D}$ and the corresponding spaces will be denoted by $V_{r,s}$. The corresponding norms are introduced in the obvious way. Theorems 1.4 and 1.8 remain true for differential forms, and all following results will be valid for spaces of functions or forms.

§2 Behaviour of A_0-operators on L^p-spaces

Now we want to analyse the behaviour of A_λ-operators with respect to different function spaces. We recall that $P(x,y) = \rho^2(x,y) + 2r(x)r(y)$ and $\Phi(x,y)$ is the extended Levi polynomial of D as constructed in Chapter IV, (5.33). Furthermore we will assume that ds^2 is a normalised Levi metric as defined in Chapter VI, §2, such that we have (cf. Chapter VI, Proposition 2.19)

$$P = 2|\Phi|^2 + \mathcal{E}_2.$$

In order to streamline the necessary estimates we first prove a technical lemma.

Lemma 2.1 $n \geq 2$. Let j, m, p, $\delta^\#$ and $\delta + 1 > 0$ be nonnegative real numbers. Then

$$\sup_{y \in D} \int\limits_D \frac{\rho(x,y)^j |r(x)|^\delta |r(y)|^{\delta^\#} \, dV(x)}{P(x,y)^p |\Phi(x,y)|^m} < \infty$$

if either one of the following cases holds

(α) $\delta + \delta^\# \geq m - 2$ and $2p - 2n + m - j < \delta + \delta^\#$,
(β) $\delta + \delta^\# < m - 2$ and $p - n + m - j/2 - 1 < \delta + \delta^\#$,
(γ) $\delta \leq 0$, $0 < \delta + \delta^\# \leq m - 5/2$ and $p - n + m - j/2 - 1 = \delta + \delta^\#$,
(δ) $0 < -\delta = \delta^\# < 1$, $0 \leq m < 2$, $j = 0$, $p \leq n - \frac{m}{2}$
      or $0 < -\delta = \delta^\# < 1$, $m > 2$, $j = 0$, $p \leq n + 1 - m$,
(ε) $\delta = p = 0$, $m = n + 1 + j/2 + \alpha$, and $\delta^\# > \alpha \geq 0$ or $\delta^\# = \alpha > 0$.

Remark (δ) covers the special admissible kernels $\mathcal{A}_0$.

Proof By passing to local coordinate patches with the appropriate coordinates from Chapter III, Lemma 5.26, we obtain

$$y = 0, \quad x = (t_1, \ldots, t_{2n}), \quad \text{Im } \Phi(x,y) = t_2, \quad -r(x) = t_1,$$

$$|\text{Re } \Phi(x,y)| \geq |r(y)| + |t_1| + |t_2| + R^2, \quad P(x,y) \geq (|t_1| + |t_2| + R)^2$$

with $R^2 = t_3^2 + \cdots + t_{2n}^2$. Set $\Theta = |r(y)|$.

In spherical coordinates with respect to $(t_3, t_4, \ldots, t_{2n})$ we need to estimate the following integral from above uniformly in Θ

$$\int\limits_{t_1, t_2, R} \frac{(|t_1| + |t_2| + R)^j |t_1|^\delta \Theta^{\delta^\#} R^{2n-3} \, dt_1 dt_2 dR}{(\Theta + |t_1| + |t_2| + R)^{2p} (\Theta + |t_1| + |t_2| + R^2)^m}.$$

Since terms $|t_1|$ and $|t_2|$ in the numerator give always rise to better estimates it suffices to majorise

$$\int\limits_{t_1, t_2, R} \frac{R^{2n-3+j} |t_1|^\delta \Theta^{\delta^\#} \, dt_1 dt_2 dR}{(\Theta + |t_1| + |t_2| + R)^{2p} (\Theta + |t_1| + |t_2| + R^2)^m}.$$

Here t_1, t_2, R vary over some fixed cube in $\mathbb{R}^3$.

Now we have to distinguish between several cases.

$0 \leq m < 1$. Estimates are trivial for $p < \frac{1}{2}$. So assume $p \geq \frac{1}{2}$. Then it suffices to estimate

$$\int\limits_{t_1,t_2,R} \frac{R^{2n-3+j}|t_1|^\delta \Theta^{\delta^\#}\, dt_1 dt_2 dR}{(\Theta + |t_1| + |t_2| + R)^{2p+m-1}(\Theta + |t_1| + |t_2| + R^2)}$$

$$\leq c_1 + c_2 \int\limits_{t_1,R} \frac{R^{2n-3+j}|t_1|^\delta \Theta^{\delta^\#}|\log(\Theta + |t_1|)|\, dt_1 dR}{(\Theta + |t_1| + |t_2| + R)^{2p+m-1}}.$$

Since $\delta + \delta^\# > -1 > m - 2$ we are in case (α) and assume $2p - 2n + m - j < \delta + \delta^\#$. For $2p - 2n + m - j < -1$ the estimates are obvious. So let $2p + m - 2n - j \geq 1$. Then it suffices to majorise

$$\int\limits_{t_1} \frac{|t_1|^\delta |\log(\Theta + |t_1|)|\, dt_1}{(\Theta + |t_1|)^\sigma},$$

with $\sigma = 2p + m + 1 - 2n - j - \delta^\# < 1 + \delta$. Boundedness is obvious for $\delta \geq 0$. If $\delta = -\nu < 0$ a change of coordinates $\tau = |t_1|^{1-\nu}$ leads to

$$\int\limits_\tau \frac{|\log(\Theta^{1-\nu} + |\tau|)|\, d\tau}{(\Theta^{1-\nu} + |\tau|)^{\sigma/(1-\nu)}} < \infty,$$

since $\frac{\sigma}{1-\nu} < 1$.

$2 > m \geq 1$. After integration with respect to t_2 it suffices to estimate for an arbitrary small constant $\varepsilon > 0$

$$I = \int\limits_{t_1,R} \frac{R^{2n-3+j}|t_1|^\delta \Theta^{\delta^\#}\, dt_1 dR}{(\Theta + |t_1| + R)^{2p+\varepsilon}(\Theta + |t_1|)^{m-1}}.$$

We first consider the case $\delta \geq 0$. This is part of (α).

Subcase $2p \geq 2 - m + \delta + \delta^\#$. Let $0 < \varkappa, \varepsilon$ be so small that $\sigma = 2p - 2n + m - j + \varepsilon + \varkappa < \delta + \delta^\#$. Then

$$I \leq c_1 \int\limits_{t_1,R} \frac{R^{2n-3+j}|t_1|^\delta \Theta^{\delta^\#}\, dt_1 dR}{(\Theta + |t_1| + R)^{2p-2+m+\varepsilon+\varkappa}(\Theta + |t_1|)^{1-\varkappa}}$$

$$\leq c_2 \int\limits_{t_1,R} \frac{dt_1 dR}{(\Theta + R)^{\sigma'}(\Theta + |t_1|)^{1-\varkappa}} < \infty,$$

with $\sigma' = \sigma + 1 - \delta - \delta^\# < 1$.

Subcase $2p < 2 - m + \delta + \delta^{\#}$. The estimates are trivial if $2n - 2 + j > 2p$. So let $2n - 2 + j \leq 2p$. Then we have to estimate

$$I \leq \int\limits_{t_1, R} \frac{|t_1|^{\delta} \Theta^{\delta^{\#}} dt_1 dR}{(\Theta + |t_1| + R)^{2p - 2n + 3 - j + \varepsilon}(\Theta + |t_1|)^{m-1}}$$

$$\leq \int\limits_{t_1} \frac{dt_1}{(\Theta + |t_1|)^{2p - 2n + 1 - j + \varepsilon + m - \delta - \delta^{\#}}} < \infty$$

for ε sufficiently small.

$\delta < 0$. The change of coordinates $\tau = |t_1|^{1-\nu}$, $\nu = -\delta > 0$, leads to

$$J = \int\limits_{\tau, R} \frac{R^{2n - 3 + j} \Theta^{\delta^{\#}} d\tau dR}{(\Theta + |\tau|^{1/(1-\nu)} + R)^{2p + \varepsilon}(\Theta^{1-\nu} + |\tau|)^{(m-1)/(1-\nu)}}.$$

Boundedness of J is obvious if $2p < 2n - 2 + j$. So let $2p \geq 2n - 2 + j$. Then we have to estimate

$$\int\limits_{\tau, R} \frac{\Theta^{\delta^{\#}} d\tau dR}{(\Theta + |\tau|^{1/(1-\nu)} + R)^{2p - 2n + 3 - j + \varepsilon}(\Theta^{1-\nu} + |\tau|)^{\frac{2p - 2n + 1 - j + \varepsilon + m}{1-\nu}}}$$

$$\leq c_1 + c_2 \int\limits_{\tau} \frac{\Theta^{\delta^{\#}} d\tau}{(\Theta^{1-\nu} + |\tau|)^{\frac{2p - 2n + 1 - j + \varepsilon + m}{1-\nu}}}.$$

The last integral is finite when $2p - 2n - j + m < \delta + \delta^{\#}$. This is the case for (α). In case (β) we also have $2p - 2n - j + m < 2(\delta + \delta^{\#}) - m + 2 < \delta + \delta^{\#}$.

$m \geq 2$. After an integration with respect to t_2 we have to estimate

$$J = \int\limits_{t_1, R} \frac{R^{2n - 3 + j}|t_1|^{\delta} \Theta^{\delta^{\#}} dt_1 dR}{(\Theta + |t_1| + R)^{2p}(\Theta + |t_1| + R^2)^{m-1}}.$$

Subcase $0 \leq \delta \leq \delta + \delta^{\#} < m - 2$. We obtain

$$J \leq c_1 + c_2 \int\limits_{t_1, R} \frac{R^{2n - 3 + j} dt_1 dR}{(\Theta + R)^{2p}(\Theta + |t_1| + R^2)^{m - 1 - \delta - \delta^{\#}}}$$

$$\leq c_3 + c_4 \int\limits_{R} \frac{R^{2n - 3 + j} dR}{(\Theta + R)^{2p}(\Theta + R^2)^{m - 2 - \delta - \delta^{\#}}}$$

$$\leq c_3 + c_4 \int\limits_{R} \frac{dR}{R^{\sigma}} < \infty,$$

since $\sigma = 2p - 2n - 1 + 2m - j - 2\delta - 2\delta^{\#} < 1$ according to (β).

In the case (γ) with $\delta = 0$ we require $0 < \delta^{\#} \le m - \frac{5}{2}$ and $2p - 2n + 2m - j - 2 = 2\delta^{\#}$. Here it suffices to estimate

$$\int\limits_{t_1, R} \frac{R^{2n-3+j} \Theta^{\delta^{\#}} dt_1 dR}{(\Theta + R)^{2p}(\Theta + |t_1| + R^2)^{m-1}}$$

$$\le c_1 + c_2 \int\limits_{R} \frac{R^{2n-3+j} \Theta^{\delta^{\#}} dR}{(\Theta + R)^{2p}(\Theta + R^2)^{m-2}}$$

$$\le c_1 + c_2 \int\limits_{R} \frac{R^{2m-5-2\delta^{\#}} \Theta^{\delta^{\#}} dR}{(\Theta + R^2)^{m-2}}$$

$$\le c_1 + c_2 \int\limits_{R} \frac{\Theta^{\delta^{\#}} dR}{(\Theta + R^2)^{\frac{1}{2}+\delta^{\#}}} < \infty.$$

Subcase $0 \le \delta$, $\delta + \delta^{\#} \ge m - 2$. Here

$$J \le \int\limits_{t_1, R} \frac{R^{2n-3+j} dt_1 dR}{R^{2p-\delta-\delta^{\#}+m-2}(\Theta + |t_1| + R^2)}$$

$$\le c_1 + c_2 \int\limits_{R} \frac{R^{2n-3+j} |\log R| dR}{R^{2p-\delta-\delta^{\#}+m-2}}$$

$$= c_1 + c_2 \int\limits_{R} \frac{|\log R| dR}{R^{\sigma}} < \infty,$$

since $\sigma = 2p - 2n + 1 + m - j - \delta - \delta^{\#} < 1$ according to (α).

Subcase $-1 < \delta < 0$. Set $\nu = -\delta$, $\tau = |t_1|^{1-\nu}$. Then

$$J \le c_1 \int\limits_{\tau, R} \frac{R^{2n-3+j} \Theta^{\delta^{\#}} d\tau dR}{(\Theta + R)^{2p}(\Theta + \tau^{\frac{1}{1-\nu}} + R^2)^{m-1}}$$

$$\le c_2 \int\limits_{\tau, R} \frac{R^{2n-3+j} \Theta^{\delta^{\#}} d\tau dR}{(\Theta + R)^{2p}(\Theta^{1-\nu} + \tau + R^{2(1-\nu)})^{\frac{m-1}{1-\nu}}}$$

$$\le c_3 + c_4 \int\limits_{R} \frac{R^{2n-3+j} \Theta^{\delta^{\#}} dR}{(\Theta + R)^{2p}(\Theta + R^2)^{m-2+\nu}} = c_3 + c_4 \cdot \tilde{J}.$$

According to the cases (α) and (β) $\delta^{\#} \ge m - 2 + \nu$, $\delta^{\#} < m - 2 + \nu$, we obtain $\tilde{J} < \infty$ as in the previous case.

In case (γ) with $\delta < 0$ and $2p - 2n + 2m - j - 2 = 2\delta + 2\delta^{\#}$ we obtain

$$
\begin{aligned}
J &= \int_R \frac{R^{2m-5-2\delta-2\delta^{\#}} \Theta^{\delta^{\#}} dR}{(\Theta + R^2)^{m-2-\delta}} \\
&\leq \int_R \frac{\Theta^{\delta^{\#}} dR}{(\Theta + R^2)^{\frac{1}{2}+\delta^{\#}}} < \infty.
\end{aligned}
$$

It remains to prove (δ). We consider at first the case $0 \leq m < 2$, $j = 0$, $p = n - \frac{m}{2}$. If p is smaller the assertion follows from this. Since $2n - m > 2n - 2$ we have for $0 < \varepsilon < 1$

$$
\begin{aligned}
&\int_{t_1,t_2,R} \frac{R^{2n-3} dt_1 dt_2 dR}{(\Theta + |t_1| + |t_2| + r)^{2n-m}(\Theta + |t_1| + |t_2| + R^2)^m |t_1|^{\varepsilon}} \\
&\lesssim \int_{t_1,t_2,R} \frac{dt_1 dt_2 dR}{(\Theta + |t_1| + |t_2| + R)^{3-m}(\Theta + |t_1| + |t_2|)^m |t_1|^{\varepsilon}} \\
&\lesssim C_1 + I_1
\end{aligned}
$$

with

$$
\begin{aligned}
I_1 &\lesssim \int_{t_1,t_2} \frac{dt_1 dt_2}{(\Theta + |t_1| + |t_2|)^2 |t_1|^{\varepsilon}} \\
&\lesssim C_2 + \int_{t_1} \frac{dt_1}{(\Theta + |t_1|)|t_1|^{\varepsilon}} \\
&\lesssim \Theta^{-\varepsilon},
\end{aligned}
$$

with the usual transformation $|t_1| = \tau^r$, with $r = \frac{1}{1-\varepsilon} > 1$.

Now let $m > 2$, $j = 0$, $p = n + 1 - m$. This will also include the cases when p is smaller. Then

$$
\begin{aligned}
&\int_{t_1,t_2,R} \frac{R^{2n-3} dt_1 dt_2 R}{(\Theta + |t_1| + |t_2| + R)^{2n+2-2m}(\Theta + |t_1| + |t_2| + R^2)^m |t_1|^{\varepsilon}} \\
&\lesssim C_1 + \int_{t_1,R} \frac{R^{2n-3} dt_1 dR}{(\Theta + |t_1| + R)^{2n+2-2m}(\Theta + |t_1| + R^2)^{m-1} |t_1|^{\varepsilon}}.
\end{aligned}
$$

Denote the latter integral by J.

Subcase $2 < m \leq \frac{5}{2}$. Then $2n + 2 - 2m \geq 2n - 3$ and consequently

$$J \lesssim \int\limits_{t_1,R} \frac{dt_1 \, dR}{(\Theta + |t_1| + R)^{5-2m}(\Theta + |t_1| + R^2)^{m-1}|t_1|^{\varepsilon}}$$

$$\lesssim \int\limits_{t_1,R} \frac{dt_1 \, dR}{(\Theta + R)^{5-2m}(\Theta + |t_1| + R^2)^{m-1}|t_1|^{\delta}}$$

$$\lesssim \int\limits_{\tau,R} \frac{d\tau \, dR}{(\Theta + R)^{5-2m}(\Theta + \tau^r + R^2)^{m-1}}$$

with $|t_1| = \tau^r, r = \frac{1}{1-\varepsilon} > 1$

$$\lesssim \int\limits_{\tau,R} \frac{d\tau \, dR}{(\Theta + R)^{5-2m}(\Theta^{1/r} + \tau + R^{2/r})^{r(m-1)}}$$

$$\lesssim C_1 + \int\limits_{R} \frac{dR}{(\Theta + R)^{5-2m}(\Theta^{1/r} + R^{(2/r)})^{r(m-1)-1}}$$

$$\lesssim C_1 + \int\limits_{R} \frac{dR}{(\Theta + R)^{5-2m}(\Theta + R^2)^{m-2+\varepsilon}}.$$

We decompose the interval of R in three subintervals

I) $0 \leq R \leq \varepsilon$. Since then $\Theta + R \leq 2(\Theta + R^2)$ we have to estimate

$$\int\limits_{0}^{\delta} \frac{dR}{(\Theta + R)^{1+\varepsilon}} \lesssim \Theta^{-\varepsilon}.$$

II) $\varepsilon < R < \sqrt{\varepsilon}$. Since then we can use $1/(\Theta + R^2) \lesssim 1/\Theta$ we majorise by

$$\int\limits_{\varepsilon}^{\sqrt{\varepsilon}} \frac{\Theta^{2-m-\varepsilon} \, dR}{(\Theta + R)^{5-2m}} \lesssim \Theta^{m-2-\varepsilon} + \Theta^{2-m-\varepsilon}\Theta^{m-2}$$

$$\lesssim \Theta^{-\varepsilon}.$$

III) $\sqrt{\varepsilon} < R < 1$. Then the transformation $R = \tau^s, s = \frac{1}{2m-4}$ leads to

$$\int\limits_{0}^{1} \frac{d\tau}{(\Theta + \tau^{2s})^{m-2+\varepsilon}} \lesssim \int\limits_{0}^{1} \frac{d\tau}{(\Theta^{1/2s} + \tau)^{2s(m-2+\varepsilon)}}.$$

Now $2s(m - 2 + \varepsilon) = 1 + \frac{\varepsilon}{m-2} > 1$. Thus the above integral is majorised by

$$C_2 + \frac{1}{\Theta^{\varepsilon/2s(m-2)}} = C_2 + \Theta^{-\varepsilon} \lesssim \Theta^{-\varepsilon}.$$

Let finally $m > \frac{5}{2}$. Then $2n + 2 - 2m < 2n - 3$ and therefore

$$I \lesssim \int\limits_{t_1,R} \frac{R^{2m-5} \, dt_1 \, dR}{(\Theta + |t_1| + R^2)^{m-1}|t_1|^\varepsilon}$$

$$\lesssim \int\limits_{t_1,R} \frac{dt_1 \, dR}{(\Theta + |t_1| + R^2)^{3/2}|t_1|^\varepsilon}$$

$$\lesssim C_1 + \int\limits_{t_1,R} \frac{dt_1 \, dR}{(\Theta^{1/2} + |t_1|^{1/2} + R)^3 |t_1|^\varepsilon}$$

$$\lesssim C_2 + \int\limits_{t_1} \frac{dt_1}{(\Theta + |t_1|)|t_1|^\varepsilon}$$

$$\lesssim \Theta^{-\varepsilon}.$$

We finally prove (ε). The case $\delta^\# > \alpha$ is contained in case (β), so let $\delta^\# = \alpha > 0$ and consider, as in (β), the integral

$$J = \int\limits_{t_1,R} \frac{R^{2n-3+j}\Theta^\alpha}{(\Theta + |t_1| + R^2)^{n+j/2+\alpha}} \, dt_1 \, dR$$

Integration with respect to t_1 gives

$$J \lesssim \int\limits_0^1 \frac{R^{2n-3+j}\Theta^\alpha}{(\Theta + R^2)^{n+j/2+\alpha-1}} \, dR \quad = \quad \Theta^\alpha J_1 \, .$$

Now substitute in J_1

$$R = \Theta^{\frac{1}{2}}\rho.$$

Then

$$J_1 \lesssim \Theta^{\frac{1}{2}(2n-3+j+1)}\Theta^{-n-j/2-\alpha+1} \int\limits_0^\infty \frac{\rho^{2n-3+j}}{(1+\rho^2)^{n-1+j/2+\alpha}} \, d\rho$$

$$= \Theta^{-\alpha}J_2 \, ,$$

with J_2 finite.   □

Lemma 2.2 *Let $0 \leq \delta < \min(1, \delta^\#)$. If A_0 is an admissible kernel of type ≥ 0 then*

$$\sup_{y \in D} \int\limits_D |A_0(x,y)||r(x)|^{-\delta}|r(y)|^{\delta^\#} \, dV(x) < \infty.$$

The same estimate holds if A_0 is special admissible of type ≥ 0 and $0 < \delta = \delta^\# < 1$.

Proof   Locally we have since $\Phi - \Phi^* = \mathcal{E}_3$

$$|\mathcal{A}_0| \leq \frac{|\mathcal{E}_j||r(x)|^k|r(y)|^l}{P(x,y)^{t_0}|\Phi(x,y)|^t} \leq \frac{|\mathcal{E}_j|}{P(x,y)^{t_0}|\Phi(x,y)|^\tau}$$

with

$$2n + j + \min(2, \tau) - 2(t_0 + \tau) \geq 0, \; j \geq 0, \; t_0 \geq 0.$$

Since $P(x,y) \geq C(r(x)^2 + r(y)^2)$ we can cancel $|r(x)|$ or $|r(y)|$ against $\sqrt{P(x,y)}$ if $k + l > t$. Therefore we can assume $\tau \geq 0$. Then the assertion follows easily from Lemma 2.1. $\qquad\square$

Corollary 2.3   *Let $\delta > 0$ and $\mathcal{A}_0$ an admissible kernel of type ≥ 0. Then*

$$\sup_{x \in D} \int_D |\mathcal{A}_0(x,y)||r(y)|^\delta \, dV(y) < \infty.$$

Proof   Trivial consequence of Lemma 2.2 $\qquad\square$

Theorem 2.4   *Let $1 < p \leq \infty$, $0 \leq \delta < \min(\delta^\#, 1, p - 1)$. If $\mathcal{A}_0$ is an admissible kernel of type ≥ 0 then $\mathcal{A}_0$ defines a bounded linear operator*

$$\mathcal{A}_0 : L^{p,\delta} \longrightarrow L^{p,\delta^\#}.$$

If $\mathcal{A}_0$ is special admissible, then it defines a bounded linear operator

$$\mathcal{A}_0 : L^{p,\delta} \longrightarrow L^{p,\delta}$$

in the following two cases:

   i) $0 < \delta < 1, \delta + 1 \leq p \leq \infty$
   ii) $\delta = 0, 1 < p < \infty$

Proof   a) $p = \infty$.   Let $f \in L^{\infty,\delta}$. Since $\delta < \delta^\#$ one has

$$\left|(\mathcal{A}_0 f)(y)|r(y)|^{\delta^\#}\right| \lesssim \int_D |\mathcal{A}_0||f(x)||r(x)|^\delta |r(x)|^{-\delta}|r(y)|^{\delta^\#} \, dV(x)$$

$$\lesssim \|f\|_{L^{\infty,\delta}} \int_D |\mathcal{A}_0||r(x)|^{-\delta}|r(y)|^{\delta^\#} \, dV(x)$$

$$\lesssim \|f\|_{L^{\infty,\delta}}$$

by Lemma 2.2 with a constant independent of y. – The same proof works if $\mathcal{A}_0$ is special admissible and $0 < \delta = \delta^\#$.

b) $1 < p < \infty$. Let first $\mathcal{A}_0$ be admissible of type ≥ 0. Let q the conjugate exponent to p such that $1/p + 1/q = 1$. We set

$$g(y) = \int\limits_D \mathcal{A}_0(x,y) f(x) \, dV(x);$$

then

$$|g(y)| \leq \int\limits_D \{|f(x)||r(x)|^{\delta/p}|\mathcal{A}_0|^{1/p}\}\{|\mathcal{A}_0|^{1/q}|r(x)|^{-\delta/p}\} \, dV(x)$$

and from Hölder's inequality

$$|g(y)|^p \leq \left[\int\limits_D |f(x)|^p |r(x)|^\delta |\mathcal{A}_0(x,y)| \, dV(x)\right] \cdot \left[\int\limits_D |\mathcal{A}_0(x,y)||r(x)|^{-\delta q/p}\right]^{p/q} dV(x).$$

Hence, if $\varepsilon > 0$ is such that $\delta + \varepsilon < \delta^{\#}$, one obtains

$$\|g\|^p_{L^p,\delta^{\#}} = \int\limits_D |g(y)|^p |r(y)|^{\delta^{\#}} \, dV(y) \tag{2.5}$$

$$\leq \int\limits_D \left[\int\limits_D |f(x)|^p |r(x)|^\delta |\mathcal{A}_0(x,y)||r(y)|^\varepsilon \, dV(x)\right]$$

$$\cdot \left[\int\limits_D |\mathcal{A}_0(x,y)||r(x)|^{-\delta q/p}|r(y)|^{(\delta^{\#}-\varepsilon)q/p} \, dV(x)\right]^{p/q} dV(y).$$

We may of course assume $\delta^{\#} < p - 1$; this implies $\delta^{\#} q/p < 1$ and

$$0 \leq \delta q/p < (\delta^{\#} - \varepsilon)\frac{q}{p} < 1.$$

So Lemma 2.2 can be applied to the second factor in (2.5) leading to the estimate

$$\|g\|^p_{L^p,\delta^{\#}} \lesssim \int\limits_D \int\limits_D |f(x)|^p |r(x)|^\delta |\mathcal{A}_0||r(y)|^\varepsilon \, dV(x) dV(y). \tag{2.6}$$

We integrate (2.6) first with respect to y and apply Corollary 2.3 to obtain

$$\|g\|^p_{L^p,\delta^{\#}} \lesssim \int\limits_D |f(x)|^p |r(x)|^\delta \, dV(x) = \|f\|^p_{L^p,\delta}.$$

c) $0 < \delta < 1, \delta \leq p - 1 < \infty$, $\mathcal{A}_0$ special admissible of type ≥ 0.

Choose ε with $1 - \delta < \varepsilon(p-1) < 1$ and set

$$\varkappa = \frac{1}{p-1} - \varepsilon.$$

Then
$$0 < \varkappa < 1, \quad 0 < \frac{\varkappa p}{q} < \delta, \quad 0 < \delta - \frac{\varkappa p}{q} < 1.$$

This implies by starting as in (b):

$$|g(y)|^p \leq \left[\int_D |f(x)|^p |r(x)|^{\frac{\varkappa p}{q}} |\mathcal{A}_0(x,y)| dV(x) \right] \left[\int_D |\mathcal{A}_0(x,y)| |r(x)|^{-\varkappa} dV(x) \right]^{\frac{p}{q}}.$$

The latter integral can be majorized by using Lemma 2.2 by

$$C|r(y)|^{-\varkappa}.$$

Therefore we obtain

$$\|g\|^p_{L^{p,\delta}} \leq C \int_D \int_D |f(x)|^p |r(x)|^{\delta} |\mathcal{A}_0(x,y)| |r(x)|^{-(\delta - \frac{\varkappa p}{q})} |r(y)|^{-(\delta - \frac{\varkappa p}{q})} dV(y) \, dV(x).$$

By integrating first with respect to y Lemma 2.2 yields the boundedness of $\mathcal{A}_0$ in $L^{p,\delta}$.

d) $\delta = 0, 1 < p < \infty$, $\mathcal{A}_0$ special admissible of type ≥ 0 (see also [PhS 77]).

Choose $0 < \varkappa < 1$ such that $\varkappa p < q$. The same approach as in (b,c) yields

$$|g(y)|^p \leq C \int_D |f(x)|^p |\mathcal{A}_0(x,y)| |r(x)|^{\frac{\varkappa p}{q}} |r(y)|^{-\frac{\varkappa p}{q}} dV(x).$$

and analogously by Lemma 2.2

$$\|g\|^p_{L^p} \leq \tilde{C} \|f\|^p_{L^p}.$$

This concludes the proof.   □

§3 Regularity Properties of A_1-operators

We need a lemma on the behaviour of the kernels $\mathcal{A}_\lambda$ with respect to differentiation.

Lemma 3.1 *Let $\mathcal{A}_\lambda$ be an admissible kernel of type $\geq \lambda$ and X a C^∞ vector field on $\overline{D}$ acting in x or in y. Then $X\mathcal{A}_\lambda$ is admissible of type $\geq \lambda - 2$. If X is allowable then $X\mathcal{A}_\lambda$ is of type $\geq \lambda - 1$.*

If $\mathcal{A}_\lambda$ is of commutator type $\geq \lambda$ and X a C^∞ vector field on $\overline{D}$ then there exists a kernel $\tilde{\mathcal{A}}_{\lambda-2}$ of commutator type $\geq \lambda - 2$ and a kernel $\tilde{\mathcal{A}}_{\lambda-1}$ of commutator type $\geq \lambda - 1$ such that

$$X\mathcal{A}_\lambda = \tilde{\mathcal{A}}_{\lambda-2} + \tilde{\mathcal{A}}_{\lambda-1}.$$

If moreover X is allowable then there exists a kernel $\tilde{\mathcal{A}}_{\lambda-1}$ of commutator type $\geq \lambda - 1$ and a kernel $\tilde{\mathcal{A}}_\lambda$ of commutator type $\geq \lambda$ such that

$$X\mathcal{A}_\lambda = \tilde{\mathcal{A}}_{\lambda-1} + \tilde{\mathcal{A}}_\lambda.$$

Proof   We have in the general case

$$Xr = \mathcal{E}_0, \; Xr^* = \mathcal{E}_0, \; X\mathrm{P} = \mathcal{E}_1 + \mathcal{E}_0 r + \mathcal{E}_0 r^*, \; X\Phi = \mathcal{E}_0.$$

Therefore $X\mathcal{A}_\lambda$ is as a sum of admissible kernels admissible. So a differentiation decreases the type by 2 in the type formula of Definition 5.49, Chapter IV.

Let X be allowable. Then

$$Xr = \mathcal{E}_0 r, \; Xr^* = \mathcal{E}_0 r,$$
$$X\mathrm{P} = \mathcal{E}_1 + \mathcal{E}_0 rr^* + \mathcal{E}_0 r^2$$
$$= \mathcal{E}_1 + \mathcal{E}_0 r + \mathcal{E}_0 r^*.$$

Since near the boundary we have with respect to local holomorphic coordinates

$$\Phi(x,y) = \sum_{i=1}^{n} \frac{\partial r(x)}{\partial x_i}(x_i - y_i) - r(x) + \mathcal{E}_2(x,y), \tag{3.2}$$

it follows

$$X\Phi = \mathcal{E}_1 + \mathcal{E}_0 r. \tag{3.3}$$

This results in a decrease of the type by 1.

Now let $\mathcal{A}_\lambda$ be of commutator type. Let x, y vary in a small neighbourhood of a boundary point of D. There we have

$$\mathcal{A}_\lambda(x,y) = \mathcal{E}_j \mathrm{P}^{-t_0} \Phi^{t_1} \overline{\Phi}^{t_2} \Phi^{*t_3} \overline{\Phi}^{*t_4} r^k r^{*l}.$$

For the proof we shall call a kernel *pure* if $t_3 = t_4 = 0$ and $t_1 t_2 \leq 0$. We want to decompose $\mathcal{A}_\lambda$ into a pure component of the same commutator type and an error term of type $\geq \lambda + 1$. Since pure kernels are of commutator type and stay pure under differentiation the assertion of the lemma will be clear by the first part of the proof.

In order to obtain the decomposition we have to eliminate Φ^{*t_3} and $\overline{\Phi}^{*t_4}$ in the above decomposition. We show it for Φ^{*t_3}. The other case is analogous.

So let $t_3 \neq 0$. If $t_3 > 0$ then $t_1 \geq 0$. Then

$$\Phi^{t_1} \Phi^{*t_3} = \Phi^{t_1}(\Phi + \mathcal{E}_3')^{t_3}$$

$$= \Phi^{t_1 + t_3} + \sum_{i=1}^{t_3} \Phi^{t_1 + t_3 - i} \mathcal{E}_{3i}',$$

where $\mathcal{E}_j'$ stands for a C^∞ function of vanishing order j if $x = y$. Therefore we obtain

$$\mathcal{A}_\lambda(x,y) = \mathcal{E}_j \Phi^{t_1 + t_3} \overline{\Phi}^{t_2} \overline{\Phi}^{*t_4} r^k r^{*l} + \mathcal{A}_{\lambda+1}',$$

where $\mathcal{A}_{\lambda+1}'$ is admissible of type $\geq \lambda + 1$.

If $t_3 < 0$ then $t_1 \leq 0$. In this case we proceed as follows

$$
\Phi^{t_1}\Phi^{*t_3} = \frac{1}{\Phi^{-t_1}(\Phi^*)^{-t_3}}
$$

$$
= \frac{1}{\Phi^{-t_1}(\Phi + \mathcal{E}'_3)^{-t_3}}
$$

$$
= \frac{\Phi^{t_1+t_3}}{\left(1 + \frac{\mathcal{E}'_3}{\Phi}\right)^{-t_3}}.
$$

Now if the neighbourhood is sufficiently small we obtain

$$
\frac{1}{1 + \frac{\mathcal{E}'_3}{\Phi}} = 1 - \frac{\mathcal{E}'_3}{\Phi}\frac{1}{1 + \frac{\mathcal{E}'_3}{\Phi}} = 1 + \frac{\mathcal{E}''_3}{\Phi}.
$$

So also in this case the above decomposition of $\mathcal{A}_\lambda$ is possible. Applying the same trick to $\overline{\Phi}^{*t_4}$ we obtain the desired decomposition

$$
\mathcal{A}_\lambda(x,y) = \mathcal{E}_j \Phi^{t_1+t_3}\overline{\Phi}^{t_2+t_4} r^k r^{*l} + \tilde{\mathcal{A}}_{\lambda+1}
$$

At last it remains to show that the error terms are also of commutator type. In the case $t_3 > 0$ we had $t_1 \geq 0$. Here the error terms add positive exponents to t_1 and Φ^* has vanished. In this case since $t_2 t_4 \geq 0$, $(t_1 + t_3)(t_2 + t_4) \leq 0$, we have $t_2 \leq 0$, $t_4 \leq 0$. So the decomposition method applied to $\overline{\Phi}^{*t_4}$ will produce error terms which add negative exponents to t_2 and $\overline{\Phi}^*$ vanishes from the picture.

If $t_3 < 0$ the same reasoning applies whith the role of t_3 and t_4 reversed. That means that the error terms occuring are pure and therefore of commutator type. $\qquad\square$

In the following we have to look more closely at the behaviour of $\mathcal{A}_\lambda$ operators with respect to differentiation in different directions.

Let the neighbourhood U of a boundary point be so small such that there exists a C^∞ orthogonal frame $\omega_1,\ldots,\omega_n$ of $(1,0)$ vector fields with $\omega_n = \partial r$. We denote the dual frame by $L_1,\ldots,L_n$.

Since $L_i r = 0$ for $i < n$ $L_1,\ldots,L_{n-1}$ are allowable vector fields which span the holomorphic tangent space of the level hyper surfaces $\{r = \text{const}\}$.

The missing tangential directions are $\overline{L}_1,\ldots,\overline{L}_{n-1}$ and $Y = L_n - \overline{L}_n$ because $L_n r = 1$. Since

$$
Y\Phi = 1 + \mathcal{E}_1 \tag{3.4}
$$

Y is the 'bad' tangential direction causing a decrease of the operator type by 2. The normal direction is given by

$$
N = L_n + \overline{L}_n.
$$

We begin our analysis with the following theorem.

Theorem 3.5 *Let A_1 be an admissible operator of type ≥ 1. Then A_1 defines bounded linear operators between the following spaces:*

i. $A_1 : L^p \to L^s$ *for* $1 \leq p, s \leq \infty$, *with* $\frac{1}{s} > \frac{1}{p} - \frac{1}{2n+2}$.

ii. $A_1 : L^{\infty,\delta} \to L^{\infty,\delta^\#-\frac{1}{2}}$ *for* $\frac{1}{2} \leq \delta^\# < 1$, $\delta > -1$, $\delta + \delta^\# > 0$.

iii. $A_1 : L^{p,\delta} \to S_1^{p,\delta^\#}$ *for* $1 < p \leq \infty$, $0 \leq \delta < \min(\delta^\#, 1, p-1)$.

iv. $A_1 : L^{\infty,\delta} \to \Gamma_{\alpha, \frac{\alpha}{\alpha+\delta^\#}}$ *for* $0 \leq \delta < \frac{1}{2}$, $\delta < \delta^\#$, $\alpha = \frac{1}{2} - \delta > 0$.

v. $A_1 : L^{\infty,\delta} \to L_1^{\infty,\delta+1/2}$ *for* $0 \leq \delta < \frac{1}{2}$.

vi. $A_1 : L^p \to L^p$, $1 < p \leq \infty$ *is a compact operator.*

vii. $A_1 : \Lambda_\alpha \to L_1^{\infty,\delta}$ *for* $0 < \alpha < 1$ *and* $2\delta > 1 - \alpha$ *and* A_1 *of commutator type.*

Proof   *i)* follows from a generalisation of Young's inequality (III.5.35). once we have shown the following two estimates for $1 \leq \gamma < (2n+2)/(2n+1)$:

$$\sup_{y \in D} \int_D |\mathcal{A}_1(x,y)|^\gamma dV(x) < \infty$$

$$\sup_{x \in D} \int_D |\mathcal{A}_1(x,y)|^\gamma dV(y) < \infty.$$

But these are direct consequences of Lemma 2.1.

ii) Here one has to show that

$$\sup_{y \in D} \int_D |\mathcal{A}_1(x,y)||r(x)|^{-\delta}|r(y)|^{\delta^\#-\frac{1}{2}} dV(x) < \infty.$$

This is also settled by Lemma 2.1.

iii) Let W be an allowable vector field. By Lemma 3.1 $W\mathcal{A}_1$ is of type ≥ 0. Therefore $W A_1 : L^{p,\delta} \to L^{p,\delta^\#}$ is bounded by Theorem 2.4 for the indicated $p, \delta, \delta^\#$. This concludes the proof of *(iii)*.

iv) *(iii)* implies boundedness of

$$A_1 : L^{\infty,\delta} \to S_1^{\infty,\delta^\#}$$

for $0 \leq \delta < \delta^\# < 1$. Let us show that for $0 \leq \delta < \frac{1}{2}$

$$A_1 : L^{\infty,\delta} \to L_1^{\infty,\delta+\frac{1}{2}}$$

is bounded. Then Theorem 1.8 gives the conclusion. In order to show this we have to prove
that for every C^∞ vector field X on $\overline{D}$

$$XA_1 : L^{\infty,\delta} \to L^{\infty,\delta+\frac{1}{2}}$$

defines a bounded operator. In other words we have to show

$$\sup_{y \in D} \int_D |X^y A_1(x,y)||r(x)|^{-\delta}|r(y)|^{\delta+\frac{1}{2}} \, dV(x) < \infty.$$

If X^y acts on a factor in $\mathcal{A}_1$ such that the type is only decreased by 1 the assertion is clear
from Theorem 2.4. So the only remaining cases are when X^y falls on $r(y)$, Φ, Φ^* or $\overline{\Phi}^*$.

After using $\Phi - \Phi^* = \mathcal{E}_3$ and simplifying we are left with the term

$$\sup_{y \in D} \int_D \frac{\rho(x,y)^j |r(x)|^{-\delta}|r(y)|^{\delta+\frac{1}{2}} \, dV(x)}{\mathrm{P}(x,y)^p |\Phi(x,y)|^{m+1}},$$

with $2n + j + \min(2,m) - 2(p+m) \geq 1$. If $m \leq 1$ the boundedness of the last integral
follows from Lemma 2.1, (α) and (β). We assume $m \geq 2$.

Then $-\delta + (\delta + \frac{1}{2}) \leq m + 1 - \frac{5}{2}$ and $p - n + (m+1) - \frac{j}{2} - 1 \leq \frac{1}{2} = -\delta + (\delta + \frac{1}{2})$. This
is covered by Lemma 2.1, (β) or (γ).

$vi)$ For $p = \infty$ this follows from $iv)$ and the Arzelà-Ascoli theorem. Let $\mathcal{A}_1 = \mathcal{N}/\mathcal{D}$ be a
kernel of type ≥ 1 where

$$\mathcal{D} = \mathrm{P}^{\alpha_0} \Phi^{\alpha_1} \Phi^{*\alpha_2} \overline{\Phi}^{\alpha_3} \overline{\Phi}^{*\alpha_4}$$

denotes its denominator and $\mathcal{N}$ its numerator. Set $s = \alpha_0 + \alpha_1 + \alpha_2 + \alpha_3 + \alpha_4$. We write
$\mathcal{D}$ as

$$\prod_{i=1}^{s} F_i,$$

where the F_i are taken from the set of factors $\mathrm{P}, \Phi, \Phi^*, \overline{\Phi}, \overline{\Phi}^*$ which occur in $\mathcal{D}$. We set for
$\varepsilon > 0$

$$\mathcal{D}^\varepsilon = \prod_{i=1}^{s} (F_i + \varepsilon)$$

and

$$\mathcal{A}_1^\varepsilon = \frac{\mathcal{N}}{\mathcal{D}^\varepsilon}.$$

Since $\mathcal{A}_1^\varepsilon$ defines an operator A_1^ε which is compact from L^p to L^p it suffices to show that
the operator norm of $A_1^\varepsilon - A_1$ tends to 0 if ε tends to 0.

Now

$$\left|\frac{1}{\mathcal{D}^\varepsilon} - \frac{1}{\mathcal{D}}\right| \le \text{const} \sum_{i=1}^s \frac{\varepsilon}{|\mathcal{D}||F_i + \varepsilon|}$$

$$\le \text{const}\, \varepsilon^\varkappa \sum_{i=1}^s \frac{1}{|\mathcal{D}||F_i|^\varkappa}$$

with a small $\varkappa > 0$.

Let $f \in L^p$. Since

$$|A_1^\varepsilon f(y) - A_1 f(y)| \le \text{const}\, \varepsilon^\varkappa \sum_{i=1}^s \int_D \frac{|f(x)||\mathcal{N}(x,y)|}{|\mathcal{D}(x,y)||F_i(x,y)|^\varkappa} dV(x)$$

it suffices to show that the kernel $|A_1|/|F_i|^\varkappa$ defines an operator which maps L^p to L^p contiuously. In order to show this it suffices to show

$$\sup_{y \in D} \int_D \frac{|A_1(x,y)| dV(x)}{|F_i(x,y)|^\varkappa} < \infty$$

$$\sup_{x \in D} \int_D \frac{|A_1(x,y)| dV(x)}{|F_i(x,y)|^\varkappa} < \infty$$

for a sufficiently small $\varkappa > 0$. But this follows from lemma 2.1.

$vii)$ Let X be a vectorfield; we show

$$\|X(f, A_1)\|_{L^{\infty,\delta}} \lesssim \|f\|_{\Lambda^\alpha}$$

if $2\delta > 1 - \alpha$. The critical case arises for

$$|X^y A_1| \lesssim \frac{\mathcal{E}_1}{|\Phi|^3 P^{n-1}};$$

all other cases are either easier or can be reduced to the above case. If $f \in \Lambda^\alpha$ then

$$(f, X^y A_1) = f(y) \int X^y A_1(x,y) + \int (f(x) - f(y)) X^y A_1(x,y)$$

The first term is reduced, by the commutator relation 3.7 below and partial integration, to

$$f(y) \int A_0(x,y)$$

and is thus dominated by

$$\|f\|_{L^\infty} |r^*|^{-\delta} \qquad \text{for any } \delta > 0.$$

The second integral is dominated by

$$I(y) = \|f\|_{\Lambda^{\alpha}} \int \frac{\rho(x,y)^{1+\alpha}}{|\Phi|^3 P^{n-1}} dV(x).$$

Lemma 2.1.β yields

$$I(y)|r(y)|^{\delta} \lesssim 1$$

for $2\delta > 1 - \alpha$.   □

In order to study regularity properties of A_{λ} operators involving higher derivatives we have to study commutator relations with vector fields. It will turn out that A_{λ} operators have better properties in holomorphic tangential directions. *From now on all A_1-operators are assumed to be of commutator type.*

Theorem 3.6 *Let A_1 be an admissible operator of commutator type ≥ 1 and X a C^{∞} tangential vector field (i.e.$X|_{bD}$ is a section of the complexified tangent bundle to bD). Then*

$$X^y A_1 = -A_1 \tilde{X}^x + A_1^{(0)} + \sum_{v=1}^{l} A_1^{(v)} W_v^x,$$

where $\tilde{X}$ is the adjoint vector field of X (see below), the W_v are allowable vector fields and the $A_1^{(v)}$ are admissible operators of commutator type ≥ 1.

Remark: The *adjoint vector field* $\tilde{X}$ is characterised by the following equation obtained by partial integration for smooth forms f and g:

$$(\tilde{X}f, g) = (f, Xg).$$

If X is allowable $\tilde{X}$ is allowable too.

Proof of 3.6. By using a partition of unity we can assume that X has arbitrarily small support on a given coordinate patch around a boundary point where we have the orthogonal frames $L_1, \ldots, L_n$ (defined at the beginning of this section).

Then we have with smooth compactly supported functions a, b, a_j, b_j the decomposition

$$X = \sum_{j=1}^{n-1} a_j L_j + \sum_{j<n} b_j \overline{L}_j + aY + bN.$$

We prove the theorem for each of the summands.

Case 1. $X = a_j L_j$ or $b_j \overline{L}_j$ or aY, $j \leq n - 1$. Since trivially

$$X^y A_1 = -X^x A_1 + (X^x + X^y) A_1,$$

an integration by parts gives

$$
\begin{aligned}
X A_1 f = X(f, A_1)_D &= (f, X^y A_1)_D \\
&= -(f, X^x A_1)_D + (f, (X^x + X^y) A_1)_D \\
&= -(\tilde{X} f, A_1)_D + (f, (X^x + X^y) A_1)_D \\
&= -A_1(\tilde{X} f) + (f, (X^x + X^y) A_1)_D.
\end{aligned}
$$

Note that X is tangential. Now we have in our case

$$
\begin{aligned}
(X^x + X^y)\mathcal{E}_j &= \mathcal{E}_j, \\
(X^x + X^y)\mathrm{P} &= \mathcal{E}_2, \\
(X^x + X^y)\Phi &= \mathcal{E}_1,
\end{aligned}
$$

since $(X^x + X^y)r = 0$, $(X^x + X^y)(\zeta - z) = \mathcal{E}_1$. Let

$$
A_1(x, y) = \mathcal{E}_j \mathrm{P}^{-t_0} \Phi^{t_1} \overline{\Phi}^{t_2} \Phi^{*\,t_3} \overline{\Phi}^{*\,t_4} r^k r^{*\,l}
$$

an admissible A_1 kernel of commutator type ≥ 1 according to Definition 5.49, Chapter IV.

In the proof of (3.1) we have shown that $A_1 = A_1' + A_2'$, where the kernels are *pure* of the indicated type. Therefore we can assume that A_1 is pure ($t_3 = t_4 = 0$).

Only when $X^x + X^y$ falls on a Φ-term the resulting term for $(X^x + X^y)A_1$ is not obviously of type A_1. Since all the occurring subcases can be treated in an analogous way we can assume that $(X^x + X^y)$ falls on Φ^{t_1}, with $t_1 \neq 0$. We obtain the term

$$
B := \mathcal{E}_{j+1} \mathrm{P}^{-t_0} \Phi^{t_1-1} \overline{\Phi}^{t_2} r^k r^{*\,l},
$$

with

$$
2n + j + \min(2, t - k - l) - 2(t_0 + t - k - l) \geq 1
$$

and $t = -(t_1 + t_2) \geq 0$.

Therefore if $t - k - l \leq 1$ B is of type ≥ 1.

So let $t - k - l \geq 2$, then $-(t_1 + t_2 - 1) = 1 + t \geq 3 + k + l \geq 3$. We set

$$
B = \Phi^{\tau_1} \overline{\Phi}^{\tau_2}(\mathcal{E}_{j+1} \mathrm{P}^{-t_0} r^k r^{*\,l}),
$$

with $-(\tau_1 + \tau_2) \geq 3 + k + l$, $\tau_1 = t_1 - 1$, $\tau_2 = t_2$.

If $\tau_2 \geq 0$ we expand $\overline{\Phi}^{\tau_2} = (-\Phi + (\Phi + \overline{\Phi}))^{\tau_2}$ and obtain a sum of terms ($0 \leq i \leq \tau_2$)

$$
\Phi^{\tau_1+i}(\Phi + \overline{\Phi})^{\tau_2-i} \mathcal{E}_{j+1} \mathrm{P}^{-t_0} r^k r^{*\,l}.
$$

If $\tau_1 \geq 0$ we obtain analogously a sum of terms ($i \leq \tau_1$)

$$
\overline{\Phi}^{\tau_2+i}(\Phi + \overline{\Phi})^i \mathcal{E}_{j+1} \mathrm{P}^{-t_0} r^k r^{*\,l}.
$$

It remains the case $t_1 < 1$, $t_2 < 0$. $t_1 < 0$ is impossible since $t_1 t_2 \leq 0$ from Definition 5.49. So $t_1 = 0$. But this case was excluded from the very beginning of the proof.

We conclude that B can be written as a sum of terms B_r such that B_r or $\overline{B}_r$ is of the form

$$\Phi^r (\Phi + \overline{\Phi})^{\tau_1 + \tau_2 - r} \mathcal{E}_{j+1} \mathrm{P}^{-t_0} r^k r^{*\,l},$$

with an integer r such that $\tau_1 \leq r \leq \tau_1 + \tau_2$ or $\tau_2 \leq r \leq \tau_1 + \tau_2$. Since $\tau_1 + \tau_2 \leq -3$ we have $r \leq -3$.

If the domain of definition of the vector fields L_i is sufficiently small we can assume because of $Y\Phi = 1 + \mathcal{E}_1$, $Y\overline{\Phi} = -1 + \mathcal{E}_1$ that $|Y\Phi| \geq \frac{1}{2}$, $|Y\overline{\Phi}| \geq \frac{1}{2}$. The following calculations are meant for B_r of the above form. $\overline{B}_r$ can be treated in an analogous way. It follows

$$
\begin{aligned}
B_r &= \Phi^r (\Phi + \overline{\Phi})^{\tau_1 + \tau_2 - r} \mathcal{E}_{j+1} \mathrm{P}^{-t_0} r^k r^{*\,l} \\
&= \frac{Y(\Phi^{r+1})}{(r+1) Y(\Phi)} (\Phi + \overline{\Phi})^{\tau_1 + \tau_2 - r} \mathcal{E}_{j+1} \mathrm{P}^{-t_0} r^k r^{*\,l} \\
&= Y \left[\Phi^{r+1} (\Phi + \overline{\Phi})^{\tau_1 + \tau_2 - r} \mathcal{E}_{j+1} \mathrm{P}^{-t_0} r^k r^{*\,l} \right] \\
&\quad + \Phi^{r+1} (\tau_1 + \tau_2 - r)(\Phi + \overline{\Phi})^{\tau_1 + \tau_2 - r - 1} \mathcal{E}_{j+2} \mathrm{P}^{-t_0} r^k r^{*\,l} \\
&\quad + \Phi^{r+1} (\Phi + \overline{\Phi})^{\tau_1 + \tau_2 - r} \mathcal{E}_j \mathrm{P}^{-t_0} r^k r^{*\,l} \\
&\quad + \Phi^{r+1} (\Phi + \overline{\Phi})^{\tau_1 + \tau_2 - r} \mathcal{E}_{j+1} \mathrm{P}^{-t_0 - 1} r^k r^{*\,l}.
\end{aligned}
$$

If we expand the powers of $(\Phi + \overline{\Phi})$ we can easily verify that

$$B_r = Y \mathcal{A}_2^r + \mathcal{A}_1^r,$$

where $\mathcal{A}_2^r$ is of type ≥ 2 and $\mathcal{A}_1^r$ of commutator type ≥ 1. We can therefore conclude that

$$(X^x + X^y)\mathcal{A}_1 = Y \mathcal{A}_2 + \mathcal{A}_1',$$

with admissible kernels $\mathcal{A}_2$ of commutator type ≥ 2 and $\mathcal{A}_1'$ of commutator type ≥ 1.

The remaining part of the proof deals with $Y \mathcal{A}_2$ and uses a feature of the Levi form.

The commutators $[L_j, \overline{L}_k] = L_j \overline{L}_k - \overline{L}_k L_j$, $j, k \leq n - 1$, cannot all vanish in a given point because this would imply Levi flatness contradicting the strict pseudoconvexity of the boundary (for more details see [Bog 91]).

Since $[L_j, \overline{L}_k]$ is a tangential vector field it has a nowhere vanishing Y component if the coordinate patch is sufficiently small. That means that there exist allowable vector fields W_1, W_2, W_3 and a smooth function φ with

$$Y = \varphi [W_1, W_2] + W_3.$$

So we obtain

$$
\begin{aligned}
Y \mathcal{A}_2 &= \varphi [W_1, W_2] \mathcal{A}_2 + W_3 \mathcal{A}_2 \\
&= [W_1, W_2](\varphi \mathcal{A}_2) + \mathcal{A}_1',
\end{aligned}
$$

with A_1' of commutator type ≥ 1.

Now one integration by parts yields

$$(f, [W_1, W_2](\varphi A_2)) = (f, W_1 W_2(\varphi A_2) - W_2 W_1(\varphi A_2))$$
$$= (\widetilde{W}_1 f, W_2(\varphi A_2)) - (\widetilde{W}_2 f, W_1(\varphi A_2)).$$

Since $\widetilde{W}_1$, $\widetilde{W}_2$ are allowable vector fields and $W_1(\varphi A_2)$, $W_2(\varphi A_2)$ are of commutator type ≥ 1 we are done.

Case 2. $X = \mathcal{E}_0 r N$, $\mathcal{E}_0$ smooth.

Now we have

$$r(x)(N^x + N^y)\mathcal{E}_j = r(x)\mathcal{E}_j$$
$$r(x)(N^x + N^y)P = r(x)r(y)\mathcal{E}_0 + r(x)^2 \mathcal{E}_0 + r(x)\mathcal{E}_2$$
$$r(x)(N^x + N^y)\Phi = r(x)\mathcal{E}_0.$$

It follows

$$\begin{aligned}
X A_1 f &= X(f, A_1) \\
&= (f, X^y A_1) \\
&= (\mathcal{E}_0 r f, N^y A_1) \\
&= (-\mathcal{E}_0 r f, N^x A_1) + (f, \mathcal{E}_0 r (N^x + N^y) A_1) \\
&= (-\tilde{N}^x(\mathcal{E}_0 r f), A_1) + (f, \mathcal{E}_0 r (N^x + N^y) A_1).
\end{aligned}$$

Integration by parts is possible because of the presence of the factor $r(x)$. Now

$$\tilde{N}^x(\mathcal{E}_0 r f) = \mathcal{E}_0 f + \mathcal{E}_0 r \tilde{N}^x f,$$

with $\mathcal{E}_0 r \tilde{N}^x$ an allowable vector field. That $r(N^x + N^y)A_1$ is of commutator type ≥ 1 is easily seen. We are done. $\square$

We also need a result on general vector fields. Since integration by parts is no longer possible it has the following weaker form.

Theorem 3.7 *Let X be an arbitrary C^∞ vector field and A_1 an admissible operator of commutator type ≥ 1. Then*

$$X A_1 = A_0^{(0)} + \sum_{v=1}^{l} A_0^{(v)} W_v,$$

where the operators $A_0^{(v)}$, $v \geq 0$, are of commutator type ≥ 0, and the W_v are allowable fields.

Proof   Let $\mathcal{A}_1$ be a kernel of commutator type ≥ 1. We consider $X = N$.

$$\mathcal{A}_1 = \mathcal{E}_j \mathrm{P}^{-t_0} \Phi^{t_1} \overline{\Phi}^{t_2} r^k r^{*\,l}.$$

If $t - k - l \leq 1$ $N^y \mathcal{A}_1$ is of type $\mathcal{A}_0$. So let $t - k - l \geq 2$, that means $t_1 + t_2 + k + l \leq -2$. If N^y falls on $\mathcal{E}_j \mathrm{P}^{-t_0}$ or r^k the type will be ≥ 0. So

$$N^y \mathcal{A}_1 = \mathcal{E}_j \mathrm{P}^{-t_0} r^k N^y (\Phi^{t_1} \overline{\Phi}^{t_2} r^{*\,l}) + \mathcal{A}_0.$$

Now

$$
\begin{aligned}
N^y (\Phi^{t_1} \overline{\Phi}^{t_2} r^{*\,l}) &= \mathcal{E}_0 \Big(t_1 \Phi^{t_1 - 1} \overline{\Phi}^{t_2} r^{*\,l} \\
&\quad + t_2 \Phi^{t_1} \overline{\Phi}^{t_2 - 1} r^{*\,l} + l \Phi^{t_1 + t_2} \overline{\Phi}^{t_2} r^{*\,(l-1)} \Big) \\
&=: (C_1 + C_2 + C_3) \mathcal{E}_0.
\end{aligned}
$$

If $t_1 - 1 < 0$, $t_2 < 0$ it will follow $t_1 = 0$, so $C_1 = 0$. Analogously if $t_1 < 0$ and $t_2 - 1 < 0$ it will follow $t_2 = 0$, so $C_2 = 0$. So in each of the C_1, C_2, C_3 one Φ-exponent is nonnegative. Therefore we always can decompose C_i or $\overline{C}_i$ as a sum of terms

$$\Phi^r (\Phi + \overline{\Phi})^{\tau_1 + \tau_2 - r} r^{*\,\tau_3},$$

with

$$\tau_3 \geq 0, \quad r \leq \tau_1 + \tau_2 \leq -2.$$

Now

$$\Phi^r = \frac{1}{(r+1) Y \Phi} Y(\Phi^{r+1}) = \mathcal{E}_0 Y(\Phi^{r+1}),$$

will give as in the foregoing proof that

$$N^y \mathcal{A}_1 = Y \mathcal{A}_1 + \mathcal{A}_0.$$

Writing $Y = \varphi[W_1, W_2] + W_3$ with allowable vector fields W_i and proceeding as before will achieve the proof since $W_i \mathcal{A}_1 = \mathcal{A}_0$. $\square$

Definition 3.8 *Let k be a nonnegative integer. The tangential weighted Sobolev norm of order k is defined as*

$$\|f\|_{W_k^{p,\delta}} = \sup\{\|X_1 \dots X_j f\|_{L^{p,\delta}} \mid X_\lambda \text{ tangential}, \|X_\lambda\|_{C^1} \leq 1, j \leq k\}.$$

We denote the corresponding Sobolev space by

$$W_k^{p,\delta} = \{f \in L^{p,\delta} \mid \|f\|_{W_k^{p,\delta}} < \infty\}.$$

Remark 3.9 *Note that $L_k^{p,\delta} \subset W_k^{p,\delta} \subset S_k^{p,\delta}$ with continuous inclusions.*

Now we are ready to formulate our higher regularity result for A_1 operators.

Theorem 3.10 *Suppose $1 < p \leq \infty$ and $0 \leq \delta < \min(\delta^\#, 1, p-1)$. Let A_1 and B_1 be admissible operators of commutator type ≥ 1. Then*

    i. $A_1 : S_k^{p,\delta} \to S_{k+1}^{p,\delta^\#}$, $k = 0, 1, 2, \ldots,$

    ii. $A_1 B_1 : L^{p,\delta} \to L_1^{p,\delta^\#}$,

    iii. $A_1 B_1 : W_k^{p,\delta} \to W_{k+1}^{p,\delta^\#}$,

    iv. $A_1 B_1 : L^\infty \to \Gamma_{\beta,2\beta}$, $\beta < 1,$

are bounded linear operators.

Proof   *i*) The case $k = 0$ is settled by Theorem 3.5, *iii*. Let $W_1, \ldots, W_{k+1}$ be allowable vector fields with $\|W_v\|_{C^1} \leq 1$. Repeated application of Theorem 3.6 gives for $k > 0$ (note that $\widetilde{W_v}$ is also allowable)

$$W_{k+1} W_k \ldots W_1 A_1 f = W_{k+1} \sum_{v=0}^{l} A_1^{(v)} g_v,$$

where the $A_1^{(v)}$ are admissible of commutator type ≥ 1 and g_v is obtained from f by applying at most k allowable vector fields of uniformly bounded C^1 norms. Hence

$$\|g_v\|_{L^{p,\delta}} \leq C \|f\|_{S_k^{p,\delta}},$$

with C independent of f. Now the case $k = 0$ achieves the proof.

ii) Let X be an arbitrary vector field and $f \in L^{p,\delta}$. By Theorem 3.7 we have

$$X A_1 B_1 f = A_0^{(0)} B_1 f + \sum_{v=1}^{l} A_0^{(v)} W_v B_1 f.$$

Choose $\delta < \delta_1 < \delta^\#$. Then

$$\|B_1 f\|_{S_1^{p,\delta_1}} \lesssim \|f\|_{L^{p,\delta}}$$

because of Theorem 3.5, *iii*. Then Theorem 2.4 gives the result.

iii) Let $X_1, \ldots, X_k$ be tangential vector fields of C^1-norm ≤ 1. By repeatedly applying (3.6) we obtain

$$X_k \ldots X_1 A_1 B_1 f = \sum_{v,\mu} A_1^{(v)} B_1^{(v,\mu)} g_{v,\mu},$$

where the operators on the right are of commutator type ≥ 1, and the functions $g_{v,\mu}$ arise from f by application of at most k tangential vector fields of bounded C^1-norm. So

$$\|g_{v,\mu}\|_{L^{p,\delta}} \lesssim \|f\|_{W_k^{p,\delta}}.$$

Now part ii) gives

$$\|X_k \ldots X_1 A_1 B_1 f\|_{L_1^{p,\delta\#}} \lesssim \|f\|_{W_k^{p,\delta}},$$

which is slightly stronger than claim iii).

iv) Part ii) with $p = \infty$ and Theorem 1.4 imply

$$\|A_1 B_1 f\|_{\Lambda_\beta} \lesssim \|f\|_{L^\infty}$$

for all $\beta < 1$. So we have to show

$$\|A_1 B_1 f\|_{\Gamma_{2\beta}} \lesssim \|f\|_{L^\infty},$$

that is

$$\|W A_1 B_1 f\|_{\Gamma_\gamma} \lesssim \|f\|_{L^\infty} \qquad\qquad (*)$$

for all $\gamma < 1$ and all allowable vector fields W of C^1-norm ≤ 1.

As usual (3.6) and the remark thereafter yield

$$W A_1 B_1 f = \sum_{v=0}^{l} A_1^{(v)} g_v,$$

where, in view of 3.5, iii

$$\|g_v\|_{L^{\infty,\delta}} \lesssim \|B_1 f\|_{S_1^{\infty,\delta}} \lesssim \|f\|_{L^\infty}$$

for $0 < \delta < 1$. Now choose $0 < \delta < \frac{1}{2}$ and $\delta < \delta^\# < 1$ and set $\alpha = \frac{1}{2} - \delta$. Then 3.5, iv gives

$$\|W A_1 B_1 f\|_{\Gamma_{\alpha/(\alpha+\delta\#)}} \lesssim \|g_v\|_{L^{\infty,\delta}} \lesssim \|f\|_{L^\infty}.$$

Letting $\delta^\#$, δ be sufficiently small, one obtains $(*)$.

The proof of the theorem is now complete. $\square$

We need a last regularity result in terms of Sobolev norms which is based on the following remark:

Proposition 3.11 *Let A_1 be a kernel of type 1, and let X be any vectorfield (operating on x or y). Then the operator B with kernel*

$$\mathcal{B} = |r^*|^{1/2} X A_1$$

sends L^2 continuously into itself.

Proof   We have

$$A_1 = \mathcal{E}_j P^{-t_0} \Phi^{t_1} \overline{\Phi}^{t_2} \Phi^{*\,t_3} \overline{\Phi}^{*\,t_4} r^k r^{*\,l}$$

with integers $j, t_0, \ldots, t_4, k, l$ satisfying

$$j, t_0, k, l \geq 0, \qquad -t = t_1 + t_2 + t_3 + t_4 \leq 0$$
$$2n + j + \min(t - k - l, 2) - 2(t_0 + t - k - l) \geq 1$$

Application of X to $\mathcal{E}_j$ or P^{-t_0} reduces the type by 1, multiplication with $|r^*|^{1/2}$ then gives a $\mathcal{Z}$-kernel as described in VII.4.5. To study the application of X to the Φ's and r's we distinguish two cases:

a) $t - k - l \leq 1$. Then differentiation again reduces the type only by 1, and we are back in the previous cases.

b) $t - k - l \geq 2$. This is the most complicated situation. Namely, differentiating one of the Φ-powers or r-powers yields admissible kernels of type ≥ -1. Using, if necessary,

$$\Phi^* = \Phi + \mathcal{E}_3, \qquad |r|, |r^*| \lesssim |\Phi|, |\Phi^*|, \qquad |\mathcal{E}_j| \lesssim P^{j/2},$$

we can decompose these kernels into a sum of type-0 kernels and an admissible kernel of type ≥ -1 which can be estimated by

$$\mathcal{B}(x, y) = \frac{1}{P^{t_0 - j/2} |\Phi|^b},$$

where

$$2n + 2 - 2(t_0 - j/2 + b) \geq -1$$

and – this is crucial! –

$$b \geq 3.$$

The last inequality holds because $t - k - l \geq 2$. The type-0 kernels can be dealt with as before. Now the kernel $|r^*|^{1/2} |\mathcal{B}|$ can be estimated as

$$|r(y)|^{1/2} |\mathcal{B}(x, y)| \leq \frac{1}{P^{t_0 - j/2} |\Phi|^{b-1/2}},$$

and can therefore be further treated, concerning L^2-boundedness, like a special admissible kernel of type ≥ 0. $\qquad \Box$

From here we easily deduce

Theorem 3.12 *An admissible operator A_1 of type ≥ 1 is continous from L^2 into the Sobolev space $L^2_{1/2}$.*

Proof Let $F(y) = A_1 f(y) = \int_\Omega f(x) A_1(x, y)$. Denoting by grad the gradient, we will show

$$\||r|^{1/2} \operatorname{grad} F\|^2_{L^2} \leq \operatorname{const} \|f\|^2_{L^2}.$$

In view of [JeK 95] this gives our result. Now

$$\||r|^{1/2}\nabla F\|_{L^2}^2 = \int\limits_{y\in\Omega} |r(y)||\nabla^y F(y)|^2 dy,$$

$$= \int\limits_{y\in\Omega} |\int\limits_{x\in\Omega} f(x)|r(y)|^{1/2}\nabla^y\mathcal{A}(x,y)dx|^2 dy$$

$$\leq \mathrm{const}\|f\|_{L^2}^2$$

because of 3.11 and the L^2-boundedness of Z-operators. □

§4   Regularity Properties of E_{1-2n}-operators

Let the kernel $\mathcal{E}_{1-2n}(x,y)$ be given as

$$\mathcal{E}_{1-2n}(x,y) = \frac{\mathcal{E}_m(x,y)}{\rho(x,y)^{2k}},$$

with $m - 2k \geq 1 - 2n$ and $\mathcal{E}_m$ a smooth function which vanishes of order $m \geq 0$ for $x = y$. We shall denote the corresponding operator by E_{1-2n}.

Theorem 4.1 E_{1-2n} is a bounded operator

$$E_{1-2n} : L^p(D) \longrightarrow L^q(D)$$

for any $1 \leq p \leq q \leq \infty$ with $\frac{1}{q} > \frac{1}{p} - \frac{1}{2n}$.

Proof   The proof is analogous to those of Theorem (3.5), i, by showing that

$$\sup_{y\in D} \int\limits_D |\mathcal{E}_{1-2n}(x,y)|^\gamma dV(x) < \infty,$$

$$\sup_{x\in D} \int\limits_D |\mathcal{E}_{1-2n}(x,y)|^\gamma dV(y) < \infty,$$

if $\gamma < \frac{2n}{2n-1}$. But this is trivial since $\rho(x,y)^{-\gamma(2n-1)}$ is uniformly integrable. □

For a smooth tangential vector field T the operator TE_{1-2n} is defined in the distributional sense. But the following results will give more precise information. We first prove

Theorem 4.2 E_{1-2n} is a bounded operator

$$E_{1-2n} : L^\infty(D) \longrightarrow \Lambda_\alpha(D)$$

for all $\alpha < 1$.

Proof Since $|\mathcal{E}_{1-2n}(x,y)|$ is uniformly integrable it suffices to show

$$|E_{1-2n}(f)(y_1) - E_{1-2n}(f)(y_2)| \leq C_\alpha \|f\|_{L^\infty} \rho(y_1, y_2)^\alpha$$

for all $\alpha < 1$. Without restriction of generality we can assume that $D \subset\subset \mathbb{C}^n$. Moreover it suffices to show

$$\int\limits_D \left| \frac{\mathcal{E}_m(x, y_1)}{\rho(x, y_1)^{2k}} - \frac{\mathcal{E}_m(x, y_2)}{\rho(x, y_2)^{2k}} \right| dV(x) \leq C_\alpha |x - y|^\alpha$$

for all $\alpha < 1$. So let $y_1 \neq y_2$. Taylor expansion yields

$$Y := |\mathcal{E}_m(x, y_1)\rho^{2k}(x, y_2) - \mathcal{E}_m(x, y_2)\rho^{2k}(x, y_1)|$$

$$\leq C \sum_{l=1}^{\max(m, 2k)} \rho^{m+2k-l}(x, y_1)\rho^l(y_1, y_2).$$

Thus it suffices to estimate for $l \geq 1$.

$$\int\limits_D \frac{\rho^l(y_1, y_2)\, dV(x)}{\rho^{l-m}(x, y_1)\rho^{2k}(x, y_2)}.$$

Case $\rho(x, y_1) \leq \rho(x, y_2)$. The case $\rho(x, y_2) \leq \rho(x, y_1)$ is analogous.

Since the hessian form of $\rho^2(x, y)$ is positive definite on the diagonal of $\overline{D} \times \overline{D}$ we have

$$0 < C_1 \leq \frac{\rho(x, y)}{|x - y|} < C_2 < \infty$$

with two constants C_1, C_2 independent of $x, y \in \overline{D}$. Therefore

$$\rho(y_1, y_2) \leq C_2 |y_1 - y_2| \leq C_2 |x - y_1| + C_2 |x - y_2|$$

$$\leq \frac{C_2}{C_1}(\rho(x, y_1) + \rho(x, y_2)).$$

This implies

$$\rho^{l-1}(y_1, y_2) \leq C \sum_{j=0}^{l-1} \rho^{l-1-j}(x, y_1)\rho^j(x, y_2)$$

so that we must estimate $(0 \leq j \leq l - 1 < 2k)$

$$\rho(y_1, y_2) \int\limits_D \frac{dV(x)}{\rho^{1+j-m}(x, y_1)\rho^{2k-j}(x, y_2)}.$$

We decompose the domain of integration into two parts.

A)   Let $\rho(y_1, y_2) \leq \rho(x, y_1)$. Then we have to estimate

$$\rho(y_1, y_2) \int_{\substack{D \\ \rho(x,y_1) \geq \rho(y_1,y_2)}} \frac{dV(x)}{\rho^{2k+1-m}(x, y_1)}$$

$$\leq C' \rho(y_1, y_2) \int_{\substack{D \\ \rho(x,y_1) \geq \rho(y_1,y_2)}} \frac{dV(x)}{\rho^{2n}(x, y_1)}$$

$$\leq C'' |y_1 - y_2| \int_{\substack{D \\ |x-y_1| \geq C''' |y_1-y_2|}} \frac{dV(x)}{|x - y_1|^{2n}}$$

$$\leq C^{IV} |y_1 - y_2| (C^V + C^{VI} \log|y_1 - y_2|).$$

B)   Now let $\rho(y_1, y_2) \geq \rho(x, y_1)$. Since

$$\rho(y_1, y_2) \leq \frac{C_2}{C_1}(\rho(x, y_1) + \rho(x, y_2)) \leq \frac{2C_2}{C_1}\rho(x, y_2)$$

and $0 \leq j < 2k$ it suffices to estimate

$$\rho(y_1, y_2) \int_{\substack{D \\ \rho(x,y_1) \leq \rho(y_1,y_2)}} \frac{dV(x)}{\rho^{2k-m}(x, y_1)\rho(x, y_2)}$$

$$\leq C \int_{\substack{D \\ \rho(x,y_1) \leq \rho(y_1,y_2)}} \frac{dV(x)}{\rho^{2n-1}(x, y_1)}$$

$$\leq \tilde{C}\rho(y_1, y_2).$$

Therefore if we add the two cases (A) and (B) we obtain for every $0 < \delta < 1$ a constant $K_\delta < \infty$ with

$$|E_{1-2n}(f)(y_1) - E_{1-2n}(f)(y_2)| \leq K_\delta \|f\|_{L^\infty} \rho^{1-\delta}(y_1, y_2).$$

$$\square$$

The next theorem contains, in particular, Lemma 4.3 from Chapter IV.

Theorem 4.3 *Let T be a tangential C^∞ vector field on $\overline{D}$. Then TE_{1-2n} defines bounded operators*

$$TE_{1-2n} : \Lambda_\alpha(D) \longrightarrow \Lambda_{\alpha'} D$$

for all $0 < \alpha' < \alpha < 1$.

Proof   Let $g(x)$ be C^∞ on $\overline{D}$. Since T is tangential it follows by an integration by parts

$$TE_{1-2n}g = \int_D g(x)(T^y + T^x)\mathcal{E}_{1-2n}(x, y)\, dV(x) + \int_D h(x)\widetilde{\mathcal{E}}_{1-2n}(x, y)\, dV(x),$$

where $\widetilde{\mathcal{E}}_{1-2n}$ is an isotropic kernel of type $\mathcal{E}_{1-2n}$ and $h(x)$ is a smooth function.

This can be easily seen by the following observations.

$$\mathcal{E}_{1-2n}(x,y) = \frac{\mathcal{E}_m(x,y)}{\rho^{2k}(x,y)}, \qquad m, k \geq 0, \, m - 2k \geq 1 - 2n.$$

We suppose that $k > 0$. Let $\varepsilon > 0$ and pose

$$\mathcal{E}^\varepsilon_{1-2n}(x,y) = \frac{\mathcal{E}_m(x,y)}{\rho^{2k}(x,y) + \varepsilon}$$

and denote the corresponding operator by E^ε_{1-2n}. Then

$$TE^\varepsilon g(y) = \int_D g(x)(T^y + T^x)\mathcal{E}^\varepsilon_{1-2n}(x,y)\,dV(x) - \int_D g(x)T^x\mathcal{E}^\varepsilon_{1-2n}(x,y)\,dV(x).$$

Since T^x is tangential it is clear that there exist $h \in C^\infty(\overline{D})$ and a kernel $\widetilde{\mathcal{E}}^\varepsilon_{1-2n}(x,y)$, which is built up similarly as $\mathcal{E}^\varepsilon_{1-2n}(x,y)$, such that we have

$$TE^\varepsilon g(y) = \int_D g(x)(T^y + T^x)\mathcal{E}^\varepsilon_{1-2n}(x,y)\,dV(x) + \int_D h(x)\widetilde{\mathcal{E}}^\varepsilon_{1-2n}(x,y)\,dV(x).$$

Furthermore a small calculation implies the existence of a constant $C > 0$, which is independent of x, y and ε, such that

$$|\Delta^\varepsilon(x,y)| = |(T^y + T^x)(\mathcal{E}^\varepsilon_{1-2n}(x,y) - \mathcal{E}_{1-2n}(x,y))| \leq \frac{C\rho^{m+2}\varepsilon}{\rho^{2k+2}(\rho^{2k} + \varepsilon)}.$$

Therefore

$$TE^\varepsilon g(y) = \int_D g(x)(T^y + T^x)\mathcal{E}_{1-2n}(x,y)\,dV(x) + \int_D h(x)\widetilde{\mathcal{E}}^\varepsilon_{1-2n}(x,y)\,dV(x)$$

$$+ \int_D g(x)\Delta^\varepsilon(x,y)\,dV(x).$$

This implies that on each compact subset $K \subset\subset D$ $TE^\varepsilon g$ converges uniformly to

$$\int_D g(x)(T^y + T^x)\mathcal{E}_{1-2n}(x,y)\,dV(x) + \int_D h(x)\widetilde{\mathcal{E}}_{1-2n}(x,y)\,dV(x)$$

where $\widetilde{\mathcal{E}}_{1-2n}$ is obtained from $\widetilde{\mathcal{E}}^\varepsilon_{1-2n}$ by posing $\varepsilon = 0$.

Since $E^\varepsilon g \to Eg$ in the distributional sense it follows $TE^\varepsilon g \to TEg$ in the distributional sense. Therefore the claimed formula at the beginning of the proof holds.

Because $(T^y + T^x)\mathcal{E}_{1-2n}(x,y)$ is an isotropic kernel of order $(1-2n)$ Theorem 4.2 implies $TE_{1-2n}g \in \Lambda_\alpha(D)$, for all $\alpha < 1$. Now let $f \in \Lambda_\alpha(D)$. By a partition of unity argument we can assume that $D \subset\subset \mathbb{C}^n$. It is easy to see because of the existence of the right hand side of the following equation that we have in the sense of functions

$$TE_{1-2n}(f)(y) = \int_D f(x)(T^y + T^x)\mathcal{E}_{1-2n}(x,y)\,dV(x)$$

$$- \int_D (f(x) - f(y))T^x\mathcal{E}_{1-2n}\,dV(x)$$

$$- f(y)\int_D (T^y + T^x)\mathcal{E}_{1-2n}(x,y)\,dV(x) + f(y)TE_{1-2n}(1).$$

More precisely, introducing E^ε as above, the formula holds for the E^ε. But $E^\varepsilon f \to Ef$ in the sense of distributions (for $\varepsilon \to 0$), and the corresponding right-hand sides converge to the right hand side of the above formula. As Ef and the right-hand side are continuous functions, the claim follows from classical distribution theory [Sch 66]. So in view of Theorem 4.2 we only have to show that the operator

$$Kf(y) = \int_D (f(x) - f(y))T^x\mathcal{E}_{1-2n}(x,y)\,dV(x)$$

maps $\Lambda_\alpha(D)$ into $\Lambda_{\alpha'}(D)$. $K : \Lambda_\alpha(D) \to L^\infty(D)$ is obvious.

$$\mathcal{K}(x,y) = T^x\mathcal{E}_{1-2n}(x,y) = \frac{\mathcal{E}_m(x,y)}{\rho^{2k}(x,y)},$$

with $m, k \geq 0$, $m - 2k \geq -2n$. It suffices to show that for $D \subset\subset \mathbb{C}^n$, $y_1, y_2 \in D$

$$\int_D |f(x) - f(y_1)\mathcal{K}(x,y_1) - (f(x) - f(y_2))\mathcal{K}(x,y_2)|\,dV(x) \leq C\|f\|_{\Lambda_\alpha}\rho^{\alpha'}(y_1,y_2).$$

Let $\rho(x,y_1) \leq \rho(x,y_2)$. The case $\rho(x,y_1) \geq \rho(x,y_2)$ is analogous. The estimates will follow along the same lines as in the proof of Theorem 4.2.

$$|(f(x) - f(y_1))\mathcal{K}(x,y_1) - (f(x) - f(y_2))\mathcal{K}(x,y_2)|$$
$$\leq |(f(x) - f(y_1))(\mathcal{K}(x,y_1) - \mathcal{K}(x,y_2))| + |(f(x) - f(y_2))\mathcal{K}(x,y_2)|.$$

A)   Let $\rho(y_1, y_2) \leq \rho(x, y_1)$. Then it suffices to estimate

$$A_1 = \int_{\substack{D \\ \rho(x,y_1) \geq \rho(y_1,y_2)}} |x - y_1|^\alpha |\mathcal{K}(x,y_1) - \mathcal{K}(x,y_2)|\,dV(x)$$

and

$$A_2 = \int_{\substack{D \\ \rho(x,y_1) \geq \rho(y_1,y_2)}} |y_1 - y_1|^\alpha |\mathcal{K}(x,y_2)|\,dV(x).$$

We have with a constant $\varkappa > 0$

$$A_2 \leq C \int\limits_{\substack{D \\ \rho(x,y_1) \geq \rho(y_1,y_2)}} \frac{|y_1 - y_2|^\alpha}{\rho^{2k-m}(x,y_2)} \, dV(x)$$

$$\leq C' \int\limits_{|x-y_1| \geq \varkappa |y_1-y_2|} \frac{|y_1 - y_2|^\alpha}{|x - y_1|^{2n}} \, dV(x)$$

$$\leq C_1 |y_1 - y_2|^\alpha (1 + |\log|y_1 - y_2||)$$

$$\leq C_2 |y_1 - y_2|^{\alpha'}.$$

As in the proof of Theorem 4.2, case (A), we majorise A_1 by a sum of terms $(0 \leq j < 2k)$

$$\int\limits_{\substack{D \\ \rho(x,y_1) \geq \rho(y_1,y_2)}} \frac{|x - y_1|^\alpha \rho(y_1, y_2)}{\rho^{1+j-m}(x, y_1) \rho^{2k-j}(x, y_2)} \, dV(x)$$

$$\leq C \int\limits_{\substack{D \\ \rho(x,y_1) \geq \rho(y_1,y_2)}} \frac{|x - y_1|^\alpha \rho^\alpha(y_1, y_2)}{\rho^{2k-m+\alpha}(x, y_1)} \, dV(x)$$

$$\leq C_1 |y_1 - y_2|^\alpha \int\limits_{|x-y_1| \geq \varkappa |y_1-y_2|} \frac{dV(x)}{|x - y_1|^{2n}}$$

$$\leq C_2 |y_1 - y_2|^\alpha (1 + |\log|y_1 - y_2||)$$

$$\leq C_3 |y_1 - y_2|^{\alpha'}.$$

B) Now let $\rho(y_1, y_2) \geq \rho(x, y_1)$. We have to estimate

$$B_1 = \int\limits_{\substack{D \\ \rho(x,y_1) \leq \rho(y_1,y_2)}} |x - y_1|^\alpha |\mathcal{K}(x, y_1) - \mathcal{K}(x, y_2)| \, dV(x)$$

and

$$B_2 = \int\limits_{\substack{D \\ \rho(x,y_1) \leq \rho(y_1,y_2)}} |y_1 - y_2|^\alpha |\mathcal{K}(x, y_2)| \, dV(x)$$

As in case (B) of the proof of Theorem 4.2 we have

$$\frac{C_1}{2C_2} \rho(y_1, y_2) \leq \rho(x, y_2), \quad \rho(x, y_1) \leq \rho(x, y_2)$$

Therefore, with a constant $\varkappa > 0$,

$$B_2 \leq C \int\limits_{\substack{D \\ \rho(x,y_1) \geq \varkappa \rho(y_1,y_2)}} \frac{|y_1 - y_2|^\alpha}{\rho(x,y_1)^{2n}} \, dV(x)$$

$$\leq C_1 |y_1 - y_2|^\alpha (1 + |\log|y_1 - y_2||)$$

$$\leq C_2 |y_1 - y_2|^{\alpha'}.$$

B_1 can be majorised by a sum of terms ($0 \leq j < 2k$)

$$\int\limits_{\substack{D \\ \rho(x,y_1) \leq \rho(y_1,y_2)}} \frac{|x - y_1|^\alpha \rho(y_1,y_2)}{\rho^{1+j-m}(x,y_1) \rho^{2k-j}(x,y_2)} \, dV(x)$$

$$\leq C \int\limits_{\substack{D \\ \rho(x,y_1) \leq \rho(y_1,y_2)}} \frac{|x - y_1|^\alpha \rho(y_1,y_2)}{\rho^{2k-m}(x,y_1) \rho(x,y_2)} \, dV(x)$$

$$\leq C' \int\limits_{\substack{D \\ \rho(x,y_1) \leq \rho(y_1,y_2)}} \frac{1}{\rho(x,y_1)^{2n-1}} \, dV(x)$$

$$\leq C'' \int\limits_{\substack{D \\ \rho(x,y_1) \leq \rho(y_1,y_2)}} \frac{1}{|x - y_1|^{2n-1}} \, dV(x)$$

$$\leq C''' |y_1 - y_2|.$$

The theorem is proven. □

Theorem 4.4 *Let X be a C^∞ vector field on $\overline{D} \subset\subset X$ and $1 < p < \infty$. Then*

$$X E_{1-2n} : L^p(D) \longrightarrow L^p(D)$$

defines a bounded linear operator.

Proof   By using a partition of unity we can assume that $D \subset\subset \mathbb{C}^n$. Moreover let $\varepsilon > 0$ be a constant and $\chi : \mathbb{R} \to \mathbb{R}$ a C^∞ function with $\chi(x) = 1$ for $x \leq 0$ and $\chi(x) = 0$ for $x \geq \varepsilon$. Set

$$E'_{1-2n} f = \int\limits_D f(x) \chi(r(x)) \mathcal{E}_{1-2n}(x,y) \chi(r(y) - \varepsilon) \, dV(x)$$

$$= \int\limits_{\mathbb{C}^n} f(x) \chi(r(x)) \mathcal{E}_{1-2n}(x,y) \chi(r(y) - \varepsilon) \, dV(x).$$

Then if ε is sufficiently small and $f : D \to \mathbb{C}$ is extended trivially to $f : \mathbb{C}^n \to \mathbb{C}$ by $f(x) = 0$ for $x \notin D$ we obtain

$$E'_{1-2n}(f)|_D = E_{1-2n}(f).$$

So it suffices to show that $XE'_{1-2n} : L^p(\mathbb{C}^n) \to L^p(\mathbb{C}^n)$ is continuous. If X^y falls on $\chi(r(y) - \varepsilon)$ in the kernel the result follows from Theorem 4.1.

Step 1. We show that $XE'_{1-2n} : L^2(\mathbb{C}^n) \to L^2(\mathbb{C}^n)$ is bounded. Since for $y \notin \text{supp}(f)$ $XE'_{1-2n}(f)(y)$ is obviously given by

$$\int\limits_{\mathbb{C}^n} f(x)\chi(r(x))X^y(\mathcal{E}_{1-2n}(x,y)\chi(r(y) - \varepsilon))\, dV(x),$$

we can apply Theorem 3 in [Ste 93] (p. 294), if the following two conditions are verified.

A) *i.* $|X^y\mathcal{E}'_{1-2n}(x,y)| \leq A\rho^{-2n}(x,y)$,

   ii. $|X^y\mathcal{E}'_{1-2n}(x,y) - X^y\mathcal{E}'_{1-2n}(x',y)| \leq A\frac{\rho(x,x')}{\rho(x,y)^{2n+1}}$ if $|x - x'| \leq \frac{|x-y|}{2}$,

   iii. $|X^y\mathcal{E}'_{1-2n}(x,y) - X^y\mathcal{E}'_{1-2n}(x,y')| \leq A\frac{\rho(y,y')}{\rho(x,y)^{2n+1}}$ if $|y - y'| \leq \frac{|x-y|}{2}$

B) Let $\Phi \in C_c^\infty \mathbb{C}^n$ be the family of test functions with support in the unit ball with

$$\sup_{x \in \mathbb{C}^n} \max(|\Phi(x)|, \|\text{grad } \Phi(x)\|) \leq 1.$$

Set for $x_0 \in \mathbb{C}^n, R > 0$ $\Phi^{R,x_0}(x) = \Phi\left(\frac{x-x_0}{R}\right)$.

Then there exists a constant $A \geq 0$ independent of Φ^{R,x_0} such that

$$\|XE'_{1-2n}(\Phi^{R,x_0})\|_{L^2} \leq A \cdot R^n, \tag{*}$$

$$\|X\left(E'_{1-2n}\right)^*(\Phi^{R,x_0})\|_{L^2} \leq A \cdot R^n. \tag{**}$$

Proof of (A). It is clear that we can replace in *i.*, *ii.*, *iii.* $\mathcal{E}'_{1-2n}(x,y)$ by $\mathcal{E}_{1-2n}$ and assume that $r(x) < \varepsilon, r(y) < 2\varepsilon$. *i.* is obvious since

$$\mathcal{E}_{1-2n}(x,y) = \frac{\mathcal{E}_m(x,y)}{\rho^{2k}(x,y)},$$

with $m, k \geq 0$, $m - 2k \geq 1 - 2n$. *iii.* follows from *ii.* since $\rho(x,y) = \rho(y,x)$. We show *ii.*

If $r(x) < \varepsilon, r(x') \geq \varepsilon$ one has

$$|\mathcal{E}_{1-2n}(x,y)| = \left|\frac{\mathcal{E}_m(x,y)\rho(x,x')}{\rho^{2k}(x,y)\rho(x,x')}\right|$$

$$\leq 2\left|\frac{\mathcal{E}_m(x,y)\rho(x,x')}{\rho^{2k+1}(x,y)}\right|$$

$$\leq A\frac{\rho(x,x')}{\rho^{2n+1}(x,y)}.$$

So let $r(x) < \varepsilon, r(x') < \varepsilon, r(y) < 2\varepsilon$.

$$X^y\mathcal{E}_{1-2n}(x,y) = \frac{\mathcal{E}_m(x,y)}{\rho^{2k}(x,y)},$$

with $m - 2k \geq -2n$. Now

$$
\begin{aligned}
Y &:= |\mathcal{E}_m(x,y)\rho^{2k}(x',y) - \mathcal{E}_m(x',y)\rho^{2k}(x,y)| \\
&= |\mathcal{E}_m(x,y)\left(\rho^{2k}(x',y) - \rho^{2k}(x,y)\right) + (\mathcal{E}_m(x,y) - \mathcal{E}_m(x',y))\rho^{2k}(x,y)| \\
&\leq C[\rho^m(x,y)|\rho^{2k}(x',y) - \rho^{2k}(x,y)| + |\mathcal{E}_m(x,y) - \mathcal{E}_m(x',y)|\rho^{2k}(x,y)] \\
&\leq C'\left[\sum_{l=1}^{2k} \rho^{m+2k-l}(x,y)\rho^l(x,x') + \sum_{l=1}^{n} \rho^{m+2k-l}(x,y)\rho^l(x,x')\right].
\end{aligned}
$$

Therefore

$$
\left|\frac{\mathcal{E}_m(x,y)}{\rho^{2k}(x,y)} - \frac{\mathcal{E}_m(x',y)}{\rho^{2k}(x',y)}\right| \leq C \sum_{l=1}^{\max(m,2k)} \frac{\rho^{m+2k-l}(x,y)\rho^l(x,x')}{\rho^{2k}(x,y)\rho^{2k}(x',y)}.
$$

Since $|y - x'| \geq |x - y| - |x - x'| \geq \frac{1}{2}|x - y|$ we can majorise each summand by

$$
\frac{\rho^{m+2k-1}(x,y)\rho(x,x')}{\rho^{4k}(x,y)} \leq C\frac{\rho(x,x')}{\rho^{2n+1}(x,y)}.
$$

Proof of (B). Since the right hand side of the following equation makes sense as a pointwise (non distributional) equation it follows easily

$$
X E'_{1-2n}\left(\Phi^{R,x_0}\right)(y) = \int \Phi^{R,x_0}(x)\left(X^y + X^x\right)\mathcal{E}'_{1-2n}(x,y) + \int \psi^{R,x_0}(x)\mathcal{E}'_{1-2n}(x,y),
$$

where ψ^{R,x_0} is obtained from Φ^{R,x_0} by an integration by parts with respect to X^x. So

$$
|\psi^{R,x_0}(x)| \leq \frac{C}{R}.
$$

Since

$$
\int_{|x-x_0|\leq R} |\mathcal{E}'_{1-2n}(x,y)|\, dV(x) \leq C \int_{|x-x_0|\leq R} \frac{dV(x)}{|x-y|^{2n-1}} \leq C' \cdot R
$$

we get

$$
\left\|\int \psi^{R,x_0}(x)\mathcal{E}'_{1-2n}(x,y)\, dV(x)\right\|_{L^2} \leq C' \cdot R\|\psi^{R,x_0}\|_{L^2} \leq AR^n
$$

since $\operatorname{supp}\psi^{R,x_0} \subset B(x_0, R)$. The fact that $(X^x + X^y)\mathcal{E}'_{1-2n}(x,y)$ is an isotropic $\mathcal{E}_{1-2n}$ kernel finishes the proof of Step 1.

Step 2. We show that $X E'_{1-2n} : L^p(D) \to L^p(D)$ is continuous for $1 < p < 2$. Since the adjoint operator has the same properties the theorem will be shown for all $1 < p < \infty$.

In order to apply Theorem 3 in [Ste 93], p. 19, we have to show that there exists constants $A \geq 0, c > 1$ with

$$
\int_{\mathbb{C}^n \backslash B(y,c\delta)} |X^y\mathcal{E}'_{1-2n}(x,y) - X^y\mathcal{E}'_{1-2n}(x,y')|\, dV(x) \leq A,
$$

for all $y, y' \in \mathbb{C}^n$, $\delta > 0$ with $|y - y'| < \delta$.

When we proceed as in the proof of Step 1 we obtain for $r(y) < 2\varepsilon$, $r(y') < 2\varepsilon$

$$|X^y \mathcal{E}_{1-2n}(x,y) - X^y \mathcal{E}_{1-2n}(x,y')| \le C \sum_{l=1}^{\max(m,2k)} X_l,$$

with

$$X_l = \frac{\rho^{m+2k-l}(x,y)\rho^l(y,y')}{\rho^{2k}(x,y)\rho^{2k}(x,y')},$$

and $m - 2k \ge 1 - 2n$.

Now $K \ge |x - y| \ge c\delta$, $|x - y'| \ge |x - y| - |y - y'| \ge \frac{c-1}{c}|x - y|$. Therefore $\rho(x,y) \gtrsim \delta$, $\rho(x,y') \gtrsim \rho(x,y) \gtrsim |x - y|$, $\rho(y,y') \lesssim \rho(x,y)$ with positive constants independent of x, y, y', δ. Thus

$$\int\limits_{|x-y| \ge c\delta} X_l \, dV(x) \le c\delta \int\limits_{|x-y| \ge c\delta} \frac{dV(x)}{|x-y|^{2k-m+1}} \le c'\delta \int\limits_{K \ge |\tilde{x}| \ge c\delta} \frac{dV(\tilde{x})}{|\tilde{x}|^{2n}} \le A.$$

Note that $|\tilde{x}| \le K$ since $r(y) < 2\varepsilon$, $r(x) < \varepsilon$.

When $r(y) < 2\varepsilon$, $r(y') \ge 2\varepsilon$ or $r(y) \ge 2\varepsilon$, $r(y') < 2\varepsilon$ the estimates are obvious since for example if $r(x) < \varepsilon$

$$\varepsilon \le |r(x) - r(y')| \le C|x - y'| \le C(|x - y| + \delta) \le C'|x - y|.$$

$\square$

Corollary 4.5 *i) Let $k = 1, 2, \ldots$ and $1 < p < \infty$. Then*

$$E_{k-2n} : L^p(D) \longrightarrow L_k^p(D)$$

is bounded.

ii) For $\alpha > 0$ and $\delta > 0$,

$$E_{1-2n} : \Lambda_\alpha \longrightarrow L_1^{\infty,\delta}(D)$$

is bounded.

Proof *i)* Obvious since $X_1 X_2 \ldots X_{k-1} \mathcal{E}_{k-2n}(x,y) = \mathcal{E}_{1-2n}(x,y)$ and Theorem 4.4.

ii) We verify for an arbitrary vectorfield X the inequality

$$\|X E_{1-2n} f\|_{L^{\infty,\delta}} \lesssim \|f\|_{\Lambda_\alpha}, \qquad \alpha > 0, \delta > 0.$$

Let f be in Λ_α and extend f to a compactly supported $\tilde{f}$ on a larger domain $\tilde{\Omega} \supset\supset \Omega$ such that

$$\|\tilde{f}\|_{\Lambda_\alpha(\tilde{\Omega})} \lesssim \|f\|_{\Lambda_\alpha(\Omega)}.$$

The existence of $\tilde{f}$ follows from [See 64] – see also [LiR 80]. Likewise extend the kernel $\mathcal{E}_{1-2n}(x,y)$ to a kernel $\tilde{\mathcal{E}}_{1-2n}(x,y)$ on $\tilde{\Omega} \times \tilde{\Omega}$ and X to a smooth vectorfield $\tilde{X}$ on $\tilde{\Omega}$ with compact support. Then

$$X E_{1-2n} f = \tilde{X}(\tilde{f}, \tilde{\mathcal{E}}_{1-2n})_{\tilde{\Omega}} - \tilde{X}(\tilde{f}, \tilde{\mathcal{E}}_{1-2n})_{\tilde{\Omega}-\Omega}$$

The first term is bounded in $\Lambda_{\alpha'}$-norm by $\|\tilde{f}\|_{\Lambda_\alpha(\tilde{\Omega})}$ in view of theorem 4.3; the second term satisfies, for $y \in \Omega$,

$$|\tilde{X}(\tilde{f}, \tilde{\mathcal{E}}_{1-2n})(y)| \lesssim \|\tilde{f}\|_{L^\infty} \log|r(y)|$$
$$\lesssim \|f\|_{\Lambda_\alpha} |r|^\delta, \qquad \delta > 0.$$

$\qquad\qquad\qquad\qquad\qquad\qquad\qquad\qquad\qquad\qquad\qquad\qquad\qquad\qquad\qquad\qquad\qquad\qquad\square$

Corollary 4.6 *Let X be a C^∞ vector field on $X \supset\supset D$ and $1 < p < \infty$, $0 \le \delta < \min(1, \delta^{\#})$. Then*

$$X E_{1-2n} : L^{p,\delta}(D) \longrightarrow L^{p,\delta^{\#}}(D)$$

is a bounded linear operator.

Proof   For $\delta = 0$ we can even choose $\delta^{\#} = 0$. So let $\delta > 0$.

Since the right hand side of the following equation exists in the pointwise sense, it follows

$$|r(y)|^{\delta^{\#}/p}(X E_{1-2n})(f)(y) =$$
$$\int_D \left(f(x)|r(x)|^{\delta/p} \right) |r(x)|^{-\delta/p} (X^y \mathcal{E}_{1-2n}(x,y)) \left(|r(y)|^{\delta^{\#}/p} - |r(x)|^{\delta^{\#}/p} \right) dV(x)$$
$$+ X \int_D f(x)|r(x)|^{\delta^{\#}/p} \mathcal{E}_{1-2n}(x,y) \, dV(x).$$

Theorem 4.4 implies

$$\left\| X \int f(x)|r(x)|^{\delta^{\#}/p} \mathcal{E}_{1-2n}(x,y) \, dV(x) \right\|_{L^p} \le C \| f |r|^{\delta^{\#}/p} \|_{L^p} \le C' \|f\|_{L^{p,\delta}}.$$

For the first term on the right hand side we choose local coordinates $(t_1, \ldots, t_{2n})$ with $t_1(x) = r(x)$. (Only near the boundary $\{r(x) = 0\}$ the estimates are not obvious.) We obtain

$$|r(x)|^{-\delta/p} |X^y \mathcal{E}_{1-2n}(x,y)| \left| |r(y)|^{\delta^{\#}/p} - |r(x)|^{\delta^{\#}/p} \right| \le \frac{C}{|t_1(x)|^{\delta/p}|t(x) - t(y)|^{2n-\delta^{\#}/p}}.$$

Changing the first coordinate by $\tau = |t_1|^{1-\delta/p}$ we can see as in the proof of Lemma 2.1 that

$$\sup_{y \in D} \int |X^y \mathcal{E}_{1-2n}(x,y)| |r(x)|^{-\delta/p} \left| |r(x)|^{\delta^{\#}/p} - |r(y)|^{\delta^{\#}/p} \right| dV(x) < \infty. \qquad (*)$$

Analogously one shows $(*)$ with the role of x and y interchanged. This gives the claim. $\qquad\square$

The analogous result to Theorem 3.6 for isotropic operators is the following result.

Theorem 4.7 *Let E_{1-2n} be an isotropic operator and T a C^∞ tangential vector field on $\overline{D}$. Then there exist isotropic operators $E_{1-2n}^{(\nu)}$, $\nu \geq 0$, of type $\mathcal{E}_{1-2n}$, with*

$$TE_{1-2n} = E_{1-2n}^{(0)} + \sum_{\nu=1}^{k} E_{1-2n}^{(\nu)} T_\nu,$$

with tangential C^∞ vector fields T_ν. If T is allowable the T_ν are also allowable.

Proof   The proof is now obvious and combines the fact $(T^y + T^x)\mathcal{E}_{1-2n} = \mathcal{E}_{1-2n}$ with an integration by parts. – cf proof of 4.3. $\square$

A direct consequence of Theorems 2.4, 3.5, 3.6, 3.7, 3.10, 4.4 and Corollaries 4.5, 4.6 is

Corollary 4.8 *Let $B = E_{1-2n} \circ A_1$ or $B = A_1 \circ E_{1-2n}$ where A_1 denotes an admissible operator of commutator type ≥ 1. Then B defines bounded linear operators*

  i. $B : S_k^p(D) \longrightarrow S_{k+2}^p(D)$,

  ii. $B : L^p(D) \longrightarrow L_1^{p,\delta}(D)$,

for $k = 0, 1, 2, \ldots,$ $0 < \delta$, $0 < p < \infty$.

§5   Notes

As mentioned before, the basic ideas for estimating admissible operators are due to many mathematicians, starting with Grauert/Lieb, Henkin and Kerzman; very influential contributions are due to Stein, see [PhS 77], [Kra 76], [FoS 74]. The important role of allowable vectorfields was pointed out by Stein. We have essentially followed [LiR 86₂] in §§2–4 and expressed the relevant estimates in the framework of weighted L^p-spaces. The estimates are not optimal as may be guessed by the frequent occurrence of conditions like $\delta < \delta^\#$ (see e. g. 3.10); optimal estimates (admitting $\delta = \delta^\#$) are much harder to establish and have only recently been proved by Phong and Stein in their theory of singular Radon transforms [PhS 86]. The discussion of function spaces in §1 is based on Krantz' work [Kra 76].

Chapter VIII

Regularity of the $\overline{\partial}$-Neumann Problem and Applications

In this chapter we exploit most of the previous results in order to establish estimates for the operators arising in the $\overline{\partial}$-Neumann problem, in particular for the canonical solution operator $\overline{\partial}^* N$, in various function spaces. As the case of Sobolev spaces is well known and can in fact be dealt with by easier methods, we lay the stress on Hölder and C^k-estimates. These latter estimates in particular are based on the detailed study of the $\overline{\partial}$-Neumann problem which we have done in the preceding chapters. The information we obtain leads to interesting consequences for the function theory on a strictly pseudoconvex domain, consequences which have actually motivated the development of the theory (see [Ran 86]). In fact, the approximation and decomposition theorems of the last paragraphs can be considered as quantitative versions of well-known results in the classical theory of Stein spaces, in analogy to the vanishing theorems for (e. g.) L^p-cohomology spaces which are variants of Cartan's Theorem B.

§1 The Basic Hölder Estimate

1. We start from the basic integral representations VI.1.26 and 3.16 which we restate in the language of Z-operators keeping only the essential information. We denote the Bergman projector by P and by N the Neumann operator; Ω is as usual strictly pseudoconvex. Then we have

Theorem 1.1 (Basic integral representation) *i) There are Z-operators of type ≥ 1 such that for $f \in L^2_{00}(\Omega) \cap \mathrm{dom}\,\overline{\partial}$,*

$$f - Pf = Z_1 \overline{\partial} f + Z_1 (f - Pf) \tag{1.2}$$

ii) There are Z-operators of type ≥ 1 such that for $f \in L^2_{0q}(\Omega) \cap \mathrm{dom}\,\overline{\partial} \cap \mathrm{dom}\,\overline{\partial}^*$ (and $q \geq 1$),

$$f = Z_1 \overline{\partial} f + Z_1 \overline{\partial}^* f + Z_1 f \tag{1.3}$$

Each of the above Z_1-operators has the structure

$$Z_1 = A_1 + E_{1-2n} \tag{1.4}$$

with A_1 of commutator type.

From (1.4) and the regularity theorems VII.3.5, 4.1 and 4.2 it is immediately clear that

$$Z_1 : L^\infty \to C^{1/2}$$

continuously, and that a suffiently high power of Z_1, say Z_1^m, is continuous from $L^2 \to C^{1/2}$:

$$Z_1^m : L^2 \to C^{1/2}. \tag{1.5}$$

Now, iterating (1.2) several times we arrive at

$$f - Pf = Z_1\bar{\partial}f + Z_1^m(f - Pf). \tag{1.6}$$

Similarly, (1.3) yields via iteration

$$f = Z_1\bar{\partial}f + Z_1\bar{\partial}^*f + Z_1^m f. \tag{1.7}$$

(Recall that Z_1 stands for many operators, all of type ≥ 1).

Let us now suppose that $f \in L_{00}^2$ and $\bar{\partial}f \in L_{01}^\infty$. Then $Z_1\bar{\partial}f \in C^{1/2}(\Omega)$, and so is $Z_1^m(f - Pf)$. So (1.6) yields, using the conituity of the operators,

$$\|f - Pf\|_{C^{1/2}} \leq \text{const} \left(\|\bar{\partial}f\|_{L^\infty} + \|f\|_{L^2}\right)$$

(because $\|f - Pf\|_{L^2} \leq \|f\|_{L^2}$).

From (1.7) we deduce in the same way,

$$\|f\|_{C^{1/2}} \leq \text{const} \left(\|\bar{\partial}f\|_{L^\infty} + \|\bar{\partial}^*f\|_{L^\infty} + \|f\|_{L^2}\right).$$

We restate this as

Theorem 1.8 (Basic Hölder estimate)   *There are constants C such that*

$$\|f - Pf\|_{C^{1/2}} \leq C \left(\|\bar{\partial}f\|_{L^\infty} + \|f\|_{L^2}\right) \tag{1.9}$$

$$\|f\|_{C^{1/2}} \leq C \left(\|\bar{\partial}f\|_{L^\infty} + \|\bar{\partial}^*f\|_{L^\infty} + \|f\|_{L^2}\right) \tag{1.10}$$

(for $f \in L_{0q}^2(\Omega)$ with $q = 0$ in (1.9), $q > 0$ in (1.10)). Ω is strictly pseudoconvex, and for (1.10) *the metric is normalised Levi.*

Remarks.   It is tacitly assumed that the right-hand side is ∞ if $\bar{\partial}f$ or $\bar{\partial}^*f$ are not defined or not bounded in L^∞-norm. *The constants C do not depend on f.* It can be shown, by a careful analysis of all the above constructions, that *they can be chosen invariant under small perturbations of $b\Omega$.* Their behaviour under large perturbations is unknown: this is one of the main open problems whose solution would have important applications.

2.   An immediate consequence of Theorem 1.8 is

Theorem 1.11 *Let f be a smooth $\overline{\partial}$-exact $(0,q)$-form on Ω, and suppose f is bounded. Then the solution u of minimal L^2-norm of $\overline{\partial}u = f$ is smooth and satisfies*

$$\|u\|_{C^{1/2}} \leq const \, \|f\|_{L^\infty}.$$

The constant does not depend on f.

In fact, since Pu respectively $\overline{\partial}^* u$ vanishes, (1.9) or (1.10) can be applied to u to give the above result. The interior regularity of u is well known — see [FoK 72]. Note that we only have to require that f is $\overline{\partial}$-closed if Ω is contained in a Stein manifold. Then the canonical solution operator $\overline{\partial}^* N$ immediately provides us with a uniformly Hölder continuous solution (a fortiori: a bounded solution) if the right-hand side of the Cauchy-Riemann system is bounded. The construction of bounded solutions in strictly pseudoconvex domains of $\mathbb{C}^n$ was one of the fundamental questions which led to the development of our theory. – Finally, on a general manifold, we only have to assume that $f \in L^2_{0q}$ is $\overline{\partial}$-exact, $f = \overline{\partial}v$, where v is a distribution solution on Ω (not neccesarily in L^2). Arguing as in VI §4 one sees that under these circumstances there is also an L^2-solution.

The recent progress in the theory of finite type convex domains described in Chapter III allows to construct Hölder continuous solutions even in this case: the following result is due to Diederich/Fornæss/Fischer [DFF 99] and — by a different approach — to Cumenge [Cum 97], [Cum 2001].

Theorem 1.12 *Let Ω be a smoothly bounded convex domain of finite type in $\mathbb{C}^n$. Then there are $\varepsilon > 0$, $C > 0$, and a continuous operator $S_{q-1} : L^\infty_{0q} \to \Lambda_{\varepsilon;0q-1}$ with*

$$\overline{\partial}S_{q-1}f = f \ \text{ for } \ \overline{\partial}f = 0$$

and

$$\|S_{q-1}f\|_{\Lambda_\varepsilon} \leq C\|f\|_{L^\infty}.$$

In fact, the operator S_{q-1} mentioned in Chapter III has the required properties. ε can be expressed in terms of the type of the domain: Moreover, ε also depends on q — the precise statement can be found in [HefT 2002].

§2   The Basic Sobolev Estimate

1.   It is now immediate to obtain Kohn's basic estimate:

Theorem 2.1 *Let Ω be strictly pseudoconvex and choose a Levi metric on the manifold $X \supset\!\supset \Omega$. Then there are constants C_q such that for $q \geq 1$ and $f \in L^2_{0q}(\Omega) \cap \mathrm{dom}\,\overline{\partial} \cap \mathrm{dom}\,\overline{\partial}^*$,*

$$\|f\|_{L^2_{1/2}} \leq C_q \left(\|f\|_{L^2} + \|\overline{\partial}f\|_{L^2} + \|\overline{\partial}^* f\|_{L^2}\right)$$

For $q = 0$ and arbitrary metrics we have

$$\|f - Pf\|_{L^2_{1/2}} \leq C_0 \left(\|f\|_{L^2} + \|\bar{\partial}f\|_{L^2} \right),$$

with P as Bergman projector.

Proof   By (1.4),

$$f = Z_1\bar{\partial}f + Z_1\bar{\partial}^*f + Z_1f,$$

and by the regularity theorems of Chapter VII, Z_1-operators are continuous from L^2 into the Sobolev space $L^2_{1/2}$. The case $q = 0$ is handled in the same way. $\square$

2.   The estimate relies, in our approach, on the special integral representation (1.4) and on the integral estimates which, in particular in the case of isotropic operators, are quite subtle in this problem. A more direct approach only using partial integration is Kohn's original solution of the $\bar{\partial}$-Neumann problem – see [FoK 72]. Anyway, for the approach we have chosen we get the estimate also if the metric is arbitrary: according to Sweeney [Swe 76] one knows that if a problem admits the basic estimate 2.1 for one metric then it does so for all metrics. Consequently we can state

Theorem 2.1′   *The estimate 2.1 holds for any metric.*

Note that $\bar{\partial}^*$ and the boundary conditions depend on the metric - Sweeney's result is highly nontrivial.

3.   It is in the scale of Sobolev estimates that our knowledge of pseudoconvex domains is most complete. In fact, combining the geometric results on finite type pseudoconvex domains obtained by Diederich, Fornæss, D'Angelo a. o. with the analytic methods underlying Kohn's proof of Theorem 2.1, Catlin proved [Cat 83], [Cat 87]

Theorem 2.2   *The following conditions for a pseudoconvex domain Ω with smooth boundary are equivalent:*

  i.   *Ω is of finite type.*

  ii.   *There are constants $\varepsilon > 0$, $C_q > 0$, such that for $q \geq 1$ and $f \in L^2_{0q}(\Omega) \cap \operatorname{dom} \bar{\partial} \cap \operatorname{dom} \bar{\partial}^*$,*

$$\|f\|_{L^2_\varepsilon} \leq C_q(\|f\|_{L^2} + \|\bar{\partial}f\|_{L^2} + \|\bar{\partial}^*f\|_{L^2}),$$

   and for $q = 0$:

$$\|f - Pf\|_{L^2_\varepsilon} \leq C_0(\|f\|_{L^2} + \|\bar{\partial}f\|_{L^2}),$$

   with P as Bergman projector.

This theorem marks the endpoint of a long development starting with Kohn's 1962 solution of the $\bar{\partial}$-Neumann problem in the strictly pseudoconvex case. — The only available proof is in Catlin's papers [Cat 83] and [Cat 87]. — It is still not known how to find the largest ε (except in special cases — for $\varepsilon = \frac{1}{2}$ we are in the strictly pseudoconvex case).

We shall not make use of this paragraph in the sequel.

§3   The Basic C^k-Estimate

1.   We arrive at one of the essential results of the theory.

Theorem 3.1 (Basic C^k-estimate) *Let $\Omega \subset\subset X$ be a strictly pseudoconvex domain in a hermitian manifold X.*

i) If P is the orthogonal projection from $L^2_{00}(\Omega)$ onto the L^2-bounded holomorphic functions, then for any $f \in L^2_{00}(\Omega)$,

$$\|f - Pf\|_{C^{k+1/2}} \leq c_0 \left(\|f\|_{L^2} + \|\overline{\partial}f\|_{C^k} \right).$$

ii) Let the metric be a normalized Levi metric and $\overline{\partial}^*$ the Hilbert space adjoint of $\overline{\partial}$. For each $f \in L^2_{0q}(\Omega)$, $q \geq 1$,

$$\|f\|_{C^{k+1/2}} \leq c_q \left(\|f\|_{L^2} + \|\overline{\partial}f\|_{C^k} + \|\overline{\partial}^* f\|_{C^k} \right).$$

The constants $c_q, q \geq 0$, depend on k and Ω.

As before, we set the right-hand side equal to infinity, if $\overline{\partial}f$ or $\overline{\partial}^* f$ is not defined or not bounded in the corresponding norms; so $f \in \mathrm{dom}\,\overline{\partial} \cap \mathrm{dom}\,\overline{\partial}^*$ if the statement is not void.

The most important consequence is

Corollary 3.2 *Let X be Stein. Then, for $f \in L^2_{0q}(\Omega), q \geq 1$, with $\overline{\partial}f = 0$, the canonical solution $u = \overline{\partial}^* N f$ of the CR equations satisfies*

$$\|u\|_{C^{k+1/2}} \leq c_{q-1}\|f\|_{C^k}$$

with c_{q-1} a constant independent of f. If X is arbitrary, the same statement holds under the additional assumption $f \perp \mathbb{H}^q$.

In fact, we have $Pu = 0$ in case $q = 1$, $\overline{\partial}^* u = 0$ in case $q > 1$, so either 3.1i or 3.1ii can be applied to u. – The metric can be arbitrary if $q = 1$ and has to be a Levi metric in case $q > 1$.

This implies in particular

Corollary 3.3 *Let f be a $\overline{\partial}$-closed $(0, q)$-form on Ω with finite C^k-norm. If there is an L^2-solution u to $\overline{\partial}u = f$, then there is also a solution v with $\overline{\partial}v = f$ and*

$$\|v\|_{C^{k+1/2}} \leq c_{q-1}\|f\|_{C^k}$$

(where again c_{q-1} is independent of f).

2.   Let us now turn to the proof of 3.1. The basic integral formula gives, for $q \geq 1$,

$$f = Z_1 \bar{\partial} f + Z_1 \bar{\partial}^* f + Z_1 f \tag{3.4}$$

with

$$Z_1 = A_1 + E_{1-2n} \tag{3.5}$$

and the A_1-operators of commutator type. In the sequel, we will express the commutator theorem VII 3.6 and 4.7 as: A_1 *essentially commutes with a tangential vector field* T, namely

$$T A_1 = \sum A_1^j T^j, \tag{3.6}$$

where the A_1^j are A_1-operators and T^j are differntial operators of order 0 or 1 (tangential in the latter case). Similarly for E_{1-2n}-operators.

Definition 3.7   *A differential operator D is tangential of order k, if it is the sum of terms*

$$T_\nu \circ T_{\nu-1} \circ \cdots \circ T_1, \quad 0 \leq \nu \leq k,$$

with tangential vector fields $T_1, \ldots, T_\nu$.

$k = 0$ is allowed – D is then a 0-order operator.

The commutator relations yield: *tangential differential operators of order k essentially commute with Z_1-operators of the form* (3.5):

$$D Z_1 = \sum Z_1^j D^j, \tag{3.8}$$

with D^j again tangential of order at most k.

Lemma 3.9   *Let T be a tangential vector field and $Z_3 = Z_1^1 \circ Z_1^2 \circ Z_1^3$ a Z_3 operator with*

$$Z_1^j = A_1^j + E_{1-2n}^j,$$

the A_1^j of commutator type. Then for each $\varepsilon > 0$,

$$\|T Z_3 f\|_{L^\infty} \lesssim \|f\|_{C^\varepsilon}. \tag{3.10}$$

Proof   The operator Z_3 is a sum of terms $Z_1 \circ Z_1 \circ Z_1$, where each Z_1-operator is either A_1 or E_{1-2n}. T essentially commutes with the Z_1. If one of the Z_1-operators is isotropic, i.e. an E_{1-2n}-operator, we move the differentiation to this operator and obtain the compositum of three operators which are either continuous, from $C^\varepsilon \to C^\varepsilon$ or continuous from $C^\varepsilon \to C^\alpha$ for each $\alpha < \varepsilon$, so they (more than) satisfy (3.10). (We have used VII 3.5 and 4.3.) Now consider a term $T A_1 \circ A_1 \circ A_1$. As T essentially commutes with A_1-operators, we are reduced to considering $A_1 \circ A_1 \circ T A_1$. But, by VII 3.5$v$), A_1 maps L^∞ boundedly into $L_1^{\infty,1/2}$, and so we have a bounded map

$$T A_1 \colon L^\infty \to L^{\infty,1/2}$$

By VII 3.5ii, $L^{\infty,1/2}$ is further mapped into $L^{\infty,\varepsilon}$ for arbitrary $\varepsilon > 0$, and this space, by the next A_1-operator, into $C^{1-\varepsilon}$, see again VII 3.5iv. All maps are bounded – this again gives more than (3.10).

3. To continue our proof of Theorem 3.1 we introduce, for each $k = 0, 1, 2, \dots$ the norms

$$Q_k(f) = \|\bar{\partial}f\|_{C^k} + \|\bar{\partial}^* f\|_{C^k} + \|f\|_{L^2} \tag{3.11}$$

and prove the decisive

Lemma 3.12 *If D^k is a tangential differential operator of order k, then*

$$\|D^k f\|_{C^{1/2}} \lesssim Q_k(f)$$

(recall $f \in L^2_{0q}, q \geq 1$).

Proof The case $k = 0$ is just Theorem 1.8. So let the estimate be proved for $k \geq 0$ and consider f with $Q_{k+1}(f) < \infty$. From 3.4 we deduce by iteration

$$f = Z_1 \bar{\partial}f + Z_1 \bar{\partial}^* f + Z_3 f$$

with Z_3 as in Lemma 3.9. If $D^{(k)}$ is tangential of order k, the commutator relations give

$$D^{(k)} f = \sum Z_1^j D^j \bar{\partial}f + \sum Z_1^j D^j \bar{\partial}^* f + \sum Z_3^j D^j f \tag{3.13}$$

where Z_1^j denotes different Z_1-operators in the first two sums and D^j stands for different tangential operators of order $\leq k$ in the formula. Let us apply a tangential vector field T to (3.13) and commute in the first two sums: so we have to study

$$Z_1^j T D^j \bar{\partial}f \quad \text{and} \quad Z_1^j T D^j \bar{\partial}^* f. \tag{3.14}$$

Now

$$\|T D^j \bar{\partial}f\|_{L^\infty}, \|T D^j \bar{\partial}^* f\|_{L^\infty} \lesssim Q_{k+1}(f) \tag{3.15}$$

and so both terms in (3.14) are dominated in $C^{1/2}$-norm by $Q_{k+1}(f)$. This takes care of the first two sums in (3.13); to handle the last sum, we have to consider $T Z_3^j D^j f$, which satisfies, by (3.10) and the induction hypothesis

$$\|T Z_3 D^j f\|_{L^\infty} \lesssim \|D^j f\|_{C^{1/2}} \lesssim Q_k(f) \tag{3.16}$$

This shows: for any tangential operator of order $\leq k + 1$ one has

$$\|D^{(k+1)} f\|_{L^\infty}, \|D^{(k+1)} \bar{\partial}f\|_{L^\infty}, \|D^{(k+1)} \bar{\partial}^* f\|_{L^\infty} \lesssim Q_{k+1}(f). \tag{3.17}$$

Let us write (3.13) again, with k replaced by $k + 1$. Since the Z_1-operators all send L^∞ into $C^{1/2}$, the assertion follows from (3.17). $\square$

In the same way we obtain for $q = 0$

Lemma 3.18 *For a tangential operator $D^{(k)}$ of order k and $f \in L^2_{00}$,*

$$\|D^{(k)}(f - Pf)\|_{C^{1/2}} \lesssim \|\bar{\partial}f\|_{C^k} + \|f\|_{L^2}$$

4.   To get the required estimates for the normal derivative we use the ellipticity of the operator $\bar{\partial} \oplus \bar{\partial}^*$. To express the required algebraic relations we introduce the usual local frame at a boundary point:

$$\begin{aligned}
&\omega_1, \dots, \omega_n = \partial r \quad \text{orthonormal } (1,0)\text{-forms} \\
&L_1, \dots, L_n \quad\quad\quad \text{dual frame of vector fields.}
\end{aligned}$$

Then $L_1, L_2, \dots, L_{n-1}, \bar{L}_1, \dots, \bar{L}_{n-1}, L_n - \bar{L}_n$ is a basis for the tangential vector fields which we numerate $T_1, T_2, \dots, T_{2n-1}$, and $N = L_n + \bar{L}_n$ is a normal vector field to the boundary. We can locally write a $(0, q)$-form and its differentials as

$$\begin{aligned}
f &= \sum_{|J|=q} f_J \, \bar{\omega}^J \\
\bar{\partial}f &= \sum_{|K|=q+1} (\bar{\partial}f)_K \, \bar{\omega}^K \\
\bar{\partial}^*f &= \sum_{|L|=q-1} (\bar{\partial}^*f)_L \, \bar{\omega}^L
\end{aligned} \tag{3.19}$$

with functions $f_J, (\bar{\partial}f)_K, (\bar{\partial}^*f)_L$. Then, by a straightforward though slightly tedious calculation we get

Lemma 3.20

$$Nf_J = \sum_{jK} a_{JjK} T_j f_K + \sum_L b_{JL} f_L + \sum_M c_{JM} (\bar{\partial}f)_M + \sum_P d_{JP} (\bar{\partial}^*f)_P,$$

where the a_{JjK} etc are smooth functions and the index sets are strictly ordered of the following size: $J, K, L, M, P \subset N = \{1, \dots, n\}$, $|J| = |K| = |L| = q$, $|M| = q + 1$, $|P| = q - 1$, $j = 1, \dots, 2n - 1$.

It is now clear how Theorem 3.1 follows from Lemmas 3.12, 3.18 and 3.20. Namely, if $Q_k(f)$ is finite, then 3.12 and 3.18 give the required estimates for tangential operators. If D is a k-th order differential operator which contains the normal field N at least once, then we may commute N with tangential fields until D hat the form $D = D_1 \circ N$, D_1 of order $k - 1$, with an error term which is a sum of differential operators of order $\leq k - 1$.

But now, by Lemma 3.20, we may replace Df by terms of the form $D_1 Tf$, $D_1 N\bar{\partial}f$ and $D_1 N\bar{\partial}^*f$ with a tangential vectorfield T. The last two terms are controlled by $Q_k(f)$; in the first term we are left with one tangential field more; so, after at most k repetitions of this procedure we are back to Lemma 3.12 or 3.18, respectively.   □

Remark. Solutions of the $\overline{\partial}$-equation which satisfy the Hölder estimate

$$\|u\|_{C^{1/2}} \lesssim \|f\|_{L^\infty}$$

can be found in $\mathbb{C}^n$ (and have been found) by explicit integral formulae involving Z_1-operators:

$$u = Z_1 f$$

Our discussion shows that tangential derivatives of u can be controlled by the C^k-norm of f, but, except for functions, normal derivatives remain inaccessible. Consequently, the canonical solution satisfying the additional differential equation $\overline{\partial}^* u = 0$, has been studied in the preceding paragraphs: it is given by an asymptotic Z_1-operator, so we control its tangential derivatives, and the normal derivative is then taken care of by the ellipticity. Of course, if it could be proved that Z_1-operators map C^k into $C^{k+1/2}$ the situation would be much simpler. There is, however, no indication that Z_1-operators have this property.

5.   As a final consequence of the basic C^k-estimate we state the weaker case $k = \infty$:

Theorem 3.21 *Let $f \in C_{0q}^\infty(\overline{\Omega})$ be $\overline{\partial}$-exact (in L^2-spaces). Then the canonical solution u of $\overline{\partial} u = f$ of minimal L^2-norm is C^∞-smooth up to the boundary.*

6.   The basic C^k-estimate can also be expressed in terms of the Laplacian:

Theorem 3.22 *There are constants c_q, $q = 1, \ldots, n-2$, such that for each $\alpha < 1$, and each $f \in$ dom $\square \subset L_{0q}^2$,*

$$\|f\|_{C^{k+\alpha}} \leq c_q(\|f\|_{L^2} + \|\square f\|_{C^k}),$$

c_q *depends on k and α.*

Theorem 3.23 *For $q = 1, 2, \ldots, n-2$, the Neumann operator N_q is continuous from $C^k(\overline{\Omega})$ into $C^{k+\alpha}(\overline{\Omega})$, $k = 0, 1, \ldots$ and $0 \leq \alpha < 1$.*

Theorem 3.23 immediately follows from 3.22, so we turn to proving that inequality. At first we need

Lemma 3.24 *Let Z_2 be a Z-operator whose A_1-terms (if they occur) are of commutator type. Then, for any $\delta > 0$, we have*

$$\|Z_2 f\|_{L_1^{\infty,\delta}} \lesssim \|f\|_{L^\infty}.$$

Proof   We destinguish the cases

i)	$Z_2 = A_2$	ii)	$Z_2 = E_{2-2n}$	iii)	$Z_2 = A_1 \circ A_1$
iv)	$Z_2 = A_1 \circ E_{1-2n}$	v)	$Z_2 = E_{1-2n} \circ A_1$	vi)	$Z_2 = E_{1-2n} \circ E_{1-2n}$

and consider a vectorfield X.

$i)$ $XA_2 = A_0 : L^\infty \longrightarrow L^{\infty,\delta}$ by Ch. VII, 2.4

$ii)$ $XE_{2-2n} = E_{1-2n} : L^\infty \longrightarrow \Lambda_\alpha$ by Ch. VII, 4.1

$iii)$ is contained in Chapter VII, 3.10ii

$iv)$ $E_{1-2n} : L^\infty \longrightarrow \Lambda_\alpha$, $\alpha < 1$. Then VII, 3.5vii applies

$v),vi)$ E_{1-2n} and A_1 are continuous from L^∞ into $\Lambda_{1/2}$. Then Ch. VII, 4.5ii can be used.

$\square$

To show 3.22 we shall prove by induction with respect to k

$$\|D^k f\|_{L_1^{\infty,\delta}} \lesssim \|f\|_{L^2} + \|\square f\|_{C^k} \tag{3.25}$$

$$\|\overline{\partial} f\|_{C_t^{k+1/2}} + \|\overline{\partial}^* f\|_{C_t^{k+1/2}} \lesssim \|f\|_{L^2} + \|\square f\|_{C^k} \tag{3.26}$$

Here D^k stands for a differential operator of order k on $\overline{\Omega}$, and the norm $\|.\|_{C_t^{k+1/2}}$ in (3.26) measures the supremum respectively Hölder norm of tangential derivatives of the indicated order. (3.25) implies the theorem.

$k = 0$. We have

$$\begin{aligned}
(\overline{\partial} f, \overline{\partial} f) + (\overline{\partial}^* f, \overline{\partial}^* f) &= (\square f, f) \\
&= (\square f, N\square f + Hf) \\
&= (\square f, N\square f),
\end{aligned}$$

consequently

$$\|\overline{\partial} f\|_{L^2}^2 + \|\overline{\partial}^* f\|_{L^2}^2 \lesssim \|\square f\|_{L^2}^2 \lesssim \|\square f\|_{C^0}^2. \tag{3.27}$$

But in view of Ch. VI, 6.10 one has

$$\overline{\partial} f = Z_1 \square f + Z_1 \overline{\partial} f + Z_1 \overline{\partial}^* f, \tag{3.28}$$

$$\overline{\partial}^* f = Z_1 \square f + Z_1 \overline{\partial} f + Z_1 \overline{\partial}^* f + Z^\infty f, \tag{3.29}$$

which gives, by interation,

$$\overline{\partial} f = Z_1 \square f + Z_l \overline{\partial} f + Z_m \overline{\partial}^* f + Z_p f, \tag{3.30}$$

$$\overline{\partial}^* f = Z_1 \square f + Z_l \overline{\partial} f + Z_m \overline{\partial}^* f + Z_p f, \tag{3.31}$$

with Z_l an l-fold iteration of a Z_1-operator, etc. All A_1-operators hidden in these representations are of commutator type. (3.30), (3.31) and (3.27) prove (3.26) in case $k = 0$.

To deduce (3.25) in this case we use Ch. VI.6.12:

$$f = Z_2 \square f + Z_3 \overline{\partial} f + Z_3 \overline{\partial}^* f + Z_3 f; \tag{3.32}$$

then (3.24) shows that $f \in L_1^{\infty,\delta}$.

For the inductive step we apply a differential operator D^{k+1} of order $k + 1$ to (3.32), assuming (3.25) and (3.26) proved for k, and distinguish several cases.

Case 1 $D^{k+1} = D_t^{k+1}$ is purely tangential. Then, by the induction assumption, $f \in C^{k+\alpha}$ for $\alpha < 1$, and (3.26) holds in case k. Tangential differential operators essentially commute with Z-operators in (3.32); so we commute, after applying D_t^{k+1} to (3.32), $k + 1$ derivatives with Z_2 and k derivatives with the Z_3-operators to end up with

$$Z_2 D_t^{k+1} \Box f; \quad D_t^1 Z_3 D_t^k \overline{\partial} f; \quad D_t^1 Z_3 D_t^k \overline{\partial}^* f; \quad D_t^1 Z_3 D_t^k f \tag{3.33}$$

In view of lemma (3.24),

$$\|Z_2 D_t^{k+1} \Box f\|_{L_1^{\infty,\delta}} \lesssim \|D_t^{k+1} \Box f\|_{L^\infty} \lesssim \|\Box f\|_{C^{k+1}} \tag{3.34}$$

Moreover, the inductive hypothesis gives $D_t^k \overline{\partial} f \in \Lambda_\alpha$ for $\alpha < 1$, so lemma 3.9 gives

$$
\begin{aligned}
\|D_t^1 Z_3 D_t^k \overline{\partial} f\|_{L^\infty} &\lesssim \|D_t^k \overline{\partial} f\|_{\Lambda_\alpha} \\
&\lesssim \|f\|_{L^2} + \|\Box f\|_{C^k}, \\
&\lesssim \|f\|_{L^2} + \|\Box f\|_{C^{k+1}},
\end{aligned}
\tag{3.35}
$$

and corresponding estimates for the other terms in (3.33). This yields

$$\|D_t^{k+1} f\|_{L^\infty} \lesssim \|f\|_{L^2} + \|\Box f\|_{C^{k+1}} \tag{3.36}$$

The same argument can be applied to (3.30) and (3.31) and leads to

$$\|\overline{\partial} f\|_{C_t^{k+1+1/2}} + \|\overline{\partial}^* f\|_{C_t^{k+1+1/2}} \lesssim \|f\|_{L^2} + \|\Box f\|_{C^{k+1}} \tag{3.37}$$

Consequently we can commute also the final D_t^1 derivative with the Z_3-operators in (3.32) to obtain

$$Z_3 D_t^{k+1} \overline{\partial} f; \quad Z_3 D_t^{k+1} \overline{\partial}^* f; \quad Z_3 D_t^{k+1} f \tag{3.38}$$

Now Lemma 3.24 leads to

$$
\begin{aligned}
\|Z_3 D_t^{k+1} \overline{\partial} f\|_{L_1^{\infty,\delta}} &\lesssim \|D_t^{k+1} \overline{\partial} f\|_{L^\infty} \\
&\lesssim \|f\|_{L^2} + \|\Box f\|_{C^{k+1}},
\end{aligned}
\tag{3.39}
$$

and correspondingly for the other terms.

This proves (3.26) for $k + 1$ and (3.25) in case of tangential differential operators of order $k + 1$.

Case 2 D^{k+1} contains exactly one normal derivative D_n^1. Here we use the ellipticity of $\Box$. So let D^1 be a further derivative (which we may assume to be either normal or purely tangential), and consider $D^1 D^{k+1}$. Commuting the derivatives leads to an equation

$$D^1 D^{k+1} = D_n D_t^{k+1} + D_t^k D_n^1 D_n^1 + \tilde{D}^{k+1}, \tag{3.40}$$

where t refers to tangential, n to normal, and upper indices give the order of differentiation. Ellipticity of $\square$ shows

$$D_n^1 D_n^1 = \lambda \square + D^2, \tag{3.41}$$

where λ is a smooth funtion and D^2 a second order differential operator which contains D_n^1 just once. Inserting (3.41) into (3.40) yields

$$D^1 D^{k+1} = D_n D_t^{k+1} + D_t^k \square + \hat{D}^{k+1} \tag{3.42}$$

with new operators of the indicated kind.

The first operator is dealt with by case 1, the second by the assumptions of the theorem, and the third by the inductive hypothesis. This proves (3.25) also in this case.

Case 3 D^{k+1} contains D_n^1 many times. We then use (3.41) several times to return to case 2.

The proof of (3.22) can be combined with Lemma 3.20 to yield the analogous statement to (3.23) for the operators $\overline{\partial} N$ and $\overline{\partial}^* N$:

Theorem 3.43   *For $1 \leq q \leq n$, $k = 0, 1, 2, \ldots$, the operators $\overline{\partial}^* N_q$ and $\overline{\partial} N_q$ are continuous from $C^k(\overline{\Omega})$ to $C^{\overline{k+1/2}}(\overline{\Omega})$.*

Proof   We adopt the notations of 3.20. The reader should not confuse the normal derivative N with N_q. Let $g = \overline{\partial}^* N_q f$. Then (3.26) and Theorem 3.22 for estimating $H_q f$ yield

$$\|g\|_{C_t^{k+1/2}} \lesssim \|f\|_{C^k} + \|H_q f\|_{C^k} \lesssim \|f\|_{C^k} .$$

That means that the case $k = 0$ is settled and that we are done when all the derivatives are tangential. Now by Lemma 3.20, if $k \geq 1$,

$$N g_I = \sum_{jK} a_{IjK} T_j g_K + \sum_L b_{IL} g_L + \sum_M c_{IM} (\overline{\partial}\overline{\partial}^* N_q f)_M .$$

Thus

$$\|N g\|_{C_t^{(k-1)+1/2}} \lesssim \|g\|_{C_t^{k+1/2}} + \|\overline{\partial}\overline{\partial}^* N_q f\|_{C_t^{(k-1)+1/2}} .$$

Moreover, $\overline{\partial}\overline{\partial}^* N_q f = f - H_q f - \overline{\partial}^* N_{q+1}(\overline{\partial} f)$ implies

$$\|N g\|_{C_t^{(k-1)+1/2}} \lesssim \|f\|_{C^k} + \|\overline{\partial}^* N_{q+1}(\overline{\partial} f)\|_{C_t^{(k-1)+1/2}} .$$

Again (3.26) (or Corollary 3.2) gives

$$\|N g\|_{C_t^{(k-1)+1/2}} \lesssim \|f\|_{C^k} + \|\overline{\partial} f\|_{C_t^{k-1}} \lesssim \|f\|_{C^k} .$$

This settles the case where there is exactly one normal derivative among the k derivatives. If there are more than one normal derivatives it suffices to consider derivatives of the following type

$$D_t^{k-l} \circ N_l , \, l \geq 2,$$

where D_t^{k-l} is purely tangential. Now N^2 can be written as a linear combination of $\square$ and second order derivatives, where at most one is normal (see (3.41)). Therefore, by an induction over an increasing number of normal derivatives and which is analogous to the proof of Lemma 3.22, we can treat the general case. We do not give the obvious details. If $g = \overline{\partial} N_q f$, the proof is even slightly simpler. $\square$

Corollary 3.44 *The projection operators $\overline{\partial}\overline{\partial}^* N$ and $\overline{\partial}^* \overline{\partial} N$ are continuous from $C^k(\overline{\Omega})$ into $C^{k-\frac{1}{2}}(\overline{\Omega})$.*

7. We conclude this paragraph with another immediate consequence of the basic estimate which we already mentioned in Chapter VI.3.17.

Theorem 3.45 *Let f be harmonic, then f is smooth up to the boundary:*

$$\mathbb{H}^q \subset C_{0q}^\infty(\overline{\Omega}), \; q = 1, 2, \ldots$$

Proof This follows immediately from 3.1.ii because $\overline{\partial} f = \overline{\partial}^* f = 0$. $\square$

8. At this point we return to the question of finite type pseudoconvex domains. Considering the parallel statements between Sobolev and C^k-estimates for the strictly pseudoconvex case, and considering further Catlin's characterisation of finite type by Sobolev estimates, it is natural to state the

Main Conjecture 3.46 *For a finite type pseudoconvex domain, there are ε, $C_q > 0$ such that*

$$\|f\|_{C^{k+\epsilon}} \le C_q(\|f\|_{L^2} + \|\overline{\partial} f\|_{C^k} + \|\overline{\partial}^* f\|_{C^k})$$

(if $f \in L_{0q}^2 \cap \operatorname{dom} \overline{\partial} \cap \operatorname{dom} \overline{\partial}^*$ and $q > 0$),

$$\|f - Pf\|_{C^{k+\epsilon}} \le C_0(\|f\|_{L^2} + \|\overline{\partial} f\|_{C^k})$$

(for $f \in L_{00}^2 \cap \operatorname{dom} \overline{\partial}$).

The actual state of the conjecture is as follows: it is true in $\mathbb{C}^2$ (Fefferman and Kohn 1988 [FeK 88]); it is moreover true for certain special domains ("diagonalisable") in arbitrary dimensions [FKM 90]. Note that there is no distinguished hermitian metric on a general pseudoconvex boundary; therefore the result of Beals, Greiner and Stanton [BGS 87] on strictly pseudoconvex domains gains special importance.

§4 Dolbeault Cohomology Spaces

1. Let Ω be a relatively compact domain with smooth boundary in a hermitian manifold X. We can associate with Ω several *cohomology spaces*:

$$H_s^{pq}(\Omega) = \{f \in L_{pq}^s(\Omega) : \bar{\partial}f = 0\} \,/\, \mathrm{im}\,\{\bar{\partial} : L_{pq-1}^s(\Omega) \to L_{pq}^s(\Omega)\},\ 2 \le s \le \infty,$$
$$H_{C^k}^{pq}(\overline{\Omega}) = \{f \in C_{pq}^k(\overline{\Omega}) : \bar{\partial}f = 0\} \,/\, \mathrm{im}\,\{\bar{\partial} : C_{pq-1}^k(\overline{\Omega}) \to C_{pq}^k(\overline{\Omega})\},\ k \in \mathbb{N}_0 \cup \{\infty\},$$
$$H^{pq}(\Omega) = \{f \in C_{pq}^\infty(\Omega) : \bar{\partial}f = 0\} \,/\, \mathrm{im}\,\{\bar{\partial} : C_{pq-1}^\infty(\Omega) \to C_{pq}^\infty(\Omega)\},$$
$$\mathbb{H}^{pq} = \ker \square_{pq} \subset L_{pq}^2.$$

The relation between these spaces is in general non-trivial. But our results now give

Theorem 4.1 *If $\Omega \subset\subset X$ is strictly pseudoconvex with smooth boundary, then all the above spaces are, for $q \ge 1$, finite-dimensional and isomorphic:*

$$\mathbb{H}^q = \mathbb{H}^{0q} \cong H^{0q}(\Omega) \cong H_{C^k}^{0q}(\overline{\Omega}) \cong H_s^{0q}(\Omega)$$

In particular their dimension depends neither on the metric nor on s or k.

Proof   The solution of the Neumann problem (for any metric!) gives us the orthogonal decomposition

$$L_{0q}^2 = \mathbb{H}^{0q} \oplus \bar{\partial}\bar{\partial}^* N L_{0q}^2 \oplus \bar{\partial}^*\bar{\partial} N L_{0q}^2,$$

and this immediately implies the isomorphism

$$\mathbb{H}^{0q} \cong H_2^{0q},\ q \ge 1.$$

The right-hand side does not depend on the metric, and this gives an isomorphism between the $\mathbb{H}^{0q}$ for different metrics.

Now choose a normalized Levi metric. Then, for $f \in \mathbb{H}^{0q}, q \ge 1$,

$$f = Z_1 f.$$

Since Z_1 is a compact operator, $\mathbb{H}^{0q}$ is finite-dimensional.

For each $s \ge 2$, cohomology classes in H_s^{0q} can be represented by $\bar{\partial}$-closed forms which are smooth on $\overline{\Omega}$ – see Ch. IV, Remark 7.5.3. So the restriction maps

$$H_s^{0q} \longrightarrow H_2^{0q}, \qquad s \ge 2,$$

are surjective. If, moreover, $f = \bar{\partial}u$ with $u \in L_{0q-1}^2$, then $f = \bar{\partial}\bar{\partial}^* N f$, and $\bar{\partial}^* N f$ is smooth on $\overline{\Omega}$ by our previous estimates: this shows injectivity of the restrictions and gives the required isomorphism

$$H_s^{0q} \cong H_2^{0q} \left(\cong H_{C^\infty}^{0q}(\overline{\Omega}) \right).$$

The same argument, now using the basic C^k-estimate shows the isomorphism

$$H_{C^k}^{0q} \cong H_2^{0q}.$$

2. We still have to prove the isomorphism

$$H^{0q}_{C^\infty}(\overline{\Omega}) \overset{\sim}{\longrightarrow} H^{0q}(\Omega).$$

The natural restriction of forms on $\overline{\Omega}$ to forms on Ω obviously induces a well defined homomorphism

$$\sigma : H^{0q}_{C^\infty}(\overline{\Omega}) \longrightarrow H^{0q}(\Omega).$$

Let us show that σ is bijective.

a) **surjective**. If f is a smooth $\overline{\partial}$-closed form in $C^\infty_{0q}(\Omega)$, we find, by remark 7.5.4 of Chapter IV, $\varepsilon > 0$ and $f_\varepsilon \in C^\infty_{0q}(\Omega_\varepsilon)$ and $u \in C^\infty_{0q-1}(\Omega)$ with $\overline{\partial} f_\varepsilon = 0$ and

$$f = f_\varepsilon + \overline{\partial} u \qquad \text{on } \Omega.$$

So the class $[f_\varepsilon|\overline{\Omega}] \in H^{0q}_{C^\infty}(\overline{\Omega})$ is mapped to the class $[f] \in H^{0q}(\Omega)$.

b) **injective**. Let $f \in C^\infty_{0q}(\overline{\Omega})$ such that there exists $u \in C^\infty_{0q-1}(\Omega)$ with $\overline{\partial} u = f$. As above by remark 7.5.4 of Chapter IV we now find $\varepsilon > 0$, $f_\varepsilon \in C^\infty_{0q}(\overline{\Omega}_\varepsilon)$ and $v \in C^\infty_{0q-1}(\overline{\Omega})$ with

$$f = f_\varepsilon + \overline{\partial} v \qquad \text{on } \overline{\Omega}.$$

So $f_\varepsilon = \overline{\partial}(u - v)$ on Ω, and by the injectivity part of 3.14 of Chapter VI this implies that f_ε is $\overline{\partial}$-exact on Ω_ε; if $f_\varepsilon = \overline{\partial} w_\varepsilon$, then $f = \overline{\partial}(w_\varepsilon + v)$ on $\overline{\Omega}$ – which shows the injectivity. $\square$

§5   Regularity of the Bergman Projection

1. We now turn to proving regularity results for the Bergman projector (for an arbitrary metric)

$$P : L^2 \to L^2 \cap \mathcal{O}$$

on a strictly pseudoconvex domain Ω. To this end we use the representation

$$Pf = P_0 f + A_1 f + (A_1 + E_{1-2n})Pf \tag{5.1}$$

of Chapter VI 3.9, where P_0 is a special admissible operator of type 0 and all $\mathcal{A}_1$-kernels that occur are of commutator type. From the regularity theorems for such operators we immediately deduce

Theorem 5.2 *P restricts, respectively extends, to a bounded linear operator from L^p to $L^p \cap \mathcal{O}$ for $1 < p < \infty$.*

Proof

At first, let $2 \le p < \infty$. We iterate (5.1): for m arbitrarily chosen

$$Pf = P_0 f + Z_1 f + Z_1^m Pf, \tag{5.3}$$

where $Z_1^m = \overbrace{Z_1 \circ \ldots \circ Z_1}^{m}$ is the m-th power of the Z_1-operator $A_1 + E_{1-2n}$ in (5.1). P_0 is continuous from L^p to L^p, $1 < p < \infty$, and the other operators have even better properties.

Now let $1 < p < 2$ and $2 < q < \infty$ with $\frac{1}{p} + \frac{1}{q} = 1$. We consider for $f \in L^p$ the bounded antilinear functional $\lambda_f : L^q \to \mathbb{C}$ with

$$\lambda_f(g) = (f, Pg) = \int f \wedge *\overline{Pg}$$

For $f \in L^2$ we have $\lambda_f(g) = (Pf, g)$. Therefore λ_f defines an element of L^p which we denote by Pf, and $P : L^p \to L^p$ extends $P : L^2 \to L^2$. Moreover, since

$$\|Pf\|_{L^p} = \sup_{\|g\|_{L^q} \le 1} |(f, Pg)| \le \|P\|_{L^q} \|f\|_{L^p},$$

P is a continuous extension. As Pf is the L^p-limit of holomorphic functions we also have $Pf \in \mathcal{O}$. $\square$

A more subtle result follows from our basic C^k-estimate: suppose $f \in C^{k+1}(\overline{\Omega})$, $k \ge 0$. Then $\overline{\partial} f \in C^k(\overline{\Omega})$, and so

$$\begin{aligned}
\|Pf\|_{C^{k+1/2}} &\le \|f\|_{C^{k+1/2}} + \|Pf - f\|_{C^{k+1/2}} \\
&\lesssim \|f\|_{C^{k+1}} + \|f\|_{L^2} + \|\overline{\partial} f\|_{C^k} \\
&\lesssim \|f\|_{C^{k+1}},
\end{aligned} \tag{5.4}$$

so P *maps C^{k+1} continuously into $C^{k+1/2}$*. In particular:

Theorem 5.5 *The Bergman projector maps $C^\infty(\overline{\Omega})$ continuously into itself.*

2.   Less elementary is

Theorem 5.6 *Let $0 < \alpha < 1$ and $k = 0, 1, 2, \ldots$ Then P is a continuous map from $C^{k+\alpha}$ into $C^{k+\alpha/2}$.*

Note that 5.4 can be regarded as the natural extension of 5.6 to the case $\alpha = 1$. But more is true

Theorem 5.7 *Under the assumptions of 5.6, P is continuous from $C^{k+\alpha}$ into itself.*

We shall not prove 5.7 and refer to [AhS 78].

3. Let us now turn to the proof of 5.6. We note, first of all, that the proof of Chapter VII.3.6 can be applied to A_0-operators and yields, as in 3.6 and 3.8:

Let A_0 be an admissible operator of type ≥ 0 and kernel

$$A_0 = \frac{\mathcal{E}_j}{\bar{\Phi}^t}. \tag{5.8}$$

Then A_0 essentially commutes with tangential differential operators, i.e.

$$DA_0 = \sum_j A_0^j D_j,$$

where D is tangential of order k, the D_j are tangential operators of order $\leq k$, and the A_0^j are A_0-operators (which are, up to A_1-terms, of the form 5.8).

Lemma 5.9 *Let A_0 be admissible of type ≥ 0 and kernel (5.8). Then, for $0 < \alpha < 1$, A_0 is continuous from C^α into $C^{\alpha/2}$.*

Proof We write (5.8) as (note that j is an even number in the worst case!)

$$A_0 = \frac{\mathcal{E}_j}{\bar{\Phi}^{n+1+j/2}}$$

and apply an arbitrary vectorfield X:

$$X A_0 = \frac{\mathcal{E}_{j-1}}{\bar{\Phi}^{n+1+(j-1)/2+1/2}} + \frac{\mathcal{E}_j}{\bar{\Phi}^{n+1+j/2+1}} \tag{5.10}$$

The first kernel defines an operator which sends L^∞ continuously into $L^{\infty,1/2}$ (by VII.2.1 ϵ). Let us look at the second kernel $\mathcal{B}$. If $f \in C^\alpha$, then

$$(f, \mathcal{B})_{L^2} = \int\limits_{x \in \Omega} (f(x) - f(y)) \frac{\mathcal{E}_j(x,y)}{\bar{\Phi}^{n+1+j/2+1}} + f(y) \int\limits_{x \in \Omega} \frac{\mathcal{E}_j(x,y)}{\bar{\Phi}^{n+1+j/2+1}} = I + II.$$

Using $|f(x) - f(y)| \lesssim \rho(x,y)^\alpha \|f\|_{C^\alpha}$ and VII.2.1ϵ once again, we obtain

$$I \lesssim \|f\|_\alpha |r^*|^{\alpha/2-1}, \tag{5.11}$$

For II we use the tangential vectorfiel Y from VII.3.4 which satisfies

$$\mathcal{E}_0(x,y) Y \frac{1}{\bar{\Phi}^{n+1+j/2}} = \frac{1}{\bar{\Phi}^{n+1+j/2+1}}$$

and get by partial integration

$$II = f(y) \int \mathcal{E}_j(x,y) Y \frac{1}{\bar{\Phi}^{n+1+j/2}}$$

$$= f(y) \int \mathcal{E}_{j-1}(x,y) \frac{1}{\bar{\Phi}^{n+1+j/2}}$$

$$= f(y) \int \frac{\mathcal{E}_{j-1}(x,y)}{\bar{\Phi}^{n+1+\frac{j-1}{2}+\frac{1}{2}}},$$

which is again a kernel as the first term in 5.10. So

$$II \lesssim \|f\|_{L^\infty} |r^*|^{-1/2} \tag{5.12}$$

From 5.10 to 5.12 we get:

$$|X A_0 f(y)| \lesssim \|f\|_\alpha \, |r(y)|^{\frac{\alpha}{2}-1};$$

since A_0 is continuous from C^∞ into L^∞, this proves the desired continuity statement $A_0 : C^\alpha \to C^{\alpha/2}$. $\qquad\square$

From here we obtain part of Theorem 5.6.

Proposition 5.13 *The Bergman projector P is continuous from C^α to $C^{\alpha/2}$ for $0 < \alpha < 1$.*

Proof Since the operator Z_1 in (5.3) is continuous from L^∞ into $C^{1/2}$ and Z_1^m sends, for m sufficiently large, L^2 continuously into $C^{1/2}$, the proposition follows from 5.3 and 5.9. $\qquad\square$

4. We can now finish the proof of 5.6 by induction. Suppose $k \geq 0$ and $f \in C^{k+\alpha}$. For $k = 0$, 5.13 gives the result. Let 5.6 be proved for $k - 1$; we use 5.3 with $m = 3$:

$$Pf = \boldsymbol{P}_0 f + Z_1 f + Z_1^3 P f.$$

Let $D = T D_1$ be a tangential differential operator of order k, T a tangential vectorfield and D_1 tangential of order $k - 1$. D essentially commutes with $\boldsymbol{P}_0$ and Z_1, so we have to look at

$$\boldsymbol{P}_0 D f \qquad \text{and} \qquad Z_1 D f$$

Clearly

$$\|Z_1 D f\|_{C^{1/2}} \lesssim \|Df\|_{L^\infty} \lesssim \|f\|_{C^{k+\alpha}},$$

and, in view of 5.13,

$$\|\boldsymbol{P}_0 D f\|_{C^{\alpha/2}} \lesssim \|Df\|_{C^\alpha} \lesssim \|f\|_{C^{k+\alpha}}.$$

The last term is handled as follows:

$$DZ_1^3 Pf = TD_1 Z_1^3 Pf;$$

as D_1 essentially commutes with Z_1^3, we have to consider

$$TZ_1^3 D_1 Pf.$$

Now

$$\|D_1 Pf\|_{C^{\alpha/2}} \lesssim \|Pf\|_{C^{k-1+\alpha/2}} \lesssim \|f\|_{C^{k-1+\alpha}} \leq \|f\|_{C^{k+\alpha}}.$$

Set

$$g = D_1 Pf$$

and apply the proof of Lemma 3.9 to $TZ_1^3 g$. This proof immediately gives $TZ_1^3 g \in C^{1/2}$.

So we see: if $f \in C^{k+\alpha}$, then DPf is in $C^{\alpha/2}$ for any tangential operator D of order k, with

$$\|DPf\|_{C^{\alpha/2}} \lesssim \|f\|_{C^{k+\alpha}}.$$

Since Pf is holomorphic, normal derivatives are controlled by tangential derivatives: this gives the required estimates.

5. To conclude this paragraph we state a famous application of Theorem 5.5, namely Fefferman's mapping theorem:

Theorem 5.14 *Let f be a biholomorphic map between two strictly pseudoconvex domains with smooth boundaries (in an arbitrary complex manifold). Then f extends to a diffeomorphism of the closures.*

By a beautiful argument due to Bell and Ligocka this result can be deduced from Theorem 5.5. We do not include this argument: it is clearly exposed in [Ran 86].

§6   The L^1-theory of the $\bar{\partial}$-Neumann Problem

The asymptotic formulae for the operators arising in the $\bar{\partial}$-Neumann theory allow to extend the theory to L^p-spaces with $1 \leq p < 2$, and even to spaces of finite measures. We restrict attention to the L^p-theory and refer to [HefT 99] for the more general measure case.

Theorem 6.1 *The following L^2-bounded operators restrict, respectively extend, to bounded operators in the following cases.*

    i. $\bar{\partial}^* N_q : L^\infty \longrightarrow \Lambda_{1/2}, q \geq 1,$
    $\bar{\partial} N_q : L^\infty \longrightarrow \Lambda_{1/2}, q \geq 1.$

ii. $\bar{\partial}^* N_q : L^p \longrightarrow L^{p'}$,
$\bar{\partial} N_q : L^p \longrightarrow L^{p'}$,
$(\bar{\partial}^* N_1)^* : L^p \longrightarrow L^{p'}$ if $q \geq 1$, $\frac{1}{p'} > \frac{1}{p} - \frac{1}{2n+2}$ and either $2 \leq p < p' \leq \infty$ or $1 \leq p < p' \leq 2$.

iii. $N_q : L^\infty \longrightarrow \Lambda_\lambda$, $0 < \lambda < 1$, $q \geq 1$.

iv. $N_q : L^p \longrightarrow L^{p'}$ if $q \geq 1$, $\frac{1}{p'} > \frac{1}{p} - \frac{1}{n+1}$ and either $2 \leq p < p' \leq \infty$ or $1 \leq p < p' \leq 2$.

v. $H_q : L^1 \longrightarrow \mathbb{H}^q \subset C^\infty(\overline{\Omega})$ if $q \geq 1$.

Remark 6.2 1. In *ii.* and its proof we deal with $(\bar{\partial}^* N_1)^*$ in order to avoid $\bar{\partial} N_0$, an operator which we did not study extensively. But note that V.6.2 implies that $\bar{\partial} N_0$ exists and is L^2-bounded.

2. If $p < 2$ then $p' \leq 2$ by definition. In order to drop this condition by letting p' jump over the 2-barrier one has to study fairly delicate L^p-regularity problems for the orthogonal projection onto $\mathcal{R}(\bar{\partial})$. This we will not do, see [Mis 91].

Proof Let B denote $\bar{\partial}^* N_q$ or $\bar{\partial} N_q$, $q \geq 1$. Since B is an asymptotic Z_1-operator there exists for any $m \geq 1$ an L^2-bounded operator $K = K_m$ with

$$B = Z_1 + Z_m \circ K .$$

Now Theorems 3.5, 4.1 and 4.2 of Chapter V imply *i.* and the regularity properties of $\bar{\partial}^* N_q$ and $\bar{\partial} N_q$ in *ii.* if $2 \leq p < p' \leq \infty$. Dualising gives

$$B^* = Z_1 + K \circ Z_m .$$

Therefore $\bar{\partial} N_q, \bar{\partial}^* N_{q+1}, q \geq 1$, and $(\bar{\partial}^* N_1)^*$ extend continuously to

$$L^p \longrightarrow L^{p'}$$

if $1 \leq p < p' \leq 2$ and $\frac{1}{p'} > \frac{1}{p} - \frac{1}{2n+2}$. To achieve *ii.* it remains to show boundedness of

$$(\bar{\partial}^* N_1)^* : L^p \longrightarrow L^{p'} \text{ if } 2 \leq p < p' \leq \infty \text{ and}$$
$$\bar{\partial}^* N_1 : L^p \longrightarrow L^{p'} \text{ if } 1 \leq p < p' \leq 2$$

and $\frac{1}{p'} > \frac{1}{p} - \frac{1}{2n+2}$ for both.

Let $f \in L_{00}^p$. It is easy to see that $g = (\bar{\partial}^* N_1)^* f$ is the solution of

$$\bar{\partial}^* g = f - Pf, \ g \perp \ker \bar{\partial}^* .$$

Thus $g \in \text{im } \bar{\partial}$. The basic integral formula for g yields

$$
\begin{aligned}
(\bar{\partial}^* N_1)^* f &= T_0^*(f - Pf) + Z_1((\bar{\partial}^* N_1)^* f) + Z_2(f - Pf) \\
&= Z_1(f - Pf) + Z_1((\bar{\partial}^* N_1)^* f) .
\end{aligned}
$$

After iterating we obtain

$$(\bar{\partial}^* N_1)^* f = Z_1(f - Pf) + Z_m((\bar{\partial}^* N_1)^* f).$$

This implies the continuity of $(\bar{\partial}^* N_1)^* : L^p \to L^{p'}$ if $2 \leq p < p' \leq \infty$ and $\frac{1}{p'} > \frac{1}{p} - \frac{1}{2n+2}$. By dualising the above equation we get

$$\bar{\partial}^* N_1 f = Z_1 f - P Z_1 f + (\bar{\partial}^* N_1) Z_m f.$$

This implies bounded extendability $L^p \to L^{p'}$ for $1 \leq p < p' \leq 2$, $\frac{1}{p'} > \frac{1}{p} - \frac{1}{2n+2}$. *ii.* is shown. Since

$$N_q = (\bar{\partial}^* N_q)^* \bar{\partial}^* N_q + \bar{\partial}^* N_{q+1}(\bar{\partial}^* N_{q+1})^*$$

iv. follows. Since

$$N_q = Z_2 + Z_m \circ K$$

iii. follows from properties of Z_2-operators. We leave the details to the reader. Finally

$$H_q = Z_m H_q$$

by 1.1. Dualising yields

$$H_q = H_q Z_m.$$

Let m be so large that $Z_m : L^1 \to L^2$ is bounded. Then *v.* follows from 3.1, for example. $\square$

So far we have obtained $N_q, \bar{\partial}^* N_q, \bar{\partial} N_q$ and $(\bar{\partial}^* N_1)^*$ on L^1 by continuous extension. The following shows that the essential relations between these operators carry over from L^2 to L^1.

Lemma 6.3 *Let $f \in L^1_{0q}(\Omega)$, $q \geq 1$. Then $N_q f \in L^1_{0q}(\Omega) \cap \operatorname{dom} \bar{\partial} \cap \operatorname{dom} \bar{\partial}^*$ in the sense of Chapter V, Remark 5.6. Moreover, $\bar{\partial}(N_q f) = (\bar{\partial} N_q) f$, $\bar{\partial}^*(N_q f) = (\bar{\partial}^* N_q) f$.*

Proof   This is easily seen from Theorem 6.1 by using a sequence $f_j \in L^2_{0q}(\Omega)$ with $f_j \xrightarrow{L^1} f$. $\square$

Lemma 6.4 *Let $f \in L^1_{0q}(\Omega)$.*

   i. If $f \in \operatorname{dom} \bar{\partial} \cap L^1$ then $N_{q+1}(\bar{\partial} f) = \bar{\partial}(N_q f)$, $q \geq 1$.

   ii. If $f \in L^1(\Omega)$ and $f \in \operatorname{dom} \bar{\partial} \cap L^1$ then

$$N_1(\bar{\partial} f) = (\bar{\partial}^* N_1)^* f = \bar{\partial} N_0 f.$$

   iii. If $f \in \operatorname{dom} \bar{\partial}^* \cap L^1$, then $N_{q-1}(\bar{\partial}^* f) = \bar{\partial}^* N_q f$, $q \geq 2$.

Proof   Obvious by 6.1 and L^1-approximation with respect to graph norms. $\square$

The following proposition extends the definition of the Bergman projector P from L^2 to L^1 in important special cases preserving its essential decomposition properties.

Proposition 6.5 *Let f be a function in $L^1(\Omega)$.*

i. *Let $\overline{\partial}f \in L^1_{01}(\Omega)$ and $(f_j)_j$ be a sequence in $L^2(\Omega) \cap \operatorname{dom} \overline{\partial}$ with $f_j \overset{L^1}{\to} f$, $\overline{\partial}f_j \overset{L^1}{\to} \overline{\partial}f$. Then $Pf := \lim_{j \to \infty} Pf_j$ converges in $L^1 \cap \mathcal{O}$ and*

$$f = Pf + \overline{\partial}^*(N_1(\overline{\partial}f)) \, .$$

In particular $f = Pf$ if and only if f is holomorphic.

ii. *Let $(f_j)_j$ be a sequence in $L^2(\Omega) \cap \operatorname{dom} \overline{\partial}$ such that $f_j \overset{L^1}{\to} f$ and $Pf := \lim_{j \to \infty} Pf_j$ exists in L^1. Then*

$$f = Pf + \overline{\partial}^*((\overline{\partial}^* N_1)^* f) = Pf + \overline{\partial}^*(\overline{\partial}N_0)f \, .$$

Proof   *i.* Since

$$f_j = Pf_j + \overline{\partial}^*(N_1(\overline{\partial}f_j)) \, ,$$

$\lim_{j \to \infty} Pf_j$ exists in L^1 and does not depend on the sequence. Now a passage to the limit gives $f = Pf + \overline{\partial}^* N_1(\overline{\partial}f)$. If f is holomorphic, then $f = Pf$. Let $f = Pf$. Since then $Pf_j \to f$ ind L^1, f is holomorphic.

ii. $(\overline{\partial}^* N_1)^* f_j \to (\overline{\partial}^* N_1)^* f$ in L^1. Moreover, $(\overline{\partial}^* N_1)^* f \in \operatorname{dom} \overline{\partial}^* \cap L^1$. Since $(\overline{\partial}^* N_1)^* f_j = N_1(\overline{\partial}f_j)$, we obtain

$$f_j = Pf_j + \overline{\partial}^*(\overline{\partial}^* N_1)^* f_j \, .$$

A passage to the limit implies

$$f = Pf + \overline{\partial}^*((\overline{\partial}^* N_1)^* f) \, .$$

$\square$

Corollary 6.6   *Let $f \in L^1(\Omega)$ such that there exists a sequence $f_j \in C^1(\overline{\Omega})$ with $f_j \to f$, $Pf_j \to 0$ in L^1. Then $f \in \operatorname{im} \overline{\partial}^* \cap L^1$. More precisely*

$$f = \overline{\partial}^*((\overline{\partial}^* N_1)^* f) = \overline{\partial}^*(\overline{\partial}N_0)f \, .$$

Along the same lines one proves the following

Proposition 6.7   *Let $f \in L^1_{0q}(\Omega) \cap \operatorname{dom} \overline{\partial} \cap \operatorname{dom} \overline{\partial}^*$. Then*

$$\begin{aligned} f &= H_q f + \overline{\partial}^* N_{q+1}(\overline{\partial}f) + \overline{\partial}N_{q-1}(\overline{\partial}^* f), \, q \geq 2 \, , \\ f &= H_1 f + \overline{\partial}^* N_2(\overline{\partial}f) + \overline{\partial}(\overline{\partial}^* N_1 f), \, q = 1 \, , \end{aligned}$$

with $\overline{\partial}^* N_1 f \in L^1 \cap \operatorname{dom} \overline{\partial}$. Moreover, if $\overline{\partial}^* f = 0$ then $\overline{\partial}^* N_1 f = 0$.

Proof   Let $f_j \in L^2_{0q}(\Omega) \cap \mathrm{dom}\,\overline{\partial} \cap \mathrm{dom}\,\overline{\partial}^*$ with $f_j \to f, \overline{\partial}f_j \to \overline{\partial}f, \overline{\partial}^* f_j \to \overline{\partial}^* f$ in L^1. Then for $q \geq 2$ a passage to the limit gives the conclusion. If $q = 1$, since $f_j \to f$, $H_1 f_j \to H_1 f, \overline{\partial}^* N_2(\overline{\partial}f_j) \to \overline{\partial}^* N_2(\overline{\partial}f), \overline{\partial}^* N_1 f_j \to \overline{\partial}^* N_1 f$ in L^1, it follows from the closedness of graph $\overline{\partial}$ that $\overline{\partial}^* N_1 f \in L^1 \cap \mathrm{dom}\,\overline{\partial}$ and the second equation.

Now let $\overline{\partial}^* f = 0$ with $f \in L^1_{01}(\Omega) \cap \mathrm{dom}\,\overline{\partial} \cap \mathrm{dom}\,\overline{\partial}^*$. We set $g = \overline{\partial}^* N_1 f$. Then

$$f = \overline{\partial}g + H_1 f + \overline{\partial}^* N_2 \overline{\partial}f .$$

Let $\varphi \in C^\infty(\overline{\Omega})$. Since $\overline{\partial}^* f = 0$ we have $(\overline{\partial}\varphi, f) = 0, (\overline{\partial}\varphi, H_1 f) = 0, (\overline{\partial}\varphi, \overline{\partial}^* N_2 \overline{\partial}f) = 0$. Here $(\cdot, \cdot)$ denotes the natural extension of the L^2-scalar product if the first factor is in $C^\infty(\overline{\Omega})$ and the second in L^1. Therefore

$$(\overline{\partial}\varphi, \overline{\partial}g) = 0 .$$

Now for $\psi \in C^\infty_{01}(\overline{\Omega})$

$$\psi = H_1 \psi + \overline{\partial}\varphi_1 + \overline{\partial}^* \varphi_2 ,$$

with $\varphi_1, \varphi_2, H_1 \psi \in C^\infty_*(\overline{\Omega})$. Therefore

$$(\psi, \overline{\partial}g) = (H_1 \psi, \overline{\partial}g) + (\overline{\partial}^* \varphi_2, \overline{\partial}g) .$$

Since $g_i = \overline{\partial}^* N_1 f_i \to g, \overline{\partial}g_i = f_i - H_1 f_i - \overline{\partial}^* N_2 \overline{\partial}f_i \to \overline{\partial}g$ in L^1, we get

$$(H_1 \psi, \overline{\partial}g) = 0 , \quad (\overline{\partial}^* \varphi_2, \overline{\partial}g) = 0 .$$

Consequently,

$$(\psi, \overline{\partial}g) = 0 \;\; \text{for all } \psi \in C^\infty_{01}(\overline{\Omega}) .$$

Hence $\overline{\partial}g = 0$. But then by 6.5

$$g = Pg + \overline{\partial}^* N_1 \overline{\partial}g = Pg \in L^1 \cap \mathcal{O} .$$

Since $Pg_j = 0$ we finally obtain $g = Pg = \lim Pg_j = 0$. $\qquad\square$

We now have the following direct consequence for the solubility of the $\overline{\partial}$-equation:

Corollary 6.8 *Let $f \in L^1_{0q}(\Omega) \cap \ker\overline{\partial} \cap \ker H_q \cap \mathrm{dom}\,\overline{\partial}^*$, and $q \geq 1$. Then $\overline{\partial}^* N_q f \in L^1_{0q-1}(\Omega) \cap \mathrm{dom}\,\overline{\partial}$ and $\overline{\partial}\overline{\partial}^* N_q f = f$.*

(We do not study the question whether $\overline{\partial}f = H_q f = 0$ implies $f \in \mathrm{dom}\,\overline{\partial}^*$.)

§7   Gleason's Problem for C^k-functions

Let $D \subset\subset X$ be a strictly pseudoconvex domain in a Stein manifold. Denote by $A^k(\overline{D})$, $k = 0, 1, \ldots$, the space of holomorphic functions on D which are in $C^k(\overline{D})$. Here C^k boundary values only serve as an example, and the following theorem can be generalised to other spaces.

Theorem 7.1 *Let $w \in D$ and $f_1, \ldots, f_N \in A^k(\overline{D})$ with $\{w\} = \{z \in \overline{D} : f_1(z) = \ldots = f_N(z) = 0\}$. Moreover, let $f \in A^k(\overline{D})$ such that there exist a neighbourhood U of w and holomorphic functions $\tilde{g}_1, \ldots, \tilde{g}_N$ on U with*

$$f = \sum_{i=1}^N f_i \tilde{g}_i \quad on \ U .$$

Then there exist $g_1, \ldots, g_N \in A^k(\overline{D})$ with $f = \sum_{i=1}^N f_i g_i$.

Proof   The proof will follow the same lines as in Hörmander [Hör 67] and Øvrelid [Øvr 71$_2$]. It relies on the solvability of the $\overline{\partial}$-problem with C^k boundary values. Namely, by Theorem 7.9 of Chapter IV and 4.1 of the present chapter, $\mathbb{H}^q = 0$ for $q \geq 1$. There-fore, if $f \in C^k_{0q}(\overline{D})$, $q > 0$, is $\overline{\partial}$-closed, $\overline{\partial}^* N f \in C^{k+1/2}_{0q-1}(\overline{D})$ solves $\overline{\partial} u = f$. Since $\{z \in \overline{D} : f_i(z) \neq 0\}$, $i = 1, \ldots, N$, is a covering of $\overline{D} - \{w\}$ by relatively open sets, there exist $\varphi_i \in C^\infty(\overline{D} - \{w\})$ with $1 = \sum_{i=1}^N \varphi_i$ on $\overline{D} - \{w\}$ and supp $\varphi_i \cap \{z \in \overline{D} : f_i(z) = 0\} = \emptyset$ (recall that supp φ_i is the closure with respect to $\overline{D} - \{w\}$ of the set $\varphi_i \neq 0$). Let $\varphi_0 \in C^\infty(X)$ with supp $\varphi_0 \subset U$ and $\varphi_0 = 1$ in a small neighbouhood U_1 of w.

We set

$$g_i^* = \varphi_0 \tilde{g}_i + (1 - \varphi_0) \frac{f \varphi_i}{f_i} .$$

Then clearly $g_i^* \in C^k(\overline{D})$ and

$$f = \sum_{i=1}^N f_i g_i^* \quad on \ \overline{D} .$$

Moreover, g_i^* is holomorphic in U_1. By the following classical method one can replace the decomposition factors g_i^* by globally holomorphic ones.

We define the Koszul complex corresponding to our problem as follows. For $r \geq 0$ we set

$$L_r = \{u \in C^k_{0r}(\overline{D}) : \overline{\partial} u \in C^k_{0r+1}(\overline{D})\} ,$$

and for $s \geq 0$

$$L_r^s = L_r \otimes_{\mathbb{C}} \overset{s}{\bigwedge} \mathbb{C}^N ,$$

that is the space of linear combinations in $\bigwedge^s \mathbb{C}^N$ with L_r-coefficients. Let $e_1, e_2, \ldots, e_N$ be the canonical vector base of $\mathbb{C}^N$. We set for $I = (i_1, i_2, \ldots, i_s)$, $1 \leq i_1 < i_2 < \cdots < i_s \leq N$,

$$e^I = e_{i_1} \wedge e_{i_2} \wedge \cdots \wedge e_{i_s} .$$

Elements of L_r^s can then be written as

$$u = \sum_{|I|=s}' u_I e^I ,$$

where $\sum'$ indicates increasingly ordered index sets I and $u_I \in L_r$. We apply $\overline{\partial}$ to L_r^s coefficientwise. Then

$$\overline{\partial} : L_r^s \to L_{r+1}^s \, .$$

Clearly, if $f \in L_{r+1}^s$ is $\overline{\partial}$-closed, then there exists $u \in L_r^s$ with $\overline{\partial}u = f$. (Here we use the regularity results for the $\overline{\partial}$-operator which we have established in this book.)

Now let

$$v = \sum_{|J|=t}' v_J e^J \in L_m^t \, .$$

We set

$$u \wedge v = \sum_{|I|=s,|J|=t}' (u_I \wedge v_J) e^I \wedge e^J \, .$$

Then $u \wedge v \in L_{r+m}^{s+t}$. If we set for $j \in I$, $j = i_\nu$,

$$e_j \lrcorner e^I = (-1)^{\nu-1} e^{(i_1,\dots,i_{\nu-1},i_{\nu+1},\dots,i_s)}$$

and $e_j \lrcorner e^I = 0$ for $j \notin I$, we can define the following homological operator

$$P_f u = \sum_{j=1}^N \sum_{|I|=s}' (f_j u_I) e_j \lrcorner e^I \, .$$

$P_f : L_r^s \to L_r^{s-1}$ has the following obvious properties. $P_f^2 = 0$, P_f commutes with $\overline{\partial}$, and if $u \in L_r^s$, $v \in L_m^t$,

$$P_f(u \wedge v) = (P_f u) \wedge v + (-1)^s u \wedge (P_f v) \, .$$

If we set $g^* = \sum_{i=1}^N g_i^* e_i$, we obtain $P_f g^* = f$.

To achieve the proof of the theorem we need two auxiliary lemmata. Set for $s, r \geq 0$

$$M_r^s = \{u \in L_r^s : u|_{U_1} \equiv 0\} \, .$$

Lemma 7.2 Let $u \in M_r^s$ with $P_f u = 0$. Then there exists $\omega \in L_r^{s+1}$ with $P_f \omega = u$ and $\overline{\partial}\omega \in M_{r+1}^{s+1}$.

Proof   Let $\varphi \in C^\infty(X)$ vanish near w and be equal to one on $X - U_1$. Then

$$\omega_0 = \sum_{i=1}^N \frac{\varphi \varphi_i}{f_i} \otimes e_i \in L_0^1 \, .$$

Moreover, $P_f \omega_0 \in L_0^0$ and $P_f \omega_0 \equiv 1$ on $\overline{D} - U_1$. Therefore

$$\omega = \omega_0 \wedge u \in L_r^{s+1}$$

satisfies $P_f \omega = P_f \omega_0 \wedge u = u$ and $\overline{\partial}\omega \in M_{r+1}^{s+1}$. $\square$

Lemma 7.3 *Let $u \in M_r^s$ with $P_f u = \bar{\partial} u = 0$. Then there exists $\omega \in L_r^{s+1}$ such that $P_f \omega = u$ and $\bar{\partial}\omega = 0$.*

Proof   The proof is by induction over decreasing $s + r$. If $s > N$ or $r > N$ the claim is trivial. So let the lemma hold for larger values of $s + r$. By Lemma 7.2 there is an $\omega' \in L_r^{s+1}$ with $P_f \omega' = u$, $\bar{\partial}\omega' \in M_{r+1}^{s+1}$. Therefore, by the incuction hypothesis, there exists an $\omega'' \in L_{r+1}^{s+2}$ with $P_f \omega'' = \bar{\partial}\omega'$ and $\bar{\partial}\omega'' = 0$, since $P_f \bar{\partial}\omega' = \bar{\partial} P_f \omega' = \bar{\partial} u = 0$, $\bar{\partial}^2 \omega' = 0$. Now let $\omega''' \in L_r^{s+2}$ with $\bar{\partial}\omega''' = \omega''$ and set

$$\omega = \omega' - P_f \omega''' \in L_r^{s+1}.$$

Then $P_f \omega = P_f \omega' = u$ and $\bar{\partial}\omega = \bar{\partial}\omega' - P_f \bar{\partial}\omega''' = \bar{\partial}\omega' - P_f \omega'' = 0$. $\square$

We can now finish the proof of the theorem. Let f, f_i, $\tilde{g}_i$, g_i^* as defined. Since g_i^* is holomorphic in U_1, $\bar{\partial} g^* \in M_1^1$. Moreover, $P_f \bar{\partial} g^* = \bar{\partial} f = 0$. By Lemma 7.3 there exists $\omega \in L_1^2$ with $P_f \omega = \bar{\partial} g^*$ and $\bar{\partial}\omega = 0$. Let $k \in L_0^2$ solve $\bar{\partial} k = \omega$ and set

$$h = g^* - P_f k \in L_0^1.$$

Obviously $\bar{\partial} h = \bar{\partial} g^* - P_f \omega = 0$ and $P_f h = P_f g^* = f$. In other words, if we write

$$h = \sum_{i=1}^{N} g_i e_i,$$

then $g_i \in A^k(\overline{D})$ and $f = \sum_{i=1}^{N} f_i g_i$. $\square$

§8   Stability of Estimates for the $\bar{\partial}$-Neumann Problem

1.   We consider, in this paragraph, a family

$$\Omega_\delta = \{x : r(x) < \delta\}, \ 0 \leq \delta \leq \delta_0, \ \delta_0 > 0, \tag{8.1}$$

of strictly pseudoconvex domains in an arbitrary complex manifold X. The function r is assumed to be strictly plurisubharmonic in a neighbourhood W of the boundary $b\Omega_0$, smooth with nonvanishing gradient, and δ_0 should be sufficiently small so that $\{x : 0 \leq r(x) \leq \delta_0\} \subset W$. The function r defines a Levi metric for each Ω_δ (which is independent of δ), and we can normalize this metric on each Ω_δ (depending on δ) such as to obtain a family of normalized Levi metrics smoothly depending on δ. We note

Remark 8.2   *The $\mathcal{Z}$-kernels $\mathcal{P}_q$, $\mathcal{T}_q$, $\mathcal{S}_q$, $\mathcal{E}_{j-2n}$, $\mathcal{A}_j$ which appear in the basic integral representations IV.5.54 and VI.1.26 on Ω_δ can all be constructed to depend smoothly on δ for $0 \leq \delta \leq \delta_0$.*

Let us denote the corresponding operators on Ω_δ by P_q^δ etc. Then (8.2) implies

Remark 8.3 *The operators $Z^\delta (= P_q^\delta, T_q^\delta, \ldots)$ corresponding to the above kernels have norms in $L^p(\Omega_\delta)$, $\Lambda_\varepsilon(\Omega_\delta), \ldots$ and all the other function spaces of Chapter VII which are uniformly bounded independently of δ.*

This gives, in particular, the following

Theorem 8.4 *There is a constant C (independent of δ) such that*

i. $\|f - P_0^\delta f\|_{C^{k+1/2}(\Omega_\delta)} \leq C(\|\overline{\partial} f\|_{C^k(\Omega_\delta)} + \|f\|_{L^2(\Omega_\delta)})$ *(for $f \in \text{dom } \overline{\partial} \subset L_{00}^2(\Omega_\delta)$).*

ii. $\|f\|_{C^{k+1/2}(\Omega_\delta)} \leq C(\|\overline{\partial} f\|_{C^k(\Omega_\delta)} + \|\overline{\partial}_\delta^* f\|_{C^k(\Omega_\delta)} + \|f\|_{L^2(\Omega_\delta)})$ *(for $f \in \text{dom } \overline{\partial} \cap$ dom $\overline{\partial}_\delta^* \subset L_{0q}^2(\Omega_\delta)$, $q \geq 1$).*

Here $\overline{\partial}_\delta^*$ is of course the Hilbert space adjoint of $\overline{\partial}$ on Ω_δ with respect to the normalized Levi metric on Ω_δ, and P_0^δ is the corresponding Bergman projection.

We want to apply (8.4) in order to show

Theorem 8.5 *There is a constant C (independent of δ) such that for each δ with $0 \leq \delta \leq \delta_0$ and each f which is $\overline{\partial}$-exact on Ω_δ we have a solution u of $\overline{\partial} u = f$ on Ω_δ satisfying*

$$\|u\|_{C^{k+1/2}(\Omega_\delta)} \leq C\|f\|_{C^k(\Omega_\delta)}. \tag{8.6}$$

More precisely, let us denote by $K^\delta = \overline{\partial}_\delta^* N^\delta$ the canonical solution operator to the $\overline{\partial}$-equation on Ω_δ; then we have

Theorem 8.5′ *If f is $\overline{\partial}$-exact on Ω_δ, one has*

$$\overline{\partial} K^\delta f = f$$
$$\|K^\delta f\|_{C^{k+1/2}(\Omega_\delta)} \leq C\|f\|_{C^k(\Omega_\delta)}, \tag{8.6'}$$

where C does not depend on f or δ.

Let us prove the above results. Theorem 8.4 immediately gives, under the above assumptions,

$$\|u\|_{C^{k+1/2}(\Omega_\delta)} \leq C(\|f\|_{C^k(\Omega_\delta)} + \|u\|_{L^2(\Omega_\delta)}) \tag{8.7}$$

with a constant C independent of δ. So we still have to control the L^2-norm of u in terms of the L^2-norm of f, independently of δ. This requires some additional effort.

2.   The essential step is

Lemma 8.8 *Let δ_j be a decreasing sequence of reals converging to 0, denote Ω_{δ_j} by Ω_j, and let $\Box^j$ be the corresponding Laplacian on Ω_j (with respect to the Levi metric on Ω_j). For $q \geq 1$, let*

$$f_j \in L_{0q}(\Omega_j) \cap \operatorname{dom} \Box^j$$

be a sequence of forms satisfying

$$\|f_j\|_{L^2(\Omega_j)} \;=\; 1 \;\; and \tag{8.9}$$

$$\lim_{j \to \infty} \|\Box^j f_j\|_{L^2(\Omega_j)} \;=\; 0\,. \tag{8.10}$$

Then there is a subsequence (also denoted by f_j) such that

$$f_j|_{\Omega_0} \;\overset{L^2(\Omega_0)}{\longrightarrow}\; f\,,$$

$$\|f\|_{L^2(\Omega_0)} \;=\; 1\,,$$

$$\bar{\partial}f = 0 \quad and \quad \bar{\partial}_0^* f = 0\,.$$

Proof   From (8.10) we have

$$\lim_{j \to \infty} \left(\|\bar{\partial}f_j\|_{L^2(\Omega_j)} + \|\bar{\partial}_j^* f_j\|_{L^2(\Omega_j)}\right) = 0\,. \tag{8.11}$$

Fix j and apply our basic integral representation (1.3) to f_j:

$$f_j = Z_1^j \bar{\partial} f_j + Z_1^j \bar{\partial}_j^* f_j + Z_1^j f_j\,,$$

where the Z_1-operators depend on j. Iteration leads to

$$f_j = Z_1^j \bar{\partial} f_j + Z_1^j \bar{\partial}_j^* f_j + Z_k^j f_j \tag{8.12}$$

with an operator Z_k^j of type $\geq k$. For k sufficiently large there exists a constant C independent of j such that

$$\|Z_k^j f_j\|_{\Lambda_{1/2}(\Omega_j)} \leq C\,; \tag{8.13}$$

k can also be chosen independent of j.

Now choose $\varepsilon > 0$ and j so large that in (8.12)

$$\|Z_1^j \bar{\partial} f_j\|_{L^2(\Omega_j)} + \|Z_1^j \bar{\partial}_j^* f_j\|_{L^2(\Omega_j)} < \varepsilon$$

and so

$$\|f_j - Z_k^j f_j\|_{L^2(\Omega_j)} < \varepsilon\,. \tag{8.14}$$

This implies

$$\|f_j - Z_k^j f_j\|_{L^2(\Omega_j - \Omega_0)} < \varepsilon\,,$$

consequently from (8.13)

$$\|f_j\|_{L^2(\Omega_j - \Omega_0)} \leq C \operatorname{vol}(\Omega_j - \Omega_0) + \varepsilon\,,$$

which means

$$\lim_{j \to 0} \|f_j\|_{L^2(\Omega_j - \Omega_0)} = 0$$
$$\lim_{j \to 0} \|f_j\|_{L^2(\Omega_0)} = 1. \tag{8.15}$$

By the Ascoli/Arzelà theorem we can find a subsequence, also denoted by f_j, such that $Z_k^j f_j$ converges uniformly on Ω_0. From (8.14) we now deduce the convergence

$$\lim_{j \to 0} f_j|_{\Omega_0} = f$$

in $L^2(\Omega_0)$, and (8.15) yields $\|f\|_{L^2(\Omega_0)} = 1$. Since, by (8.11),

$$\lim_{j \to 0} \bar{\partial} f_j = 0 \text{ in } L^2(\Omega_0),$$

we have $f \in \text{dom } \bar{\partial}$ with $\bar{\partial} f = 0$. Moreover, if $g \in C^\infty_{0q-1}(X)$ then

$$|(f_j, \bar{\partial} g)_{L^2(\Omega_0)}| \leq |(f_j, \bar{\partial} g)_{L^2(\Omega_j)}| + |(f_j, \bar{\partial} g)_{L^2(\Omega_j - \Omega_0)}|$$
$$\leq C_g(\|\bar{\partial}_j^* f_j\|_{L^2(\Omega_j)} + \|f_j\|_{L^2(\Omega_j - \Omega_0)})$$

with a constant depending on g. Letting $j \to \infty$ we obtain

$$\lim_{j \to \infty} (f_j, \bar{\partial} g)_{L^2(\Omega_0)} = 0,$$

and so $f \in \text{dom } \bar{\partial}_0^*$ and $\bar{\partial}_0^* f = 0$. — This proves the lemma. $\qquad\square$

Theorem 8.16 *Under the assumptions (8.1) let $\delta_0 > 0$ be sufficiently small and denote by N^δ the Neumann operator (for $(0, q)$-forms, $q \geq 1$) in Ω_δ. Then there is a constant C independent of δ such that the operator norm*

$$\|N^\delta\|_{L^2(\Omega_\delta)} \leq C.$$

Proof Let λ_δ be the smallest positive eigenvalue of $\square^\delta$, then

$$\|N^\delta\|_{L^2(\Omega_\delta)} = \frac{1}{\lambda_\delta}.$$

Choose for each $\delta > 0$ an orthonormal base

$$e_1^\delta, \ldots, e_k^\delta$$

of the harmonic space $\{\square^\delta = 0\}$ and let f^δ be an eigenform of norm 1 for λ_δ. Let us note that, in view of VI.3.14, k does not depend on δ for $0 \leq \delta \leq \delta_0$.

Now assume that the theorem is false. Then there is a sequence $\delta \searrow 0$ with $\lambda_\delta \to 0$. Applying Lemma 8.8 several times we obtain a subsequence of δ's, say δ_j, such that

$$e_l^{\delta_j} \xrightarrow{L^2(\Omega_0)} e_l^0 \,,$$

$$f^{\delta_j} \xrightarrow{L^2(\Omega_0)} f^0 \,,$$

and such that the system

$$\{e_1^0, \ldots, e_k^0, f\}$$

is orthonormal. But — according to 8.8, the e_l^0 and f are harmonic in Ω_0, so

$$\dim \mathbb{H}^{0q}(\Omega_0) \geq k + 1 \,.$$

This contradicts the constancy of the dimension. $\square$

Remark. The surjectivity of the restriction map in cohomology is all that we needed; this result can be easily obtained by Grauert's bump method and does not depend on the existence of the Remmert reduction.

If we denote the canonical solution operator $\bar{\partial}_\delta^* N^\delta$ by K^δ, we deduce from 8.16 and from the formula

$$N^\delta f = K^\delta K^{\delta *} f + K^{\delta *} K^\delta f$$

(with orthogonal summands) the

Corollary 8.17 *There is a constant C independent of δ which satisfies*

$$\|\bar{\partial}_\delta^* N^\delta\|_{L^2(\Omega_\delta)} \leq C \quad and$$
$$\|\bar{\partial} N^\delta\|_{L^2(\Omega_\delta)} \leq C \,.$$

Let us now return to Theorem 8.5'. We have, by (8.7), for $u = K^\delta f$,

$$\|u\|_{C^{k+1/2}(\Omega_\delta)} \leq C(\|f\|_{C^k(\Omega_\delta)} + \|u\|_{L^2(\Omega_\delta)})$$

and by 8.17:

$$\|u\|_{L^2(\Omega_\delta)} \leq C\|f\|_{L^2(\Omega_\delta)} \leq C\|f\|_{C^k(\Omega_\delta)} \,.$$

So 8.5 and 8.5' are proved.

3.   In Theorem 8.16 and its corollary we can get rid of the special metrics. To be specific: let ds^2 be any hermitian metric on X and consider the corresponding Neumann operator $\widetilde{N}^\delta$ on Ω_δ and the canonical solution operators $\widetilde{K}^\delta = \bar{\partial}_\delta^* \widetilde{N}^\delta$. Then the old and the new L^2-norms on Ω_δ are equivalent, so the arising constants are independent of δ. This shows that the *new* norm of the old solution operator K^δ in 8.17 is uniformly bounded:

$$\|K^\delta\|_{L^2(\Omega_\delta)} \leq C \,.$$

But the canonical solution operator $\widetilde{K}^\delta$ has smaller (new) norm:

$$\|\widetilde{K}^\delta\|_{L^2(\Omega_\delta)} \leq \|K^\delta\|_{L^2(\Omega_\delta)} \leq C.$$

This then forces the norm of the new Neumann operators to be uniformly bounded, too, and we have (with new notations):

Theorem 8.18 *Let N^δ be the Neumann operator on Ω_δ for a fixed metric on X. Then, for $\delta_0 > 0$ sufficiently small, there is a constant $C > 0$ such that for all δ with $0 \leq \delta \leq \delta_0$:*

$$\|N^\delta\|_{L^2(\Omega_\delta)} \leq C,$$
$$\|\overline{\partial}_\delta^* N^\delta\|_{L^2(\Omega_\delta)} \leq C \text{ and}$$
$$\|\overline{\partial} N^\delta\|_{L^2(\Omega_\delta)} \leq C.$$

§9 Mergelyan's Approximation Theorem with C^k Boundary Values on Hermitian Manifolds

A classical result of 1-dimensional function theory is Mergelyan's approximation theorem: the *polynomial algebra $\mathbb{C}[z]$ is uniformly dense in the algebra $A^0(K)$ of continuous functions which are holomorphic in the interior, if K is a simply connected compact set in the plane.* Such a global result cannot be true in higher dimensions; what can be reasonably expected is uniform denseness in $A^0(K)$ of the algebra $\mathcal{O}(K)$ of germs of holomorphic functions, provided K is a suitable compact set in $\mathbb{C}^n$. Here we choose $K = \overline{\Omega}$, where Ω is a strictly pseudoconvex domain which may even be contained in an arbitrary complex manifold, and we prove an approximation theorem for all the algebras $A^k(\Omega)$ of holomorphic functions with C^k-boundary values. To be specific: we set as before $A^k(\Omega) = \{f \in C^k(\overline{\Omega}) : f \text{ holomorphic in } \Omega\}$ and analogously $A^k(\Omega_\delta)$ on Ω_δ, $0 \leq \delta \leq \delta_0$. We will prove the following theorem.

Theorem 9.1 *Let $\Omega \subset\subset X$ be a strictly pseudoconvex domain with C^∞ boundary in a hermitian manifold. Then there exist, for $\delta_0 > 0$ sufficiently small, linear operators*

$$E^\delta : A^0(\Omega) \to A^0(\Omega_\delta),$$

such that for all $k = 0, 1, 2, \ldots$

$$E^\delta : A^k(\Omega) \to A^k(\Omega_\delta)$$

is bounded. Moreover, for $f \in A^k(\Omega)$

$$\lim_{\delta \to 0} \|E^\delta f - f\|_{C^k(\overline{\Omega})} = 0.$$

Remarks. *i. An analogous result can be obtained for holomorphic L^p-functions, $p < \infty$. ii. If $f \in A^\infty(\Omega)$, then $E^\delta f \to f$ with respect to any C^k-norm on Ω.*

Proof   From here on we follow the approach as given in [Ran 86] but with the necessary modifications.

1. *Construction of a cocycle.*

There exist a finite open covering of $b\Omega$ by coordinate patches $\{U_j\}_{1 \leq j \leq N}$ and a family of biholomorphic maps

$$T_j^\tau : U_j \to T_j^\tau U_j \subset X$$

with the following properties.

i. With respect to the holomorphic coordinate map

$$x_j : U_j \to U_j' \subset \mathbb{C}^n \, ,$$

which may be defined in a neighborhood of $\overline{U}_j$ the map $x_j \circ T_j^\tau \circ x_j^{-1}$ corresponds to a translation in $\mathbb{C}^n$ by the amount τ along the inward normal to the boundary of the image set $x_j(\Omega \cap U_j)$ in $x_j(p_j)$ where $p_j \in b\Omega \cap U_j$. If $\tau > 0$ and U_j are chosen sufficiently small, one can achieve

$$T_j^\tau(U_j \cap \Omega) \subset\subset \Omega \, .$$

ii. $V^\tau = \{(T_j^\tau)^{-1}(U_j \cap \Omega)\}_{1 \leq j \leq N} \cup \{\Omega\}$ is an open covering of $\overline{\Omega}$. Therefore, if $\delta_0 > 0$ is sufficiently small, there exists for any $0 < \delta < \delta_0$ a $\tau_\delta > 0$ such that V^{τ_δ} covers $\overline{\Omega_\delta}$ and $\lim_{\delta \to 0} \tau_\delta = 0$.

If $f \in C^0(\overline{\Omega})$, then

$$f_j^\delta(x) = f(T_j^{\tau_\delta} x)$$

is continuous on $(T_j^{\tau_\delta})^{-1}(U_j \cap \Omega)$. If $f \in C^k(\overline{\Omega})$, then f_j^δ is in C^k with C^k boundary values on $(T_j^{\tau_\delta})^{-1}(b(U_j \cap \Omega))$. Finally, if f is holomorphic, f_j^δ is holomorphic, too.

iii. If $f \in C^k(\overline{\Omega})$, then

$$\lim_{\delta \to 0} \|f_j^\delta - f\|_{C^k(\overline{U_j \cap \Omega})} = 0 \, .$$

The whole setup only depends on the geometry of Ω and neither on k nor on f. Moreover, $f \mapsto f_j^\delta$ is a linear operator.

Set $V_0^\delta = \Omega$ and $f_0^\tau = f$. Now let $f \in A^k(\Omega)$. Then for $V_j^\delta = (T_j^{\tau_\delta})^{-1}(U_j \cap \Omega) \cap \Omega_\delta$, $1 \leq j \leq N$,

$$g_{ij}^\delta = f_j^\delta - f_i^\delta \, , \; 0 \leq i, j \leq N \, ,$$

defines a holomorphic cocycle ξ^δ on Ω_δ with C^k boundary values. Set

$$M_k^\delta(\xi^\delta) = \max_{0 \leq i, j \leq N} \|g_{ij}^\delta\|_{C^k(\overline{V_i^\delta \cap V_j^\delta})} \, .$$

Then $M_k^\delta(\xi^\delta) \leq C_k \|f\|_{C^k(\overline{\Omega})}$ and $\lim_{\delta \to 0} M_k^\delta(\xi^\delta) = 0$ with a constant independent of δ. The above cocycle is holomorphically solvable by definition, but we look for a holomorphic decomposition

$$g_{ij}^\delta = g_j^\delta - g_i^\delta \, , \; g_\iota^\delta \in A^k(V_\iota^\delta) \, , \iota = i, j \, ,$$

with

$$\sup_{0 \leq \iota \leq N} \|g_\iota^\delta\|_{C^k(\overline{V_\iota^\delta})} \leq C_k M_k^\delta(\xi^\delta) \, ,$$

where C_k is independent of δ.

2. *Solution of the cocycle problem.*

Let $\{\chi_j\}_{0\leq j\leq N}$ be a partition of unity in a neighbourhood of $\overline{\Omega}$ with $\mathrm{supp}\,\chi_0 \subset\subset \Omega$, $\mathrm{supp}\,\chi_j \subset\subset U_j$, $1 \leq j \leq N$. If δ_0 is small enough, this is also a partition of unity for $\overline{\Omega_\delta}$, $0 \leq \delta < \delta_0$, subordinate to $\{\Omega\} \cup \{(T_j^{\tau_\delta})^{-1}(U_j)\}_{1\leq j\leq N}$.

If we set on V_j^δ

$$v_j^\delta = \sum_{v=0}^{N} \chi_v g_{vj}^\delta \,,$$

then $v_j^\delta \in C^\infty(V_j^\delta)$ and $\|v_j^\delta\|_{C^k(\overline{V_j^\delta})} \leq C_k^j M_k^\delta(\xi^\delta)$, with constants independent of δ. Since

$$g_{ij}^\delta = v_j^\delta - v_i^\delta \quad \text{on } \overline{V_j^\delta} \cap \overline{V_i^\delta}$$

$\{\alpha^\delta\} = \{\overline{\partial} v_j^\delta\}_{0\leq j\leq N}$ defines a $\overline{\partial}$-closed $(0,1)$-form on $\overline{\Omega_\delta}$. The operator $\overline{\partial}$ only acts on the χ_j's, therefore

$$\alpha^\delta \in C_{01}^k(\overline{\Omega_\delta})\,.$$

By the same reason we have

$$\|\alpha^\delta\|_{C^k(\overline{\Omega_\delta})} \leq C_k M_k^\delta(\xi^\delta)$$

with C_k independent of δ.

Since ξ^δ is solvable, α^δ has to be a $\overline{\partial}$-exact form. We make this fact explicit. Set

$$h^\delta = \sum_{v=0}^{N} \chi_v f_v^\delta\,.$$

Then on V_j^δ

$$
\begin{aligned}
f_j^\delta - v_j^\delta &= \sum_{v=0}^{N} \chi_v(f_j^\delta - g_{vj}^\delta) \\
&= \sum_{v=0}^{N} \chi_v f_v^\delta \\
&= h^\delta\,.
\end{aligned}
$$

Therefore $\alpha^\delta = -\overline{\partial} h^\delta$. Unfortunately, h^δ is too "large". Set

$$\beta^\delta = \overline{\partial}^* N^\delta \alpha^\delta\,.$$

Then by the estimate of Theorem 8.5′

$$\|\beta^\delta\|_{C^k(\overline{\Omega_\delta})} \leq \widetilde{C}_k \|\alpha^\delta\|_{C^k(\overline{\Omega_\delta})} \leq \widetilde{C}_k C_k M_k^\delta(\xi^\delta)$$

with constants independent of δ. Now we set

$$g_j^\delta = v_j^\delta - \beta^\delta \in A^k(V_j^\delta)\,.$$

Then

$$\|g_j^\delta\|_{C^k(\overline{V_j^\delta})} \leq C_k M_k^\delta(\xi^\delta)\,,\ 0 \leq j \leq N\,,$$

with C_k independent of δ and j. Since

$$f_j^\delta - f_i^\delta = g_j^\delta - g_i^\delta\ \text{ on } V_i^\delta \cap V_j^\delta$$

$\{f_i^\delta - g_i^\delta\}_{0 \leq i \leq N}$ defines a holomorphic function in $A^k(\Omega_\delta)$, denoted by $E^\delta f$. E^δ is clearly a linear operator. Moreover, for all $0 \leq j \leq N$,

$$\|E^\delta f - f\|_{C^k(\overline{V_j^\delta \cap \Omega})} \leq \|f_j^\delta - f\|_{C^k(\overline{V_j^\delta \cap \Omega})} + \|g_j^\delta\|_{C^k(\overline{V_j^\delta})}\,.$$

Consequently

$$\lim_{\delta \to 0} \|E^\delta f - f\|_{C^k(\overline{\Omega})} = 0\,.$$

$\square$

§10 Notes

The chapter contains important applications of the theory of the Cauchy-Riemann equations to strictly pseudoconvex domains: Mergelyan's theorem 9.1, the solution to Gleason's problem 7.1, and Fefferman's mapping theorem 5.14. They are all based on the fundamental C^k-estimate 3.1 for the $\bar{\partial}$-equation and for the Bergman projector (5.6). The first regularity result to be proved (in 1963) was Kohn's basic Sobolev estimate 2.1 which contains a solution of the $\bar{\partial}$-Neumann problem   [Koh 63, Koh 64]. L^∞- and Hölder estimates came later: they were first established for special solutions of the $\bar{\partial}$-equation for $(0, 1)$-forms by Grauert and Lieb [GrL 70] and Henkin [Hen 70] in 1969, independently, using global integral formulae; Lieb [Lie 69] immediately applied this result to prove Mergelyan's theorem for continuous boundary values; this result was simultaneously obtained by Henkin [Hen 70] using a slightly different method. N. Kerzman [Ker 71] proved Hölder and L^p-estimates for $(0, 1)$-forms; he was the first to combine local integral formulae with the bump method and to work, in this context, on general manifolds instead of $\mathbb{C}^n$. Lieb [Lie 70], and later on Øvrelid [Øvr 71₁], carried the solution method over to forms of arbitrary degree. The precise $\frac{1}{2}$-Hölder estimates are due to Romanov/Henkin [RoH 71] and Range/Siu [RaS 73]. Some time later, P. Greiner and E. Stein proved (for Levi metrics and $(0, 1)$-forms) the basic Hölder estimate 3.1.i as well as its C^k-analogue [GrS 77]. C^k-estimates for special solutions of the $\bar{\partial}$-equation (for $(0, 1)$-forms) were first derived by W. Alt and Y. T. Siu, independently, in 1974 [Alt 74], [Siu 74]; the general case of $(0, q)$-forms is due to Lieb and Range [LiR 80]. They also established the basic C^k-Neumann estimates of §3 in the case of Levi metrics [LiR 86₁, LiR 86₂]; somewhat later R. Beals, P. Greiner and N. Stanton obtained the general result for arbitrary metrics using a completely different method [BGS 87]. The relation between the different cohomology groups as described in §4 is based on the preceding estimates and on Grauert's bump method; for the first cohomology one has to use, in addition, the existence of the Remmert reduction. — Except for Theorem 5.5 which is a consequence of Kohn's theory [FoK 72] all the other regularity results for the Bergman projector were obtained in the late 70s by Ch. Fefferman [Fef 74], Ahern/Schneider [AhS 78], Phong/Stein [PhS 77] and Ligocka [Lig 84]. Theorem 5.14 is due to Fefferman [Fef 74], simplified proofs based on a study of the Bergman projector were later found by Bell and Ligocka [BeL 80]; it started an intensive study of the boundary behaviour of biholomorphic and proper holomorphic maps which we do not pursue here. — Integral formulae easily allow to carry the Neumann problem over to more general spaces than L^2; even measure spaces are possible — see [HefT 99]. Gleason's problem was formulated by Gleason in 1964 [Gle 64] and almost immediately solved on the ball by Leibenzon (unpublished). Its solution on general strictly pseudoconvex domains in $\mathbb{C}^2$ is due to Kerzman and Nagel [KeN 71], in $\mathbb{C}^n$, n arbitrary, to Lieb [Lie 70], Henkin [Hen 71], and also to Øvrelid [Øvr 71$_2$] whose treatment we have carried over to Stein manifolds in §7. All these results refer to the algebra $A^0(\Omega)$. The A^k-case of the problem relies, of course, on C^k-estimates for the $\bar{\partial}$-equation or the Bergman projector; it was first handled by Ahern/Schneider [AhS 78] and Lieb/Range [LiR 80]. Mergelyan's theorem 9.1 for continuous boundary values marks the beginning of the theory of $\bar{\partial}$-estimates outside the scale of Sobolev spaces. The general case of C^k-boundary values is due to W. Alt [Alt 74] and Y. T. Siu [Siu 74] — compare also [Jak 81]. In our approach it depends on some stability results for $\bar{\partial}$-estimates on families of strictly pseudoconvex domains, which follow fairly easily from our theory of Z-operators once a uniform estimate for the Neumann operator has been proved. This latter result again uses the theory of Z-operators. An extensive study of stability of $\bar{\partial}$-estimates for Sobolev norms was given by Greene and Krantz who establish (suboptimal) stable estimates in a very general situation [GrK 82].

Bibliography

[AhS 78]   Ahern, P., Schneider, R. Holomorphic Lipschitz functions in pseudoconvex domains. *Ann. of Math.* **101**, 543–565 (1978)

[Ale 2001] Alexandre, W. Construction d'une fonction de support à la Diederich-Fornæss. *Pub. IRMA, Lille* **54,III** (2001)

[Alt 74]   Alt, W. Hölderabschätzungen von Lösungen der Gleichung $\bar{\partial}u = f$ bei streng pseudokonvexem Rand. *Man. Math.* **13**, 381–414 (1974)

[AnG 62]   Andreotti, A., Grauert, H. Théorèmes de finitude pour la cohomologie des espaces complexes. *Bull. Soc. Math. France* **90**, 193–259 (1962)

[BGS 87]   Beals, R., Greiner, P., Stanton, N. L^p and Lipschitz estimates for the $\bar{\partial}$-equation and the $\bar{\partial}$-Neumann problem. *Math. Ann.* **277**, 185–196 (1987)

[BeL 80]   Bell, S., Ligocka, E. A simplification and extension of Fefferman's theorem on biholomorphic mappings. *Inv. math.* **57**, 283–289 (1980)

[BeA 83]   Berndtsson, B., Andersson, M. Henkin-Ramírez formulas with weight factors. *Ann. Inst. Fourier* **32**, 91–110 (1983)

[BeS 39]   Behnke, H., Stein, K. Konvergente Folgen von Regularitätsbereichen und die Meromorphiekonvexität. *Math. Ann.* **116**, 204–216 (1939)

[BoS 91]   Boas, H., Straube, E. Sobolev estimates for the $\bar{\partial}$-Neumann operator on domains in $\mathbb{C}^n$ admitting a defining function which is plurisubharmonic on the boundary. *Math. Z.* **206**, 81–88 (1991)

[Boc 43]   Bochner, S. Analytic and meromorphic continuation by means of Green's formula. *Ann. Math.* **44**, 652–673 (1943)

[Bog 91]   Boggess, A. *CR manifolds and the tangential Cauchy-Riemann complex.* Studies in Advanced Mathematics. Boca Raton, CRC Press., 1991

[Car 51]   Cartan, H. Séminaire ENS 1951/52.

[Car 60]   Cartan, H. Quotients of analytic spaces. *Cont. Funct. Th. Tata Institute*, 1–16, Bombay (1960)

[Cat 83]   Catlin, D. Necessary conditions for subellipticity of the $\bar{\partial}$-Neumann problem. *Ann. Math.* **117**, 147–171 (1983)

[Cat 84]   Catlin, D. Boundary invariants of pseudoconvex domains. *Ann. Math.* **120**, 529–586 (1984)

[Cat 87]   Catlin, D. Subelliptic estimates for the $\bar{\partial}$-Neumann problem on pseudoconvex domains. *Ann. Math.* **126**, 131–191 (1987)

[ChC 88]   Chaumat, J., Chollet, A. Noyaux pour résoudre l'équation $\bar{\partial}u = v$ dans les classes indéfiniement différentiables. *C.R.A.Sci. Paris* **I.306,14**, 585–588 (1988)

[ChS 2001] Chen, S., Shaw, M. *Partial differential equations in several complex variables*. AMS Int. Press (2001)

[Cum 97]   Cumenge, A. Estimées Lipschitz optimales dans les convexes de type fini. *C. r. A. S. Paris* **325**, 1077–1080 (1997)

[Cum 2001] Cumenge, A. Sharp estimates for $\bar{\partial}$ on convex domains of finite type. *Ark. mat.* **39**, 1–25 (2001)

[DAn 82]   D'Angelo, J. Real hypersurfaces, orders of contact, and applications. *Ann. Math.* **115**, 615–637 (1982)

[DFF 99]   Diederich, K., Fischer, B., Fornæss, J. E. Hölder estimates on convex domains of finite type. *Math. Z.* **232**, 43–61 (1999)

[DiF 77$_1$] Diederich, K., Fornæss, J. E. Pseudoconvex domains: bounded strictly plurisubharmonic exhaustion functions. *Inv. math.* **39**, 129–141 (1977)

[DiF 77$_2$] Diederich, K., Fornæss, J. E. Pseudoconvex domains: an example with non-trivial nebenhülle. *Math. Ann.* **225**, 275–292 (1977)

[DiF 78]   Diederich, K., Fornæss, J. E. Pseudoconvex domains with real-analytic boundary. *Ann. Math.* **107**, 371–384 (1978)

[DiF 99]   Diederich, K., Fornæss, J. E. Support functions for convex domains of finite type. *Math. Z.* **230**, 145–164 (1999)

[Dol 56]   Dolbeault, P. Formes différentielles et cohomologie sur une variété analytique complexe I. *Ann. Math.* **64**, 83–130 (1956)

[Dol 57]   Dolbeault, P. Formes différentielles et cohomologie sur une variété analytique complexe II. *Ann. Math.* **65**, 282–330 (1957)

[Ehr 61]   Ehrenpreis, L. A new proof and an extension of Hartogs' theorem. *Bull. AMS* **67**, 507–509 (1961)

[Fef 74]   Fefferman, Ch. The Bergman kernel and biholomorphic mappings of pseudoconvex domains. *Inv. Math.* **26**, 1–65 (1974)

[FeK 88]   Fefferman, Ch., Kohn, J. J. Hölder estimates on domains of complex dimension 2 and on 3-dimensional CR manifolds. *Adv. Math.* **69**, 223–303 (1988)

[FKM 90]   Fefferman, Ch., Kohn, J. J., Machedon, M. Hölder estimates on CR manifolds with a diagonalizable Levi form. *Adv. Math.* **84**, 1–90 (1990)

[Fis 2001] Fischer, B. L^p estimates on convex domains of finite type. *Math. Z.* **236**, 401–418 (2001)

[FiL 74]   Fischer, W., Lieb, I. Lokale Kerne und beschränkte Lösungen der Cauchy-Riemannschen Differentialgleichungen auf q-konvexen Gebieten. *Math. Ann.* **208**, 249–265 (1974)

[FoK 72]   Folland, G. B., Kohn, J. J. *The Neumann problem for the Cauchy-Riemann complex.* Ann. Math. St. **75**, Princeton 1972

[FoS 74]   Folland, G. B., Stein, E. M. Estimates for the $\bar{\partial}_b$-complex and analysis on the Heisenberg group. *Comm. Pure Appl. Math.* **27**, 429–522 (1974)

[ForS 87]   Fornæss, J. E., Stensønes, B. *Lectures on counterexamples in several complex variables.* Princeton University Press, Princeton 1987

[Fri 44]   Friedrichs, K. O. The identity of weak and strong extension of differential operators. *Trans. AMS* **55**, 132–151 (1944)

[Gaf 55]   Gaffney, M. P. Hilbert space methods in the theory of harmonic integrals. *Trans. AMS* **78**, 426–444 (1955)

[Gle 64]   Gleason, A. Finitely generated ideals in Banach algebras. *J. Math. Mech.* **13**, 125–132 (1964)

[Gra 58]   Grauert, H. On Levi's problem and the embedding of real analytic manifolds. *Ann. Math.* **68**, 460–472 (1958)

[Gra 60]   Grauert, H. Ein Theorem der analytischen Garbentheorie und die Modulräume komplexer Strukturen. *Publ. IHES* **5**, 233–292 (1960)

[GrL 70]   Grauert, H., Lieb, I. Das Ramírezsche Integral und die Lösung der Gleichung $\bar{\partial}f = \alpha$ im Bereich der beschränkten Formen. *Rice Univ. St.* **56**, 29–50 (1970)

[GrR 84]   Grauert, H., Remmert, R. *Coherent analytic sheaves.* Springer, Berlin etc., 1984

[GrK 82]   Greene, R. E., Krantz, S. G. Deformations of complex structures, estimates for the $\bar{\partial}$-equation, and stability of the Bergman kernel. *Adv. Math.* **43**, 1–86 (1982)

[GrS 77]   Greiner, P., Stein, E. *Estimates for the $\bar{\partial}$-Neumann problem.* Princeton Univ. Press, Princeton, 1977

[HaP 79]   Harvey, R., Polking, J. Fundamental solutions in complex analysis. *Duke Math. J.* **46**, 253–300 (1979)

[Har 06]   Hartogs, F. Einige Folgerungen aus der Cauchyschen Integralformel bei Funktionen mehrerer komplexer Veränderlicher. *Münchner Berichte* **36**, 223–242 (1906)

[HefH 50]   Hefer, H. Zur Funktionentheorie mehrerer Veränderlichen. Über eine Zerlegung analytischer Funktionen und die Weilsche Integraldarstellung. *Math. Ann.* **122**, 276–278 (1950)

[HefT 99]   Hefer, T. Regularität von Randwerten der kanonischen Lösung der $\bar{\partial}$-Gleichung auf streng pseudokonkaven Gebieten. *Bonner Math. Schr.* **320** (1999)

[HefT 2002]   Hefer, T. Hölder and L^p estimates for $\bar{\partial}$ on convex domains of finite type depending on Catlin's multitype. *to appear in Math. Z.* (2002)

[Hen 69]   Henkin, G. M. Integral representations in strongly pseudoconvex domains and some applications. *Math. Sb.* **78**, 611–632 (1969)

[Hen 70]   Henkin, G. M. Integral representations in strongly pseudoconvex domains and applications to the $\bar{\partial}$-problem. *Math. Sb.* **82**, 300–308 (1970)

[Hen 71]   Henkin, G. M. Approximation of functions in pseudoconvex domains and a theorem of Z. L. Leibenzon. *Bull. Acad. Pol. Sci.* **19**, 37–42 (1971)

[Hen 77]   Henkin, G. M. H. Lewy's equation and analysis on pseudoconvex manifolds I. *Usp. Mat. Nauk* **82**, 57–118 (1977), II. *Mat. Sb.* **102**, 71–108 (1977)

[Hen 90]   Henkin, G. M. The method of integral representations in complex analysis. *Enc. Math. Sci.* **7**, Sev. Cpl. Var. 1, Springer, Berlin etc., 57–116 (1990)

[HeL 84]   Henkin, G. M., Leiterer, J. *Theory of functions on complex manifolds.* Birkhäuser, Boston, 1984

[HeL 88]   Henkin, G. M., Leiterer, J. *Andreotti-Grauert theory by integral formulae.* Birkhäuser, Boston, 1988

[Hod 41]   Hodge, W. D. V. *The theory and application of harmonic integrals.* Cambridge, 1941

[Hör 65]   Hörmander, L. L^2 estimates and existence theorems for the $\bar{\partial}$-operator. *Acta Math.* **113**, 89–152 (1965)

[Hör 66]   Hörmander, L. *An introduction to complex analysis in several variables.* North Holland, Amsterdam etc., 1966

[Hör 67]   Hörmander, L. Generators of some rings of analytic functions. *Bull. AMS* **73**, 943–949 (1967)

[Hua 58]   Hua, L. K. *Harmonic analysis of functions of several complex variables in classical domains.* Sci. Press, Peking, 1958

[Jak 81]   Jakóbczak, P. Approximation and decomposition theorems for the algebras of analytic functions on strictly pseudoconvex domains. *Zesz. Nauk. Univ. Jag.* **30**, 95–109 (1981)

[JeK 95]   Jerison, D., Kenig, C. The inhomogeneous Dirichlet problem in Lipschitz domains. *J. Func. Anal.* **130**, 161–219 (1995)

[Ker 71]   Kerzman, N. Hölder and L^p estimates for solutions of $\bar{\partial}u = f$ in strongly pseudoconvex domains. *Comm. Pure Appl. Math.* **24**, 301–379 (1971)

[KeN 71]   Kerzman, N., Nagel, A. Finitely generated ideals in certain function algebras. *J. Func. Anal.* **7**, 212–215 (1971)

[KeS 78]   Kerzman, N., Stein, E. The Szegö kernel in terms of Cauchy-Fantappiè ker-
            nels. *Duke Math. J.* **45**, 197–224 (1978)

[Koh 63]   Kohn, J. J. Harmonic integrals on strongly pseudoconvex domains I. *Ann.
            Math.* **78**, 112–148 (1963)

[Koh 64]   Kohn, J. J. Harmonic integrals on strongly pseudoconvex domains II. *Ann.
            Math.* **79**, 450–472 (1964)

[KoN 73]   Kohn, J. J., Nirenberg, L. A pseudoconvex domain not admitting a holomor-
            phic support function. *Math. Ann.* **201**, 265–268 (1973)

[Kop 67_1]   Koppelman, W. The Cauchy integral for functions of several complex vari-
            ables. *Bull. AMS* **73**, 373–377 (1967)

[Kop 67_2]   Koppelman, W. The Cauchy integral for differential forms. *Bull. AMS* **73**,
            554–556 (1967)

[Kra 76]   Krantz, S. G. Optimal Lipschitz and L^p regularity for the equation $\bar{\partial}u = f$
            on strongly pseudoconvex domains. *Math. Ann.* **219**, 233–260 (1976)

[Kyt 95]   Kytmanov, A. M. *The Bochner-Martinelli integral and its applications.* Birk-
            häuser, Basel etc., 1995

[Lau 75]   Laufer, H. B. On the infinite dimensionality of the Dolbeault cohomology
            groups. *Proc. AMS* **52**, 293–296 (1975)

[LTh 97]   Laurent-Thiébaut, C. *Théorie des fonctions holomorphes de plusieurs vari-
            ables.* Inter Ed., Paris, 1997

[Ler 56]   Leray, J. Fonctions de variables complexes. *Rend. Acc. Naz. Lincei* **20**, 589–
            590 (1956)

[Ler 59]   Leray, J. Le calcul différentiel et intégral sur une variété complexe III. *Bull.
            Soc. Math. France* **87**, 81–180 (1959)

[Lev 11]   Levi, E. E. Sulle ipersuperficie delle spazi a 4 dimensioni che possono essere
            frontiera del campo di esistenza di una funzione analitica di due variabili
            complesse. *Ann. Mat. Pura Appl.* **18**, 69–79 (1911)

[Lie 66]   Lieb, I. Über komplexe Räume und komplexe Spektren. *Inv. Math.* **1**, 45–58
            (1966)

[Lie 69]   Lieb, I. Ein Approximationssatz auf streng pseudokonvexen Gebieten. *Math.
            Ann.* **184**, 56–60 (1969)

[Lie 70]   Lieb, I. Die Cauchy-Riemannschen Differentialgleichungen auf streng pseu-
            dokonvexen Gebieten. Beschränkte Lösungen.*Math.Ann.***190**,6–44 (1970)

[Lie 72]   Lieb, I. Die Cauchy-Riemannschen Differentialgleichungen auf streng pseu-
            dokonvexen Gebieten. Stetige Randwerte.*Math.Ann.***199**,241–256(1972)

[Lie 84]   Lieb, I. Integral formulae in complex analysis. *Proc. Hangzhou Conf. 1981*,
            187–198 (1984)

[LiR 80]   Lieb, I., Range, R. M. Ein Lösungsoperator für den Cauchy-Riemann Komplex mit C^k-Abschätzungen. *Math. Ann.* **253**, 145–164 (1980)

[LiR 83]   Lieb, I., Range, R. M. On integral representations and a priori Lipschitz estimates for the canonical solution of the $\bar{\partial}$-equation. *Math. Ann.* **265**, 221–251 (1983)

[LiR 84]   Lieb, I., Range, R. M. Integral representations on hermitian manifolds: the $\bar{\partial}$-Neumann solution of the Cauchy-Riemann equations. *Bull. AMS* **11**, 355–358 (1984)

[LiR 86$_1$]   Lieb, I., Range, R. M. Integral representations and estimates in the theory of the $\bar{\partial}$-Neumann problem. *Ann. Math.* **123**, 265–301 (1986)

[LiR 86$_2$]   Lieb, I., Range, R. M. Estimates for a class of integral operators and applications to the $\bar{\partial}$-Neumann problem. *Inv. Math.* **85**, 415–438 (1986)

[LiR 87]   Lieb, I., Range, R. M. The kernel of the $\bar{\partial}$-Neumann operator on strictly pseudoconvex domains. *Math. Ann.* **278**, 151–179 (1987)

[LiR 93]   Lieb, I., Range, R. M. The $\bar{\partial}$-Neumann kernel for codimension-one forms on strictly pseudoconvex domains. *Sev. Cpl. Var. Princeton Math. Notes* **38**, 473–482 (1993)

[Lig 84]   Ligocka, E. The Hölder continuity of the Bergman projection and proper holomorphic mappings. *Stud. Math.* **80**, 89–107 (1984)

[Mar 38]   Martinelli, E. Alcuni teoremi integrali per le funzioni analitiche di più variabili complesse. *Mem. R. Accad. d'Italia* **9**, 269–283 (1983)

[Mic 91]   Michel, J. Integral representations on weakly pseudoconvex domains. *Math. Z.* **208**, 437–462 (1991)

[Mic 92]   Michel, J. Der Neumannoperator auf streng pseudokonkaven Gebieten und andere Anwendungen der Integralformelmethode. *Bonner Math. Schr.* **238** (1992)

[Mis 91]   Michels, I. Integraldarstellung und Regularitätseigenschaften von Projektionsoperatoren. *Bonner Math. Schr.* **216** (1991)

[Nor 60]   Norguet, F. Problèmes sur les formes différentielles et les courants. *Ann. Inst. Fourier* **11**, 1–88 (1960)

[Oka 84]   Oka, K. *Collected Papers.* Springer, New York etc., 1984

[Øvr 71$_1$]   Øvrelid, N. Integral representation formulas and L^p-estimates for the $\bar{\partial}$-equation. *Math. Scand.* **29**, 137–160 (1971)

[Øvr 71$_2$]   Øvrelid, N. Generators of the maximal ideals of $A(D)$. *Pac. J. Math.* **39**, 219–223 (1971)

[PhS 77]   Phong, D. H., Stein, E. Estimates for the Bergman and Szegö projections on strongly pseudoconvex domains. *Duke Math. J.* **44**, 695–704 (1977)

[PhS 86]   Phong, D. H., Stein, E. Hilbert integrals, singular integrals and Radon transforms II. *Inv. Math.* **86**, 75–113 (1986)

[Pom 58]   Pompeiu, D. *Œuvre mathématique.* Ed. Acad. RPR, Bucuresti, 1958

[Ram 70]   Ramírez de Arellano, E. Ein Divisionsproblem und Randintegraldarstellungen in der komplexen Analysis. *Math. Ann.* **184**, 172–184 (1970)

[Ran 82]   Range, R. M. An elementary integral solution operator for the Cauchy-Riemann equations on pseudoconvex domains in $\mathbb{C}^n$. *Trans. AMS* **274**, 809–816 (1982)

[Ran 86]   Range, R. M. *Holomorphic functions and integral representations in several complex variables.* Springer, Berlin etc., 1986

[Ran 98]   Range, R. M. *Holomorphic functions and integral representations in several complex variables,* 2nd ed., Springer, Berlin etc., 1998

[Ran 89]   Range, R. M. *Cauchy-Fantappiè formulas in multidimensional complex analysis.* Geometry and complex variables in Bologna, unpublished lecture, 1989

[RaS 73]   Range, R. M., Siu, Y. T. Uniform estimates for the $\bar{\partial}$-equation on domains with piecewise smooth strictly pseudoconvex boundaries. *Math. Ann.* **206**, 325–354 (1973)

[Rha 60]   de Rham, G. *Variétés différentiables.* Hermann, Paris, 1960

[RoH 71]   Romanov, A. V., Henkin, G. M. Exact Hölder estimates for the solution of the $\bar{\partial}$-equation. *Izv. Ak. Nauk SSR* **35**, 1171–1183 (1971)

[Rud 73]   Rudin, W. *Functional Analysis.* McGraw-Hill, New York, 1973

[Sch 66]   Schwartz, L. *Théorie des distributions,* 2ième éd., Hermann, Paris, 1966

[See 64]   Seeley, R. Extension of C^∞-functions defined in a half space. *Proc. AMS* **15**, 625–626 (1964)

[Siu 74]   Siu, Y. T. The $\bar{\partial}$-problem with uniform bounds on derivatives. *Math. Ann.* **207**, 163–176 (1974)

[Ste 70]   Stein, E. *Singular integrals and differentiability properties of functions.* Princeton Univ. Press, Princeton, 1970

[Ste 93]   Stein, E. *Harmonic analysis.* Princeton Univ. Press, Princeton, 1993

[Swe 76]   Sweeney, W. A condition for subellipticity in Spencer's Neumann problem. *J. Diff. Equ.* **21**, 316–362 (1976)

[Vek 62]   Vekua, I. N. *Generalised analytic functions.* Pergamon, Oxford, 1962

[Wei 58]   Weil, A. *Introduction à l'étude des variétés Kähleriennes.* Hermann, Paris, 1958

Notations

(See also the index and the table of contents.)

a) Differential operators

b) Operation on forms

c) Operators and kernels

d) Neumann problem

e)   Function spaces

L^p, L^2, L^∞	Lebesgue spaces	135
L_s^2	Sobolev spaces	256
$L^{p,\delta}, L_k^{p,\delta}$	weighted Lebesgue (Sobolev) spaces	256
$W_k^{p,\delta}$	tangential Sobolev spaces	284
$S_k^{p,\delta}$	Sobolev-Stein spaces	261
C^k, C^∞, C^α	spaces of differentiable functions,	
	Hölder spaces (Lipschitz spaces)	2, 6, 210
$\Lambda_\alpha = C^\alpha$	Hölder spaces	256
$\Gamma_{\alpha,\beta}$	Hölder-Stein spaces	261
C_c^k	compact support	14
E_{pq}	space of (p,q)-forms with coefficients	
	in the function space E, e. g.	
C_{pq}^k		
$\|\cdot\|_E$	norm of the function space E, e. g. $\|\cdot\|_{L^p}$	
$\mathcal{O}$	space of holomorphic functions	
Ω^p	space of holomorphic p-forms,	
	sheaf of germs of holomorphic p-forms	

f)   Cohomology spaces

$\mathbb{H}^{pq}, \mathbb{H}^{0q} = \mathbb{H}^q$	harmonic space	148, 250
H_s^{pq}	L^s-bounded cohomology space	107
$H^{pq}, H_{C^k}^{pq}$		314

g)   Miscellaneous

$\mathbb{D}$	unit ball	
Δ	diagonal in $X \times X$	
Λ	diagonal in $b\Omega \times b\Omega$	
$b\Omega$	boundary of Ω	

Index